Marine and Offshore Corrosion

Marine Engineering Series

MARINE AND OFFSHORE PUMPING AND PIPING SYSTEMS
J Crawford, CEng, FIMarE

MARINE AUXILIARY MACHINERY—5th edition
E Souchotte, CEng, FIMechE, MIMarE
David W Smith, CEng, MIMarE

MARINE DIESEL ENGINES—5th edition
Edited by C C Pounder

MARINE ELECTRICAL PRACTICE—5th edition
G O Watson, FIEE, FAIEE, MIMarE

MARINE STEAM BOILERS—4th edition
J H Milton, CEng, FIMarE, MNECIES
R M Leach, CEng, MIMechE, FIMarE

MARINE STEAM ENGINES AND TURBINES—4th edition
S C McBirnie, EEng, MIMechE

Marine and Offshore Corrosion

Kenneth A Chandler
BSc, CEng, FIM, ARSM, FICorrT

BUTTERWORTHS

London—Boston—Durban—Singapore—Sydney—Toronto—Wellington

First published 1985

© Butterworth & Co (Publishers) Ltd, 1985

British Library Cataloguing in Publication Data

Chandler, Kenneth A.
    Marine and offshore corrosion.
    1. Sea-water corrosion.
    I. Title
    620'.4162   TA462

    ISBN 0–408–01175–0

Library of Congress Cataloging in Publication Data

Chandler, Kenneth A.
    Marine and offshore corrosion.
    Bibliography:
    Includes index.
    1. Sea-water corrosion. 2. Offshore structures—
Corrosion. I. Title.
TA462.C462   1984   620.1'1223   84-7009

    ISBN 0–408–01175–0 (U.S.)

Filmset by Mid-County Press
Printed by Thetford Press Ltd.
Bound by Anchor Brendon Ltd.

Preface

'Thy needles—now rust
disus'd and shine no more'
              W Cowper (17th Century)

This book was originally intended as a revision of *Marine Corrosion* written by T Howard Rogers and published fifteen years ago. There have, however, been many developments in the field of marine corrosion in the intervening period. Corrosion control in marine situations has become a matter of increasing importance with discoveries of offshore oil and the greater demands being made on the utilisation of the sea's resources for food, fresh water and for energy requirements, e.g. by wave motion.

The original book *Marine Corrosion* was particularly concerned with shipbuilding and the marine engineering associated with it. To allow for a more detailed treatment of general engineering requirements, particularly of protective coatings for steelwork, the balance of content places less emphasis on ships although a treatment of this topic is included. Consequently, this is really a new book although elements from the original book have been included. It also has a new title to reflect its purpose.

Clearly when preparing a book, the author must have in mind his readership. This particular book has been prepared for engineers and designers who are not corrosion specialists but have to deal with marine corrosion problems as part of their day-to-day professional activities. Although the term 'engineer' covers a wide field, the book is not primarily for those who describe themselves as 'corrosion engineers'. However, because corrosion is becoming very specialised and many experts in, say, cathodic protection do not necessarily have expertise in alloy selection, it is hoped that much will be of interest to them. Because the topic—marine and offshore corrosion—covers such a large field, it would be impracticable in a book of this size to provide a detailed treatment of every aspect. The aim has been to cover the major aspects in reasonable detail but where appropriate, having drawn a matter to the attention of the reader, to indicate further sources of information.

Although general data is included in the book it is not intended as a data handbook. The aim is to indicate the principles on which good corrosion control treatments are based and to highlight some of the problems that have to be overcome to achieve economic solutions to corrosion.

Data is generally provided to illustrate a point or a principle and is not intended for design purposes. Where corrosion data is used by designers it is advisable to refer to the original publication because there may be 'caveats' regarding the use of such information in circumstances different from those in which it was obtained. Materials are often being used to near their engineering limit in marine situations and new materials will be developed to meet the requirements which are bound to increase in the next decade or so. Such materials may well be subject to localised and stress-related forms of corrosion and considerable research efforts will be required to overcome some of these difficulties and some of our present problems.

The requirements for adequately protecting steel structures from corrosion by painting will also lead to new developments in coatings, cleaning and application techniques. Those involved in marine corrosion must, therefore, maintain close contact with such developments. The basic principles of corrosion and its control are largely established and they are set out in this book but the solutions to many problems remain to be resolved.

It should be borne in mind that the aim of designers is not usually to prevent corrosion but to control it within acceptable economic limits, so they are not necessarily concerned with the most corrosion-resistant materials but the ones that have the overall properties of strength, weldability, formability, corrosion resistance and cost that will provide the most effective economic solutions to engineering problems. Corrosion control is only a part of this overall requirement but often a very important part, and so deserves the attention appropriate to the economic consequences that will arise from its neglect.

Finally, in any book there will be some chapters where the reader considers the level of treatment too low and others where it is at a level unnecessarily high for his purposes. This book is essentially practical in concept and, apart from the chapter concerned with principles of corrosion, the treatment of theoretical matters is kept to a minimum, although it is hoped that readers will—where appropriate—study matters in more detail by referring to the publications noted at the end of each chapter.

KAC

Acknowledgements

In this book reference is made to published work by other authors and investigators but because of the practical nature of the book and because it is intended for practising engineers, references have been limited to those papers and books from which data has been extracted or where the original work may, to advantage, be studied.

The knowledge of corrosion and its control is, of course, gained in many ways other than by one's own reading, research and experience. Discussions with colleagues, attendance at conferences and meetings are all important means of developing views on matters of interest. Often the original source of the information or concepts that one uses as a basis for published work is no longer clear. I, therefore, take this opportunity to thank all those with whom I have been involved on corrosion and related matters and apologise for any omissions of specific acknowledgements for contributions they may have made to my views on marine corrosion.

I wish to thank Dr Lionel Shreir, OBE, for useful comments and for his permission to use material from his many published papers on the principles and theories of corrosion. These and his work for the Department of Industry (UK) form the basis of Chapter 2 'Principles of Corrosion'.

I thank Mr Bryan Wyatt who supplied the material for the chapter on cathodic protection and Mr Hector Campbell for his advice on parts of the chapter on non-ferrous metals. Also thanks are due to the late Mr T Howard Rogers who gave permission shortly before his death for the use of material from his book *Marine Corrosion* (Butterworths, 1968) in the chapters dealing with marine environments.

Mr. John Morley is thanked for data on piling.

Finally, I wish to express my thanks to my wife, Veronica, who has painstakingly typed and checked the drafts of this book.

CONTENTS

Introduction: The control of corrosion in marine environments

1

This book is concerned with the control of corrosion in marine environments and in this chapter the general causes of corrosion and the various means of controlling it will be considered in broad terms. Other chapters deal with specific areas of corrosion and control in more detail.

Corrosion can be defined in a number of ways but 'the chemical or electrochemical reaction of a metal or an alloy with its environment' provides a reasonable explanation of the term 'corrosion'. It is one of the two common causes of metal deterioration, the other being the mechanical loss of the metal by erosion, abrasion or wear. Sometimes there is a joint action of corrosion and erosion.

Although corrosion can sometimes be prevented, the aim is usually to control it within economic limits. There are situations where no corrosion at all is acceptable, but these are few. Generally, the choice of materials is based on regular maintenance during the life of the construction. This is a sensible approach provided it is part of the design philosophy. Corrosion can be a very serious problem when it occurs unexpectedly and emergency measures have to be taken to deal with it.

Economic considerations are the essence of corrosion control procedures, but not necessarily in a straightforward way. Designers and engineers should be aware of the economics of corrosion but they cannot necessarily make the final decision in these matters. Other personnel are involved, not least accountants, and hopefully the corrosion requirements will properly be considered in relation to other factors. Nevertheless, the deciding factor may be the strength of a structure, time for construction, availability of materials and, of course, capital cost. Therefore corrosion control is a matter of options to fit in with the many other requirements to be taken into account by the design team. Often corrosion problems arise because, due to changes in design that often take place in the early stages of a project, the materials or coatings originally selected may not be suitable at a later stage, where the conditions and environments may have changed. Furthermore, on large structures and plant, different parts may be designed by different teams. If there is not close contact between them, corrosion and, incidentally, other problems may well occur.

Generally, the overall corrosion requirements can be handled in a fairly straightforward way. The solutions may be a compromise but if the materials or coatings chosen are known to have limitations, this can be taken into account in the maintenance or monitoring procedures. More serious problems arise where changes are made, of which those concerned or responsible for the corrosion elements in the design are not aware. A change to a more corrosion-resistant alloy may be advantageous for a particular section of the plant or structure. However, the new alloy may cause problems, e.g. bi-metallic corrosion, in another part of the plant. Designers must take such matters into account.

The word 'design' will appear many more times in this book. Corrosion control in marine engineering is basically a design problem, using the term in its widest sense. It is not just a question of choosing resistant materials or very durable coatings, it is a matter of fitting together all the different elements to reduce corrosion to an acceptable and economic level. Additionally it concerns the choice and selection of materials and coatings that will meet the requirements of both performance and cost.

The marine environment is a very aggressive one and new materials will be developed to provide improved properties, including corrosion resistance, for the constructions and plant that will be required as more demands are made upon the sea's resources. It is, therefore, necessary to maintain a close contact with such developments. The basic principles of corrosion and its control, however, remain largely unchanged; it is their application to new problems that will result in improvements both of an economic and technical nature.

1.1 THE CORROSION PROCESS

Corrosion is an electrochemical process and is discussed in detail in Chapter 2. To assist in the explanation of control methods, a summary of the basic reactions will be considered. The process can be broken down into anodic and cathodic reactions. If one of these reactions is controlled then the overall rate of corrosion is affected. It is important to appreciate the electrochemical nature of corrosion because the control methods are based on altering or stifling one or other, or both, of the electrode processes. If the corrosion of carbon steel is considered in very simple terms it can be explained as follows. The steel is not homogeneous and at the initiation of corrosion anodic and cathodic sites are formed on the surface of the alloy. In the presence of an electrolyte, small corrosion cells are set up on the surface and at the anodic areas iron goes into solution as ferrous ions, i.e. the steel corrodes. Various reactions can occur at the cathode. The cathodic reaction under ordinary atmospheric or immersed conditions results in the production of hydroxyl ions and the two reactions can be written as follows:

Anode: $2Fe \rightarrow 2Fe^{2+} + 4e^-$

Cathode: $O_2 + 2H_2O + 4e^- \rightarrow 4OH^-$

The two products Fe^{2+} and OH^- react together to form ferrous hydroxide which eventually is oxidised further to rust:

$$Fe^{2+} + 2OH^- \rightarrow Fe(OH)_2 \rightarrow FeOOH \quad (rust)$$

Either the anodic or cathodic process can be controlled. If rust is formed at the steel surface, it will stifle the process so, despite the availability of moisture and oxygen, the corrosion rate may be reduced. In practice, on carbon steels, the anodic process at one place is stifled and corrosion begins at new sites—hence the general nature of the corrosion. With other alloys, e.g. stainless steel, the anodic reaction may continue for some time at the original site and result in localised pitting. However, under most circumstances, there are many fewer anodic sites on stainless steel than on carbon steel, so the overall corrosion is much less.

Most corrosion is of the general type and arises from the rusting of steel, the most widely used constructional alloy. The general methods for controlling it are considered below and, of these, coating the steelwork is the most commonly used. To achieve reasonable standards of protection by coatings does not generally require a very deep knowledge of corrosion. However, particularly in marine environments, an understanding of corrosion processes does assist engineers to appreciate the importance of matters such as surface preparation of steelwork and the choice of suitable coatings.

When considering other forms of attack such as pitting and bimetallic corrosion, an appreciation of corrosion processes is important. Many simple general corrosion concepts may turn out to be either incorrect in specific situations or not quite as simple as had been anticipated. For example, it is generally considered that reinforcements in concrete do not corrode because concrete provides an alkaline environment, which is non-corrosive to steel. This is generally true, but if chloride ions—in plentiful supply in marine conditions—reach the steel, they can in fact affect its passivity, causing attack.

Again although in theory two metals joined together might be expected to lead to serious bimetallic corrosion of one of them, this may not necessarily happen in practice because of various effects such as the polarisation of the anodic and cathodic sites, the nature of the environment or the relative areas of the two metals.

1.2 CORROSION CONTROL

As corrosion results from a reaction between an alloy and its environment, the basic control methods are concerned with treating the environment or selecting suitable alloys. The methods can be grouped into four main categories, irrespective of which of the two factors—environment or material—are of major importance.

(i) Use of coatings.
(ii) Selection of materials that will resist a particular environment.
(iii) Cathodic protection.
(iv) Control or treatment of the environment.

A fifth method really covers the proper application of the other methods in practice but may be considered as a method of control, i.e. design.

If corrosion control is considered to be more than a technical approach to the problem, then other matters related to the maintenance of the structural integrity or operational efficiency must be considered. This would include requirements for access to carry out maintenance and the overall design in relation to the successful application of the control procedures.

Often more than one method of control may be employed and this may be advantageous from the economic standpoint.

1.2.1 Coatings

Coatings used to protect carbon steel in marine environments are considered in detail in other parts of the book. A few fundamental points concerning them are, however, worth considering. Coatings are the most common method by far, in terms of steel tonnage, for controlling corrosion. The basic concept is not usually electrochemical in nature, it is simply to insulate the alloy from the corrosive environment. In practice, however, many problems arise and often insufficient attention is paid to them because of the apparently simple nature of coating systems. A range of different types of coating is available, the most common of which are paints although metal coatings, plastics, waxes, greases and other materials are also used.

Organic coatings are basically of two types: those that are applied manually, e.g. by spraying or brushing, and those that are applied in works, e.g. plastisols. All organic coatings are permeable to moisture and oxygen to a greater or lesser extent and to overcome this, there has been a growing tendency to specify thicker coatings for marine situations. This has been made possible by the nature of coatings such as epoxies and urethanes which harden or cure chemically. Such coatings are basically barrier coatings to the environment.

Most paint films are not impermeable to water or oxygen. This ability of oxygen and moisture to diffuse through a paint film indicates that the protective action arises not through control of diffusion but rather to the high ionic resistance of paint films, which impede the migration of ions so reducing the corrosion current to a low value (resistance control).

Apart from metallic pigments in organic (or inorganic) binders, metals are also used as coatings. They are basically insulating or isolating barriers to the environment. Furthermore, being metals, they corrode—albeit at a lower rate than steel.

Although sprayed metal coatings are porous, even they to a considerable extent act as barrier coatings. The coatings widely used on steel, i.e. zinc, aluminium and cadmium, have potentials more negative than that of steel, i.e. they are anodic to steel so at discontinuities or damaged areas they 'protect' the steel. This is an additional advantage but they are primarily used because they corrode at a lower rate than does uncoated carbon steel. Other coating metals such as copper, nickel and chromium have corrosion potentials that are more positive than that of steel. This means that at discontinuities, the steel will corrode at an accelerated rate, so these cathodic coatings should be of a continuous nature with a minimum of defects. In practice, however, in many environments corrosion products tend to block small pores in the coating.

Probably the most resistant coating in marine environments is vitreous enamel. The fact that such coatings are virtually never used for structures illustrates the essential elements in the choice of coatings. Apart from their corrosion resistance, they must be capable of easy application to a variety of sizes and shapes of sections and must be reasonably cheap, properties not possessed by vitreous enamel.

1.2.2 Selection of materials

Generally, the approach to corrosion control is to consider carbon steel in conjunction with other control methods such as inhibitors, cathodic protection and coatings, and only to choose more resistant alloys where carbon steel would be uneconomic. Factors other than corrosion resistance may also play a role, e.g. forming, shaping, casting etc. Carbon steel is the cheapest of all constructional alloys. However, for a limited number of small components or for certain precision parts, material costs may be a minor consideration and alloys other than steel might be used. For large components and structural members, material costs are an important element in the overall economics and generally more resistant alloys would be used only in special circumstances.

The use of corrosion resistant alloys can be considered as an increase in anodic control. This arises from either the formation on the surface of an alloy of protective films or, possibly, the sluggishness of the dissolution reaction. Metals that form these protective films include Cr, Al, Ti and Ni and alloys such as stainless steels, 'Monel' and 'Inconel' containing a high proportion of such metals as alloying elements also form such films.

Sometimes thicker films which—unlike those discussed above—are visible may be formed, e.g. the patina of basic copper sulphate and basic copper salts formed on copper.

1.2.3 Cathodic protection

When different metals and alloys are joined together in an electrolyte, one of them generally corrodes at a higher rate while the other is protected. Steel generally corrodes at a greater rate under these circumstances but when joined to certain metals, notably aluminium, magnesium and zinc, steel is protected, and the coupled metal corrodes at an increased rate. In practice, a large steel structure can be protected by attaching blocks of such metals to it. The steel and the anode materials must, of course, be immersed in an electrolyte of suitable conductivity. Similar results can be obtained by impressing a current so that the steel potential is less than 0.56 V compared with a standard hydrogen electrode.

1.2.4 Control of the environment

This has only a limited, albeit important, application in marine environments. If the environment can be altered to make it less corrosive, then clearly this will reduce the corrosion of alloyrs exposed to it. This method is generally only applicable to enclosed atmospheric spaces and to closed aqueous systems.

This form of control may also be used in some open aqueous systems where the treatment is considered to be more economical than other methods of control. Apart from its use in structures, plant and buildings, the method is widely used in packaging to control corrosion of components etc, in transit.

1.2.5 Design

From the corrosion standpoint, design can be considered in relation to the proper planning and execution of a project. This covers the selection of materials and choice of suitable control measures within the overall project design.

Undoubted.y, the proper planning and design of corrosion control methods will provide a more effective solution to the problem than the 'ad

Table 1.1 **Summary of control methods in marine situations**

Control method	Immersed open	Immersed 'closed'	Atmospheric open	Atmospheric enclosed
Coatings	●	●	●	●
Inhibitors	○	●	×	●
Desiccants	×	×	×	●
Conditioning the environment	○	●	×	●
Cathodic protection	●	●	×	×
pH control	×	●	×	×
Resistant alloys	●	●	●	●
Design	●	●	●	●

●, Can be used; ○, technically feasible but not economic; ×, not used.

hoc' approach so often used. However, design can also be considered as the detailed work carried out by draughtsmen to produce working drawings to be used for the fabrication and construction of plant, buildings and structures. Most of this design work arises from the requirement to produce structurally stable and aesthetically acceptable constructions. However, during this process of producing detailed drawings, there is an opportunity to consider the influence of the design on future corrosion performance.

Obvious examples of design details that can be considered at the 'drawing board' stage include features that will trap water, crevices, siting of drainage holes and access for future maintenance. Attention to such features will improve the performance of both alloys and coatings.

1.2.6 Testing and monitoring

Although testing and monitoring cannot strictly be considered as methods of control, they are essential parts of the whole process of controlling corrosion. A great deal of the information and data on which the selection of materials is based arises from testing. It is, therefore, essential that the tests are based on sound principles and that the limitations of data emanating from them are clearly understood.

In many situations tests are devised and carried out to obtain trends in performance of coatings and alloys. Often an order of merit of coatings may be useful but the exact 'life' of a coating exposed under a particular set of conditions at a single test site may be of little value as a design criterion.

Monitoring is of growing importance. Apart from providing information on the operation and performance of plant or structures, it is also a valuable method of obtaining data under service conditions.

A summary of control methods in marine situations is given in Table 1.1.

2 Principles of corrosion

2.1 INTRODUCTION

Engineers and 'materials specialists' do not necessarily require a deep knowledge of corrosion mechanisms and processes to satisfactorily specify alloys and coatings for marine situations. However, a broad appreciation of the principles involved will be useful. Many of the corrosion problems that occur with structures, equipment and plant arise from localised geometrical and/or environmental situations. It is, therefore, important to be able to appreciate where such situations are likely to occur and this is assisted by a knowledge of the mechanisms of the processes involved. Many of the situations that can cause localised corrosion, e.g. crevices and galvanic couples, are considered in other chapters, but generally these do not include detailed explanations of the processes themselves. This policy has been adopted partly to avoid repetition and partly because, in many cases, solutions to problems have been established and can be adopted without necessarily knowing the reasons why they are used. It is not possible, however, in a book such as this to cover all situations but an understanding of the mechanism of corrosion and the process involved will often assist in the solution of such problems. In this chapter, some areas are dealt with in more detail than others; within the confines of available space it is clearly not possible to provide a full treatment of the subject. For those requiring further information, a list of references is given at the end of the chapter.

2.2 WHY METALS CORRODE

Corrosion may be defined as the chemical reaction of a metal with a non-metal (or non-metals) in the surrounding environment, with the formation of compounds which are referred to as *corrosion products*. Since metals are used for engineering constructions because of their unique

† This chapter is based on the many papers written by Dr Lionel Shreir, OBE. The author acknowledges with thanks Dr Shreir's permission to use such material and for his helpful comments.

8

mechanical properties (strength and hardness combined with ductility) this conversion of the metal into a powdery, non-adherent, friable compound, if allowed to proceed, will result in the deterioration of the metallic construction or component. The degree to which this occurs will depend on the *rate* of the corrosion reaction, which determines the *extent* of conversion of the metal into corrosion products after a given period of time.

It is not surprising that most metals corrode, and indeed it is remarkable that so many of them have such a high stability in very aggressive environments such as the strong acids. With the exception of silver, gold and the platinum family, and also to some extent copper and mercury, metals are found in nature combined with non-metals as minerals (oxides, sulphides, silicates, carbonates, etc). If the metals are to be extracted, these minerals need to be subjected to a chemical reduction process, using an element with greater affinity for the non-metal. Such a process is illustrated in *Figure 2.1* which represents the energy changes

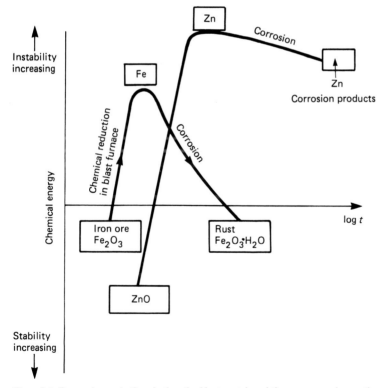

Figure 2.1 Energy changes in the reduction of oxides to metals and the reverse corrosion reaction

that take place with time t when the mineral hematite, Fe_2O_3, is reduced to metallic iron by carbon in the form of coke according to the chemical reaction

$$2Fe_2O_3 + 3C \rightarrow 4Fe + 3CO_2 \qquad (2.1)$$

and zinc oxide is reduced to Zn metal

$$2ZnO + C \rightarrow 2Zn \text{ (vapour)} + CO_2 \qquad (2.2)$$

Clearly, metallic zinc is *energetically* less stable than metallic iron, and when exposed to the atmosphere will tend to revert back to a complex mixture of zinc hydroxide, basic zinc carbonate, etc. However, whereas rust on steel is non-adherent and non-protective and allows access of water and oxygen to the surface of the metal, the corrosion product on zinc is more protective and although zinc will continue to corrode when exposed to the atmosphere it does so at a much slower rate than iron. It is for this reason that mild steel is protected in many applications by a coating of zinc, although the latter is energetically more reactive than the former.

Corrosion may take a variety of forms that range from fairly uniform wastage resulting in general loss of thickness of cross-section, to highly localised attack resulting in pitting and perforation or in cracking and fracture, the major part of the surface of the metal remaining unaffected.

Unlike the mechanical and physical properties of metals and alloys, the corrosion characteristics are not an intrinsic property and the rate and nature of the corrosion will depend upon the environment to which the alloy is exposed. Some metals and alloys are considered to be more corrosion resistant than others. For example, stainless steel is generally superior to carbon steel but there are situations where this may not be so and in some acids carbon steel corrodes at a lower rate than does stainless steel. Furthermore, under certain environmental conditions, some types of stainless steel may crack whereas carbon steels, although corroding at a higher rate, maintain their structural integrity for long periods.

The environment is an essential feature in determining the corrosion behaviour of alloys. Sometimes slight changes can have a significant effect on corrosion performance of an alloy. The chemical composition of the environment is clearly important but other factors are often of equal or greater importance. These include velocity, temperature and pressure and all can influence the local or microclimate at the alloy surface. Velocity, for example, can affect corrosion in a number of ways. It may remove protective corrosion products, alter the local environment by increasing the availability of oxygen or at high velocities may cause impingement attack—a form of pitting which can lead to the rapid perforation of thin sections of metal such as are used for heat-exchanger tubes.

2.3 RATE AND FORM OF CORROSION

Mechanical properties are expressed in terms of environment-independent constants, such as ultimate tensile stress, yield stress, hardness, elongation, etc. This is obviously not possible with corrosion, and the nearest approach to quantifying corrosion is to specify the corrosion rate *under the environmental conditions prevailing*. Since the surface area of the metal must be taken into account, the rate must be expressed as a weight loss per unit area per unit time and typical units are $mg\,dm^{-2}\,d^{-1}$ (abbreviated to mdd) and $gm^{-2}\,d^{-1}$ (gmd). Corrosion may also be expressed as a depth of penetration, which can be obtained by dividing the above units by the metal's density, expressed as mmy^{-1} or $\mu m/year$, or by the older more widely used units iny^{-1} (ipy) or mil/y ($1\,mil = 10^{-3}\,in$). These units are of value when corrosion is fairly uniform and, if it is assumed that the rate remains constant, they provide a means of predicting the *extent* of corrosion (rate × period of time, in appropriate units) after a predetermined period of time. In this way a 'corrosion allowance' can be made to the thickness of the metal. The position is quite different when corrosion is highly localised, since the location of pits and fine cracks is seldom predictable and the damage caused is usually far more serious than general attack. Thus perforation of heat-exchanger tubes (thin sections) may lead to the shut-down of an entire electricity generating station.

2.4 ELECTROCHEMICAL MECHANISM OF CORROSION

Corrosion that occurs in aqueous solutions such as sea water at ordinary temperatures is electrochemical in nature and is similar to that occurring in an electrochemical cell. The mechanism at high temperatures, i.e. oxidation, is different and will not be considered in detail here. Although the explanations given in this chapter can readily be related to situations where alloys are fully immersed in sea water, the mechanisms are basically the same for atmospheric conditions where moisture, often containing salts such as sodium chloride, is present on the surface. In simple terms, an electrochemical cell is as depicted in *Figure 2.2*. It basically consists of two electrodes immersed in an electrolyte, i.e. a liquid that conducts electricity, and joined by an external conductor, e.g. metallic wire. The anode is the electrode from which positive electric

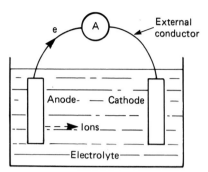

Figure 2.2 Simple electrochemical cell

current flows to the solution or where electrons flow through the external circuit in the reverse direction; the anode reaction is always one of oxidation involving the electrode itself or, if this is inert, a species in the solution. Generally, corrosion occurs at the anode. Conversely, the cathode is the electrode which receives a positive current from the solution or where electrons flow in the reverse direction. During this process the electrons liberated by the oxidation reaction at the anode are transferred through the external circuit to the cathode where they are accepted by a species in solution (an electron acceptor), which is simultaneously reduced to a lower state of valence.

The thermodynamic and kinetic approaches to corrosion are best understood in terms of the charge-transfer processes and electrode reactions that occur in an electrochemical cell having well-defined electrodes, and the well-known *Daniell cell* has been selected for this purpose. However, although corrosion occurs by a similar mechanism the electrodes are seldom so well defined as in the Daniell cell and other electrochemical cells, and in the case of uniform corrosion they cannot be physically identified, although theoretical consideration indicates that they must exist.

It is possible by means of the Daniell cell to illustrate thermodynamic concepts such as equilibrium, equilibrium e.m.f. and potentials, enthalpy and free enthalpy which form the basis of the potential-pH equilibrium diagrams that were formulated originally by Professor M Pourbaix in 1940. It should be noted that thermodynamics provides information on the nature of the species at equilibrium, and in the case of ions in solution, and gases, their concentrations and pressures respectively. It provides no information on the mechanism of the process nor the rate at which equilibrium is achieved, which are topics that are dealt with by electrode kinetics.

The Daniell cell (see *Figure 2.3*) consists of an electrode of Zn immersed in a $ZnSO_4$ solution and an electrode of Cu immersed in a $CuSO_4$

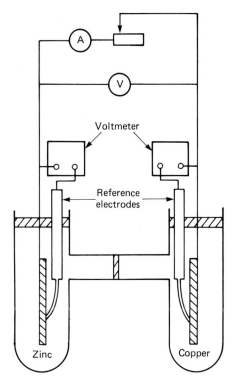

Voltmeter

Reference electrodes

Zinc

Copper

Figure 2.3 Daniell cell

solution; these electrodes may be regarded as the two half-cells of the complete cell

$$Zn \mid ZnSO_4(aq) \parallel CuSO_4(aq) \mid Cu \qquad (2.3)$$

in which $\parallel$ represents the liquid junction between the two different solutions. The cell reaction written in the direction in which it proceeds spontaneously is

$$CuSO_4 + Zn \rightarrow ZnSO_4 + Cu \qquad (2.4a)$$

but since the compounds are dissociated into ions in solution the SO_4^{2-} anion can be omitted and equation (2.4a) becomes

$$Cu^{2+}(aq) + Zn \rightarrow Zn^{2+}(aq) + Cu \qquad (2.4b)$$

in which the arrow indicates that the reaction proceeds spontaneously from left to right. In order to show the ions are hydrated the abbreviation of aqua (aq) has been inserted.

The fact that the reaction proceeds spontaneously in the direction shown can be demonstrated by the deposition of metallic copper on zinc (or on a piece of iron such as the steel blade of a penknife) when it is immersed in a solution of $CuSO_4$. The reverse reaction in which Zn is deposited from Zn^{2+}(aq) and Cu metal converted to Cu^{2+}(aq) cannot occur spontaneously, but requires the supply of energy from an external source of e.m.f.

2.5 SPONTANEITY OF REACTION

It is a thermodynamic principle that for a reaction to proceed spontaneously in a given direction the chemical energy of the reactants must be greater than that of the products, i.e. energy must be released which may appear solely as heat energy or as a combination of work energy and heat energy. For the purpose of the present discussion, which is restricted to electron transfer, work energy will be confined to electrical energy and no consideration will be given to work energy produced by changes in volume or pressure.

The absolute *heat content* or *enthalpy* (formerly referred to as *internal energy*) cannot be determined, and to overcome this difficulty the enthalpies of the elements in the form in which they normally exist at 25 °C are taken as reference states and given an enthalpy of zero. In this way it is possible to evaluate the change in enthalpy (units kJ) when an element forms compounds or ions in solution. Since the enthalpy will depend not only on the nature of the species but also upon its mass, this too must be defined, and the 'standard state' is taken as 1 mole of the species (see Appendix 1). This is denoted by the superscript $\ominus$, and $\Delta H_i^{\ominus}$ is the standard enthalpy of an ion i formed from an element (or elements) at 25 °C. It can be shown by summing algebraically $\Delta H_{Cu^{2+}(aq)}^{\ominus}$ (64.4 kJ) and $\Delta H_{Zn^{2+}(aq)}^{\ominus}$ (-152.5 kJ) that $\Delta H_{reaction}^{\ominus} = -216.9$ kJ, which means that the reaction proceeds spontaneously. This is evaluated experimentally by carrying out the reaction in a calorimeter and measuring the increase in temperature of the solution produced by the loss of heat energy during the reaction. However, a more useful criterion of the spontaneity of a reaction is the standard *free enthalpy* $\Delta G^{\ominus}$ which may be evaluated by measuring the maximum e.m.f. of the cell in which the reaction proceeds. This will be considered subsequently.

2.6 RATES OF ANODIC AND CATHODIC REACTIONS

The Daniell cell reaction (equation 2.4) consists of two half-reactions

$$Zn \rightarrow Zn^{2+}(aq) + 2e^- \tag{2.5}$$

$$Cu^{2+}(aq) + 2e^- \rightarrow Cu \tag{2.6}$$

and since these could occur spontaneously at the surface of metallic Zn, if this was immersed in $CuSO_4$ solution, equation (2.5) may be regarded as the anodic oxidation or corrosion of Zn and equation (2.6) the cathodic reduction of Cu^{2+} (aq) ions. It follows from these definitions that equation (2.4) occurs by an electrochemical cell mechanism, even though the anodic and cathodic sites of the zinc cannot be identified, as is the case when the reaction is carried out in a beaker or in a calorimeter. However, more information can be obtained on the detailed mechanism of the reaction if it is carried out in an electrochemical cell, which is a device by means of which the chemical energy of the reactants is converted into electrical energy.

A feature of this cell and circuit is that the rate of the reaction (the current I) can be controlled at a given value by means of a variable resistance (*Figure 2.3*) in the external circuit, and at the same time the e.m.f. of the cell and current can be measured with a high impedance voltmeter and an ammeter, respectively. By means of this circuit it is possible to determine the relationship between the e.m.f. of the cell and the electrochemical reaction rate k_e which is related to the current I.

The current I is the rate of charge transfer, and a rate of 1 A is equivalent to 1 Cs^{-1}. Since the charge on the electron is 1.6020×10^{-19} C, and since 1 mole (the molar mass in kg) of an element contains 6.0235×10^{23} atoms (Avogadro's number), the cathodic reduction of a univalent metal ion M^+ (aq) to metal, or the reverse anodic oxidation of M to M^+ (aq), will require

$$1.6020 \times 10^{-19} \times 6.0235 \times 10^{23} = 96\ 494 \text{ C (mol of electrons)}^{-1}$$

$$\sim 96\ 500 \text{ C mol}^{-1}$$

The above statement summarises Faraday's law of electrolysis and the constant 96 500 C mol^{-1} is the faraday, F; if the number of electrons required in one act of a reduction or oxidation reaction is n then nF C mol^{-1} will be required.

Thus the rate of charge transfer of an electrochemical reaction is given by

$$k_e = \frac{I}{nF} \text{ mol s}^{-1} \qquad (2.7)$$

or

$$k_e = \frac{IM}{nF} \text{ kg s}^{-1} \qquad (2.8)$$

where I is the current in amperes and M is the molar mass (kg mol^{-1}) of the species involved.

This can be illustrated by assuming that the current produced by the

Daniell cell is 0.193 A. Since two electrons are required for each act of the reaction, the rate of cathodic reduction of $Cu^{2+}(aq)$ to Cu and anodic oxidation of Zn to $Zn^{2+}(aq)$ expressed as mol s^{-1} must be equal and

$$k_c = \frac{0.193}{2 \times 96\ 500} = 1 \times 10^{-6} \text{ mol s}^{-1}$$

However, since the molar masses of Cu and Zn are 0.063 54 kg and 0.065 38 kg respectively, the rates expressed as a mass per unit time will differ and will be $0.063\ 54 \times 10^{-6}$ kg s^{-1} and $0.065\ 38 \times 10^{-6}$ kg s^{-1}, respectively, i.e. 0.005 49 kg/day and 0.005 65 kg/day.

It must be remembered that electrode reactions are heterogeneous and involve a solid electrode whose surface area is obviously of the greatest importance; whereas, for example, a current of 0.193 A acting on a small Zn electrode of 1 mm thickness would result in its rapid destruction, it would have little effect if the surface area was thousands of square metres.

Thus the intensity factor $i = I/S$ where i is the current density and S the geometrical surface area in appropriate units is of far greater significance than the extensive factor I, which takes no account of surface area of the metal undergoing anodic dissolution. The rate will now become

$$k_c = \frac{iM}{nF} \text{ kg m}^{-2} \text{ s}^{-1} \tag{2.9}$$

where i is the current density in A m^{-2}.

The converse of the above consideration is also applicable, and the weight gain of a cathode or the weight loss of an anode may be used to calculate the rate of the reaction in terms of I or i.

2.7 CORROSION PRODUCTS

The corrosion product is a most important factor in controlling the rate of corrosion. It has already, for example, been pointed out that while metallic zinc is energetically less stable than metallic iron, it is more resistant to corrosion in a number of environments; this is explained by the fact that the corrosion products of Zn are more protective than those of iron, and this phenomenon applies to an even greater extent in metals such as Al, Ti and Ta, which, although thermodynamically highly unstable, form very protective films of metal oxide.

The role of the corrosion product in acting as a protective barrier can be illustrated by comparing the high-temperature oxidation of steel with its corrosion in an oxygen-containing aqueous solution such as a natural water. It is well known that heating bright mild steel in air results in the formation of 'temper colours' that are initially pale straw and ultimately deep blue. This is due to the formation of a *transparent* film of oxide and to the phenomenon of 'interference' in which reflection of light from the surface of the metal and from the surface of the oxide results in the extinction of light of certain wavelengths, leaving a colour representing the remaining part of the spectrum.

The relevant equations for this 'dry' and 'wet' corrosion may be expressed as follows:

Dry corrosion: $$2Fe + 1\tfrac{1}{2}O_2 \rightarrow Fe_2O_3 \qquad (2.10)$$
(oxidation)   (ferric ocide)

Wet corrosion: $$2Fe + 1\tfrac{1}{2}O_2 + H_2O \rightarrow Fe_2O_3 . H_2O \qquad (2.11)$$
(hydrated ferric oxide or rust)

Although these two equations appear to differ only in respect to the absence or presence of water it will be seen that this leads to a very fundamental difference in mechanism.

In the case of oxidation it is apparent that the position of formation of Fe_2O_3 and the position at which the Fe reacts with oxygen must be identical (see *Figure 2.4*) and that if a continuous film of Fe_2O_3 is formed further reaction can occur only by the transport of oxygen (O^{2-}) or ferrous (Fe^{2+}) ions through the oxide film. This oxide will be *protective*, and the rate of oxidation will vary inversely with the thickness of the oxide, i.e. the rate law of oxidation will be parabolic.

On the other hand the corrosion of Fe in water, which results in the formation of *non-protective* rust, can be considered in relation to the reactions occurring at the anodic and cathodic areas

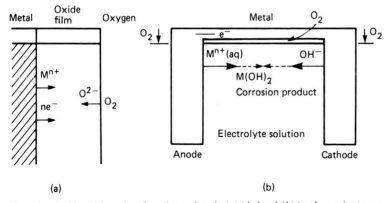

Figure 2.4 Position of formation of reaction products in (a) 'dry' and (b) 'wet' corrosion process

Anode: $2Fe \rightarrow 2Fe^{2+}(aq) + 4e^-$ (2.12)

(hydrated ferrous cation)

Cathode: $O_2 + 2H_2O + 4e^- \rightarrow 4OH^-$ (2.13)

(hydroxyl anion)

Although equations (2.10) and (2.11) represent the oxidation (gain of oxygen) of iron to iron oxide it must be appreciated that similar considerations apply to an electrically neutral atom becoming positively charged by loss of electrons as in equation (2.12). Conversely, the process of removal of oxygen or gain of electrons is a process of *reduction*. *Oxidation of one species cannot occur without the simultaneous reduction of another*, and equation (2.13) shows how dissolved oxygen accepts the electrons (oxygen is referred to as an *electron acceptor*) liberated by the oxidation of Fe, and becomes reduced to negatively charged hydroxyl ions. Equation (2.12) and equation (2.13) represent half-reactions, since if they are added together the electrons are eliminated giving the overall reaction:

$$2Fe + O_2 + 2H_2O \rightarrow 2Fe^{2+} + 4OH^-$$ (2.14)

The two sides of such an equation representing a half or complete reaction must balance one another in terms of both the nature and number of the atoms and the electric charge, e.g. equation (2.14) has the same number of Fe, O and H atoms on each side, although in different states of combination, and the four positive charges on the r.h.s. $(2 \times 2+)$ are balanced by the four negative charges due to the OH^- ions giving the same charge as that on the l.h.s. It can also be seen from equation (2.14) that the reaction products are ferrous and hydroxyl ions, which are unstable since, above a certain very small concentration, they will combine to form ferrous hydroxide, $Fe(OH)_2$, a very sparingly soluble solid. This is due to the equilibrium (as shown by the reverse arrows $\rightleftharpoons$) between ferrous and hydroxyl ions and ferrous hydroxide, which can be represented by

$$Fe(OH)_2 \rightleftharpoons Fe^{2+} + 2OH^-$$ (2.15)

and as long as some sparingly soluble $Fe(OH)_2$ is present in the system the equilibrium will be unaffected so that it can be eliminated to give

$$c_{Fe^{2+}} \times c_{OH^-}^2 = S_p$$ (2.15a)

where S_p is an equilibrium constant known as the *solubility product*. [It should be noted that the concentration of hydroxyl ions must be raised to the power of 2, since there are two present in the equilibrium $(c_{OH^-} \times c_{OH^-})$.] S_p is a constant at constant temperature and if the concentration of one of the species is increased the other must decrease, and this must result in the precipitation of $Fe(OH)_2$, i.e. the higher the pH the lower the solubility of $Fe(OH)_2$.

Thus equation (2.14) represents a situation that is thermodynamically unstable, and since $Fe(OH)_2$ is the stable phase in equilibrium with very small concentrations of $Fe^{2+}(aq)$ and OH^- ions in solution it is more appropriate to write it in the form

$$2Fe + O_2 + 2H_2O \rightarrow 2Fe(OH)_2 \qquad (2.15b)$$

However, $Fe(OH)_2$ is a powerful reducing agent and is readily oxidised by dissolved oxygen to form stable hydrated ferric oxide $Fe_2O_3 . H_2O$ or ferric hydroxide $Fe(OH)_3$

$$4Fe(OH)_2 + O_2 \rightarrow 2H_2O + 2Fe_2O_3 . H_2O \qquad (2.16)$$

Thus the *overall* reaction for the conversion of metallic iron to rust which as shown here proceeds by a series of intermediate steps may be written as

$$4Fe + 3O_2 + 2H_2O \rightarrow 2Fe_2O_3 . H_2O \qquad (2.17)$$
$$\text{(rust)}$$

The anodic oxidation of iron to Fe^{2+}, the reduction of dissolved O_2 to OH^- and the formation of $Fe(OH)_2$ may take place at different sites on the surface of the metal, and be separated by distances ranging from atomic dimensions to hundreds of metres. Furthermore, the precipitation of $Fe(OH)_2$ does not necessarily take place at the surface of the metal and may not, therefore, influence the rate of the corrosion reaction. It is for this reason that dry oxidation often results in a protective oxide whereas wet corrosion usually results in non-protective rust.

The above considerations indicate an important method of corrosion control, that of ensuring the metal readily forms a protective film; this is referred to as *passivation* and metals such as aluminium, chromium, nickel, titanium and tantalum spontaneously become passive in a variety of environments. However, mild steel is passive only under certain conditions, e.g. in alkaline solutions of a pH value > 9.5 and in neutral solutions (pH 7) containing passivating inhibitors such as sodium chromate, sodium nitrite, etc.

Another method is to alloy the steel with metals such as Cr and Ni that have a pronounced tendency to become passive, and this gives rise to the important series of alloys known as the 'stainless steels'.

2.8 EQUILIBRIUM AND EQUILIBRIUM POTENTIALS

The term 'potential' is widely used when considering corrosion reactions, particularly in relation to the electrode potentials of metals and alloys in

sea water and the protection of steelwork by cathodic protection. There is
also a theoretical EMF Series of Metals or more correctly Standard
Equilibrium Potentials. Whereas metal/solution potentials used to
ensure the effective application of cathodic protection or to predict the
probable effect of coupling different alloys together in sea water are
determined experimentally, the Standard Equilibrium Potential in the
EMF Series of Metals has a theoretical basis and is not of the same
practical value to engineers. The equilibrium potentials for a series of
metals (not alloys) is given in *Table 2.1*, based on equilibria of the type

$$M^{n+}(aq) + ne^- \rightleftharpoons M$$

arranged in order of sign and magnitude.

**Table 2.1 EMF series of metals or standard equilibrium
potentials** $E^{\ominus}_{M^{n+}(aq)/M}$ **of** $M^{n+}(aq) + ne^- \rightleftharpoons M$ **equilibria
(SHE)**

Metal	Equilibrium	$E^{\ominus}$ (V)
Gold	$Au^{3+}(aq) + 3e^- \rightleftharpoons Au$	1.5
Silver	$Ag^+(aq) + e^- \rightleftharpoons Ag$	0.799
Mercury	$Hg_2^{2+}(aq) + 2e^- \rightleftharpoons 2Hg$	0.789
Copper	$Cu^{2+}(aq) + 2e^- \rightleftharpoons Cu$	0.337
(Hydrogen)	$H^+(aq) + e^- \rightleftharpoons \frac{1}{2}H_2$	0.000
Lead	$Pb^{2+}(aq) + 2e^- \rightleftharpoons Pb$	−0.126
Tin	$Sn^{2+}(aq) + 2e^- \rightleftharpoons Sn$	−0.136
Nickel	$Ni^{2+}(aq) + 2e^- \rightleftharpoons Ni$	−0.250
Cadmium	$Cd^{2+}(aq) + 2e^- \rightleftharpoons Cd$	−0.403
Iron	$Fe^{2+}(aq) + 2e^- \rightleftharpoons Fe$	−0.440
Chromium	$Cr^{3+}(aq) + 3e^- \rightleftharpoons Cr$	−0.74
Zinc	$Zn^{2+}(aq) + 2e^- \rightleftharpoons Zn$	−0.763
Titanium	$Ti^{2+}(aq) + 2e^- \rightleftharpoons Ti$	−1.63
Aluminium	$Al^{3+}(aq) + 3e^- \rightleftharpoons Al$	−1.66
Magnesium	$Mg^{2+}(aq) + 2e^- \rightleftharpoons Mg$	−2.37

This table gives the misleading impression that any of the metals listed
when immersed in a solution of its cations at a concentration equivalent
to unit activity (see Appendix 1 for an explanation of activity) will take
up the standard equilibrium potential given in the table, and that these
potentials have been evaluated in the same way using an electrochemical
method involving the determination of an e.m.f. However, this is not the
case and it is possible to distinguish between the following different types
of equilibria:

(i) Reversible systems, e.g. $Ag^+(aq)/Ag$, $Cu^{2+}(aq)/Cu$, $Cd^{2+}(aq)/Cd$,
$Sn^{2+}(aq)/Sn$, $Zn^{2+}(aq)/Zn$, etc, where the potential is due solely to
the $M^{n+}(aq) + 2e^- \rightleftharpoons M$ equilibrium.

(ii) Irreversible systems: $Ni^{2+}(aq)/Ni$, $Fe^{2+}(aq)/Fe$, $Co^{2+}(aq)/Co$, Pt^{4+}/Pt, Au^{3+}/Au, etc, where the potential is influenced by other equilibria such as $H_3O^+ + e^- \rightleftharpoons \frac{1}{2}H_2 + H_2O$.

(iii) Systems in which the metal is coated with an oxide film: $Cr^{3+}(aq)/Cr$, $Al^{3+}(aq)/Al$, $Ti^{4+}(aq)/Ti$, etc, where the potential is not characteristic of the M^{n+}/M equilibria.

2.8.1 Reversible and irreversible systems

Before explaining what is meant by reversible and irreversible systems it is necessary to consider what happens when the metal of a reversible system is placed into a solution of its cations, and for this purpose the system $Cu^{2+}(aq)/Cu$ will be taken as being typical. Metallic Cu consists of an ordered rigid lattice of metal atoms surrounded by a cloud of mobile free electrons, whose movement through the lattice cations constitutes the electric current. If it were possible to examine the metal's structure in detail it would be found that all points within the interior would be identical and electrically neutral, but that the distribution of electrical charge at the surface would be different.

A solution of $CuSO_4$ in water will contain hydrated cupric cations, $Cu^{2+}(aq)$, hydrated hydrogen ions H^+H_2O, or H_3O^+, sulphate SO_4^{2-} and hydroxyl OH^- anions and neutral water molecules. In contrast to the atoms in the metal these ions will be mobile and randomly oriented so that within the solution there is no preferentially directed field. However, under the influence of a potential difference between two electrodes placed in the solution electrical charge will be transported by the migration of the cations and anions to the cathode and anode respectively, where the appropriate electrode reactions will occur. The transport of electrons from anode to cathode through the metallic circuit, transport of cations and anions through the solution and the electrode reactions involving the generation and consumption of electrons must occur simultaneously, and the rate of electron transfer will determine the rate of the electrode reactions.

At all interfaces there will be a redistribution of charge and, although this applies to the interface between metallic copper and air, little is known of its nature in comparison with a metal/solution interface. The system $Cu^{2+}(aq)/Cu$ is characterised by the fact that it consists of a metal in two states of valence, i.e. Cu^0 (neutral atoms in the metal lattice) and $Cu^{2+}(aq)$ (metal atoms that have lost two electrons and become hydrated copper cations). This, of course, would not be the case if, for example, Na_2SO_4 had been used in place of $CuSO_4$ and it is a characteristic of a reversible system that it involves an element in two states of valence so that an exchange of electrons results in the conversion of one form to the other. This reaction is not exclusive to metals, provided that an inert electronic conductor is present, and Pt is often used for this purpose because of its stability and its ability to facilitate electron transfer. Examples of equilibria that do not involve a metal as one of the reactants are $2H_3O^+ + 2e^- = H_2 + 2H_2O$, $Fe^{3+}(aq) + e^- = Fe^{2+}(aq)$, etc, and as will be seen the equilibrium between hydrogen ions in solution and molecular hydrogen gas is of particular importance in defining equilibrium potentials.

On immersing the Cu metal into the $CuSO_4$ solution there will be a tendency for an exchange process to occur, i.e. for the metal atoms to escape from the metal lattice and pass into the solution (anodic dissolution) and for hydrated metal ions to lose their water of hydration, escape from the solution and enter the lattice of the metal (cathodic deposition). Initially and momentarily one of these will predominate and this will determine the redistribution of charge at the interface, which could be regarded as analogous to the 'plates' of a capacitor. In the simplest case the interface between the metal and solution could be regarded as a line of excess electrons at the surface of the metal and an equal number of positive charges in the solution in contact with the metal or vice versa (*Figure 2.5*).

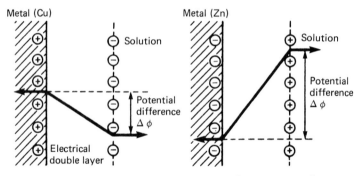

Figure 2.5 Electrochemical double layer at the interface $Cu/Cu^{2+}(aq)$ and $Zn/Zn^{2+}(aq)$ resulting in a potential difference $\Delta\phi$

This separation of charge at the metal/solution interface results in a strong electric field in the space between the charged layers and a potential difference across its extremities. It will be seen that the potential differences are comparatively small (< 1.3 V), but since the distance of separation of the plates is also very small, about 0.2 nm, the field strength will be about 10^9 V m^{-1}, and this high field strength is responsible for the high rate of electron transfer across the interface, the velocity of the reaction, when the potential difference is changed slightly.

The potential difference and its associated field prevent the passage of cations in either direction proceeding very far, and the amounts involved are so minute that they are undetectable. Thus almost immediately after immersing the Cu in the solution there will be a dynamic equilibrium at the interface in which the rates of the cathodic process and anodic process are equal, and this is the reason why the reverse arrows $\rightleftharpoons$ are used to signify that the system is at equilibrium. (Note: the concept of a dynamic equilibrium applies to most chemical processes, including the dissolution of a solid, e.g. sugar, in a liquid, e.g. water. If small amounts of sugar are added successively to a fixed volume of water at constant temperature they will dissolve, but a point will be reached when the water is 'saturated', and no further sugar will be taken into solution. At this point there will be a *dynamic equilibrium*, and the rate of dissolution of the sugar will exactly equal its rate of crystallisation.)

Adopting the convention that when the arrow head points to the right the reaction is cathodic and when to the left it is anodic, the situation at equilibrium may be summarised as follows:

At $\qquad E_{eq,I}\ \overrightarrow{i_I} = \overleftarrow{i_I} = i_{o,I}$ $\qquad\qquad\qquad\qquad\qquad$ (2.18)

where the subscript I indicates a single exchange process I, $E_{eq,I}$ is the equilibrium potential for this exchange and $i_{o,I}$ is the *equilibrium exchange current density*, whose magnitude will determine whether or not the equilibrium potential will be solely dependent on the exchange process I.

In the case of a reversible system i_o for the $M^{n+}(aq)/M$ equilibrium is so large that other exchange processes of lower i_o are insignificant and do not affect the equilibrium potential. However, in the case of irreversible systems the potential taken up by the metal is not solely due to the $M^{n+}(aq)/M$ exchange because of its low i_o, and it may be affected by other exchanges such as those due to H_3O^+/H_2, O_2 (dissolved in solution)/H_2O, redox reactions involving organic impurities, etc. Thus Ni immersed in a solution of Ni^{2+} ions will not take up the equilibrium potential of the $Ni^{2+}(aq)/Ni$ exchange but will take up a *mixed* or *compromise* potential due to both the $Ni^{2+}(aq)/Ni$ exchange and the H_3O^+/H_2 exchange.

Finally, metals such as Al, Cr, Ti, Ta, Nb give potentials that are affected by the thin oxide film present on the surface, which make the potential more electropositive than the value given in the e.m.f. series. Thus Al in a number of aqueous solutions will have a potential of about $-0.5\ V$ (SHE) compared with $E^{\ominus}_{Al^{3+}/Al} = -1.7\ V$, and many other examples could be quoted showing that reversible potentials have little relevance to the actual electrode potentials of metals in practical environments.

2.9 MEASUREMENT OF POTENTIALS OF HALF-CELLS

All devices for measuring e.m.f. or potential difference (p.d.) have two terminals, and this means that it is not possible to measure the absolute p.d. at the metal/solution interface of a single electrode unless it is connected to another to form a two-electrode cell; for this reason a single electrode is referred to as a half-cell.

It is necessary, therefore, to define an equilibrium that can be used as a standard and to give it an arbitrary single electrode potential under specified conditions, and for this purpose the equilibrium between hydrogen ions in solution and gaseous molecular H_2 has been selected

$$H_3O^+ + e^- \rightleftharpoons \tfrac{1}{2}H_2 + H_2O \qquad (2.19)$$

An electron conductor must be present to allow electron transfer and to measure the potential difference, and for this purpose platinised platinum (Pt whose surface has been activated to improve its catalytic ability) is the most appropriate metal, since it is inert in most solutions and has a high value of i_0 for the $H^+/\tfrac{1}{2}H_2$ equilibrium. The magnitude of the equilibrium potential will depend not only on the nature of the exchange process but also on the activities (see Appendix 1) of the species involved in the exchange. For any electrochemical equilibrium of the type

$$aA + bB + ne^- = cC + dD \qquad (2.20)$$

in which A, B are the reactants and C, D the products and the equilibrium is written so that the electrons are on the left-hand side, the relationship between E_{eq} and activities of reactants and products is given by

$$E_{eq} = E^{\ominus} - \frac{RT}{nF} \ln\left(\frac{a_C^c \times a_D^d}{a_A^a \times a_B^b}\right) \qquad (2.21)$$

in which the convention is that the products and reactants in equation (2.20) are placed in the numerator and denominator respectively of the activity quotient shown in the brackets in equation (2.21). Here R is the gas constant $(8.3143\,J\,mol^{-1}\,K^{-1})$, T the temperature in K, F the Faraday and a represents the activity of the ion. Using the identity $\log x = 2.303 \ln x$, and substituting for the constants, equation (2.21) becomes at 25 °C (298.16 K)

$$E_{eq} = E^{\ominus} - \frac{0.059}{n} \log\left(\frac{a_C^c \times a_D^d}{a_A^a \times a_B^b}\right) \qquad (2.22)$$

in which $E^{\ominus}$ is a constant for a given equilibrium.

It is necessary to define the *standard state* of an element, compound, gas and ion in solution. In the case of solid and liquid elements and compounds, this is defined as the pure substance (element or compound) in the form that it normally exists at 25 °C, e.g. solid Cu and Zn, liquid bromine (Br_2); liquid H_2O; Cu_2O, CuO, ZnO, $Zn(OH)_2$, liquid ethyl alcohol (C_2H_5OH), liquid benzene (C_6H_6), etc. An element or compound in its standard state is given an arbitrary activity of unity, and this will apply to the Cu and Zn electrodes irrespective of their mass. However, the activities of the individual metals in an alloy will not be unity and will be related to their molar concentrations. In the case of gases and ions in solution the above definition has to be modified to take account of the fact that the pressure of a gas and the concentration of ions in solution are variables, and the standard state is taken as the gas or ion in solution at unit activity. Any species in its standard state will have unit activity in the logarithmic activity quotient in equation (2.22), and when all the reactants and products in the equilibrium of equation (2.20) are in their standard state the logarithm of the activity quoteint becomes zero, and

$$E = E^{\ominus}$$

in which $E^{\ominus}$ is the Standard Equilibrium Potential.

Equation (2.21) is the well-known *Nernst equation* and provides the basis for the thermodynamic approach to corrosion in aqueous solutions. Applying the Nernst equation to the hydrogen equilibrium given in equation (2.19)

$$E = E^{\ominus}_{H^+/\frac{1}{2}H_2} - \frac{0.059}{1} \log \frac{a^{\frac{1}{2}}_{H_2}}{a_{H_3O^+}} \qquad (2.23)$$

and taking the H_3O^+ ions in solution and H_2 gas both to be at an $a = 1$, the equilibrium potential is given an arbitrary value of 0.00 V at 25 °C, i.e.

$$E = E^{\ominus}_{H^+/\frac{1}{2}H_2} = 0.00 \text{ V at } 25 °C \qquad (2.23a)$$

This potential of the Standard Hydrogen Electrode (SHE) forms the reference standard for defining the potentials of all single electrodes, although for convenience other reference electrodes such as $Cu/CuSO_4$ saturated, $Ag/AgCl$, Cl^-, Hg/Hg_2Cl_2, Cl^-, etc, are more widely used than the hydrogen electrode for actual determinations of single potentials (see Appendix 1). It is, of course, essential when referring to a single potential to specify the reference electrode used for its determination, although it is possible to convert from one scale to another.

Referring now to the e.m.f. series of metals it is evident that the equilibrium potentials given are really the e.m.f. of cells such as

$$Zn/Zn^{2+}(aq)\,|\,H_3O^+, H_2/Pt$$
(anode)   (cathode)

involving equilibria of the type:

$$M + 2H_3O^+ \rightleftharpoons M^{2+}(aq) + H_2 + 2H_2O \qquad (2.24)$$

However, it is necessary to adopt a convention for writing an equation representing an equilibrium, and reference to equation (2.24) shows that it has been written so that the reduced form of the metal (M) and the oxidised form of hydrogen (H_3O^+) are on the left-hand side and the oxidised form of the metal (M^{2+}) and the reduced form of hydrogen (H_2) are on the right-hand side. With this convention $E^{\ominus}_{Cu^{2+}/Cu} = +0.34$ V and $E^{\ominus}_{Zn^{2+}/Zn} = -0.76$ V. There is, of course, no reason why this equation should not have been written in the reverse direction. In which case $E^{\ominus}_{Cu/Cu^{2+}} = -0.34$ V and $E^{\ominus}_{Zn/Zn^{2+}} = +0.76$ V. The formen of these conventions is now internationally accepted.

2.10 ELECTRODE POLARISATION

Preliminary considerations have been confined to a reversible cell at equilibrium in which the net transfer of charge, or rate of reaction, approximates to zero. The short-circuited electrochemical cell that is set up when a metal is placed in a corrosive environment is not so simple, and it is necessary to consider the changes in the e.m.f. of this cell and the potentials of the electrodes when the reaction rate becomes significant. A completely non-polarisable electrode would be characterised by an equilibrium potential that was unaffected by a significant net transfer of charge in either direction. Although this is impossible to achieve fully in practice, reversible electrodes can approximate to such a constancy of potential provided that the current is very small, and it is for this reason that they are used as reference electrodes. However, when a reversible electrode produces current at an appreciable rate its potential becomes displaced from its equilibrium value, and similar considerations will also apply to the steady-state potentials of irreversible electrodes, e.g. $Ni^{2+}(aq)/Ni$ and $Fe^{2+}(aq)/Fe$.

This departure of an electrode from its reversible or its steady state potential is due to electrode polarisation, and may be illustrated by considering the spontaneous operation of the Daniell cell using the circuit shown in *Figure 2.3* to measure the polarised e.m.f. of the cell $E_{p,cell}$, the polarised potentials of the individual electrodes when they become anodic $E_{p,a}$ or cathodic $E_{p,c}$ and the rate of reaction I.

Figure 2.6 shows diagrammatically how, with increase in the rate of

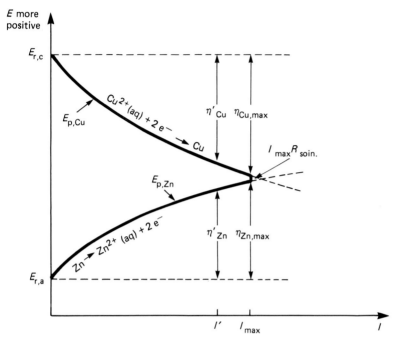

Figure 2.6 Diagrammatic representation of the E–i relationship when the Daniell cell operates spontaneously

reaction, the potential of the Cu cathode becomes more negative and the Zn anode more positive, with a consequent decrease in the e.m.f. of the cell. Polarisation can be summarised as follows:

$$E_{p,c} < E_{r,c} \quad \text{and} \quad \eta_c = E_{p,c} - E_{r,c} < 0 \tag{2.25}$$

$$E_{p,a} > E_{r,a} \quad \text{and} \quad \eta_a = E_{p,a} - E_{r,a} > 0 \tag{2.26}$$

where η is the displacement of the electrode from its equilibrium (or steady-state) potential is called the *overpotential* of the electrode, which provides a measure of the extent of the electrode polarisation. It should be noted that since η is the difference between two potentials it is not necessary to state the reference electrode used for its determination.

It can also be seen that $E_{p,cell} < E_{r,cell}$ or

$$E_{r,cell} = E_{p,cell} - \{(\eta_a + \eta_c) + (iR_e + iR_i)\} \tag{2.27}$$

where R_e is the resistance of the external circuit (electronic resistance) and R_i is the resistance of the electrolyte solution (electrolytic resistance); the term $(\eta_a + \eta_c)$ is sometimes referred to as a 'back e.m.f.', since it opposes the open-circuit e.m.f. of the cell. As the rate of reaction is increased by decreasing R_e the magnitude of the polarisation increases and $E_{p,c} \to E_{p,a}$ until a limit is reached when $R_e \to 0$ and $E_{p,cell} = I_{max} R_i$ where I_{max} is the maximum current that can be produced by the cell. In this connection it should be noted that the electrodes of corrosion cells formed on relatively massive sections of metal may be regarded as being short-circuited, i.e. R_e is negligible.

The magnitude of electrode polarisation will depend on (a) the nature of the electrode reaction (for example, a slight increase in the potential of a Ag electrode immersed in $AgNO_3$ solution—a reversible electrode with a high value of $i_{o,Ag^+/Ag}$—will result in a high rate of dissolution, whereas for Ni immersed in a $NiSO_4$ solution—an irreversible electrode with a low value of $i_{o,Ni^{2+}/Ni}$—only a small current will be produced), (b) on the nature of the anion and impurities present in the solution, and (c) on temperature.

When the electrodes are short-circuited iR_e in equation (2.27) can be neglected, and similar considerations apply to $E_{p,cell}$ in comparison with $E_{r,cell}$ providing the resistance of the solution R_i is small. Rearranging equation (2.27)

$$i = \frac{E_{r,cell} - (\eta_a + \eta_c)}{R_i} \qquad (2.27a)$$

which shows that the magnitude of the rate depends partly on the thermodynamic or reversible parameter $E_{r,cell} = E_{r,Zn} + E_{r,Cu}$, and partly on the magnitude of the overpotentials of the individual electrodes and the resistance of the solution. In view of the thermodynamic instability of most metals, polarisation is the most important natural phenomenon in favour of their preservation, and although metals such as Ti, Ta and Nb are highly unstable thermodynamically, and require considerable energy to reduce them from their oxides, their stability is such that they are unlikely to revert to their oxides during normal atmospheric exposure.

An analogy that might be helpful in appreciating electrode polarisa-tion is to consider a ship moving through the sea. The combustion of the fuel provides the chemical energy which is converted into the mechanical energy required to drive the propeller, and the speed of the ship may be regarded as being analogous to the reaction rate. The rate of progress in a given direction will be opposed by various dissipative factors (analogous to polarisation) such as the wave-making of the ship, the wind, and the frictional resistance of the hull (which will be markedly affected by the degree of fouling), and these will result in a decrease in the speed of the ship for a given consumption of fuel per unit time. It is not possible to fully discuss polarisation here but two particular terms may be encountered: 'Activation Overpotential' (also referred to as activation polarisation) and 'Transport Overpotential' (also referred to as concentration polarisation, concentration overpotential or diffusion overpotential).

Activation overpotential refers to reactions which are controlled by slow steps in the reaction sequence. The relationship between the rate of reaction and the overpotential is given by the Tafel equation

$$\eta_A = a \pm b \ln i$$

where η_A is the activation overpotential and a and b are constants for a particular electrode reaction under specified experimental conditions. The constant b is referred to as the Tafel slope.

Transport overpotential results from the inability of the cathode reactant to be transported from the bulk solution at a sufficiently high rate to avoid its concentration becoming depleted at the electrode surface by the cathodic reaction. Transport of a species to the surface of the cathode will be by diffusion and convection; also by migration if the species is a cation.

2.11 POTENTIAL-pH EQUILIBRIUM DIAGRAMS

It is possible to summarise all the different equilibria between metal, metal cations and anions and solid oxides in a diagram having $E_{eq,H}$ (subscript H indicates the potential is on the Hydrogen scale) and pH as co-ordinates. As Professor M Pourbaix was the first to construct such diagrams, they are often referred to as Pourbaix diagrams. They are constructed from calculations based on the Nernst equation and solubility data for various of the metal compounds. Pourbaix has adopted the convention that an equilibrium activity of metal ions $> 10^6$ g ion/l represents corrosion and on this basis a number of zones can be distinguished in diagrams of this type. *Figure 2.7* shows a E_H-pH diagram for the Fe-H_2O system with the following zones—corrosion, passivation and immunity, in the case of passivation and immunity the activity of metal ions is $< 10^{-6}$. In the zone of passivity a solid oxide or hydroxide of the metal is stable, and it is assumed that it forms a protective film on the metal insulating it from the aqueous environment. Metals such as Al, Cr, Ti, etc, and alloys such as stainless steel rely on passivity for their resistance to corrosion. The small zone of corrosion at high pH and low potential is due to the formation of FeO_2H^-, the anions responsible for the stress corrosion cracking of mild steel in hot solutions of alkalis. These diagrams represent equilibrium conditions and cannot be used to determine or predict the rate of corrosion. They serve a useful purpose as a means of showing how corrosion may be controlled. Steel may be prevented from corroding in near neutral solutions by lowering its potential to below -0.62 V. This can be achieved by cathodic

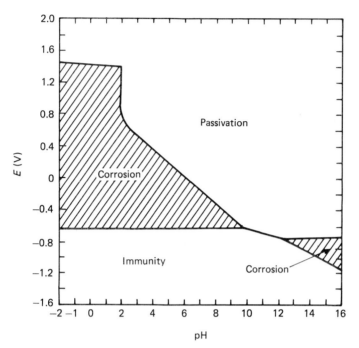

Figure 2.7 E_H–pH diagram for the Fe–H_2O system (after Pourbaix)

protection. Passivation can be achieved by raising the potential and/or the pH into the appropriate zone.

2.12 POTENTIAL-CURRENT DIAGRAMS

The graphical method of showing how corrosion rates depend on the polarisation of the electrodes of the corrosion cell was originated by U R Evans and are often called Evans Diagrams. A corroding metal is characterised by a *corrosion potential* E_{corr}, which is a mixed or compromise potential resulting from an anodic and cathodic reaction at the surface of the metal, and by a corrosion rate I_{corr} (or i_{corr}). Although the corrosion potential can be readily determined by means of an appropriate reference electrode, which must be specified, it is not possible in the case of the corrosion of a single metal to insert an ammeter in the circuit and I_{corr} must be evaluated indirectly by, for example, the rate of weight loss of metal, the rate of hydrogen evolution, etc, and by the conversion of these rates to I_{corr} by means of Faraday's law.

In the same way as E_r-pH diagrams summarise M-H_2O equilibria, the rates of corrosion and the mechanism of the individual half-reactions can be summarised by polarised potential E_p-I diagrams. These Evans diagrams present the individual electrode reactions as linear E_p-I relationships which take no account of whether the cathodic reaction is controlled by the activation energy of electron transfer or by transport overpotential resulting from slow mass transfer. The electrochemical mechanism of corrosion can be illustrated by considering the polarisation curves for the corrosion of Zn in acid. Equation (2.27a) shows that I depends on a reversible parameter $E_{r,cell}$ and the kinetic irreversible effects η_a, η_c and R_i (R_e is assumed to be negligible). The half-reactions will be

$$Zn \rightarrow Zn^{2+}(aq) + 2e^-$$

$$2H_3O^+ + 2e^- \rightarrow H_2 + 2H_2O$$

and these if added give the complete reaction

$$Zn + 2H_3O \rightarrow Zn^{2+}(aq) + H_2 + 2H_2O$$

If, for example, $a_{H_3O^+} = 10^{-1}$ in the acid and assuming that at a corroding zinc surface $a_{Zn^{2+}(aq)} = 1$, then

$$E_{r,H} = -0.059 \text{ V} \quad \text{and} \quad E_{r,Zn} = -0.76 \text{ V}$$

which defines the points on the E axis for $I=0$. Since both the h.e.r. (hydrogen evolution reaction) and $Zn \rightarrow Zn^{2+}(aq)$ will be activation controlled and conform to the Tafel equation it is appropriate to use log i rather than i for the x axis.

It is also possible to measure E_{corr} and to evaluate i_{corr} indirectly and hence define the position at which the anodic and cathodic curves intersect, but in contrast to the Daniell cell it is not possible to determine any intermediate points on the two curves. However, the full lines in *Figure 2.8* represent the extrapolated extensions of the anodic and cathodic curves, and these can be obtained experimentally by means of an external source of e.m.f. and a counter-electrode (Pt), a Luggin capillary and reference electrode. The metal at its corrosion potential is cathodically polarised at a low constant current i and the steady-state value of E_p measured; i is then increased incrementally and E_p measured at each value to give the extension of the cathodic curve. The same procedure is then adopted to determine the extension of the anodic curve, and if these two curves are extrapolated they will intersect at E_{corr} and i_{corr}, thus enabling the corrosion rate/unit area to be evaluated.

2.12.1 Types of corrosion control

The various types of corrosion control are shown in *Figure 2.8* by means of Evans-type diagrams.

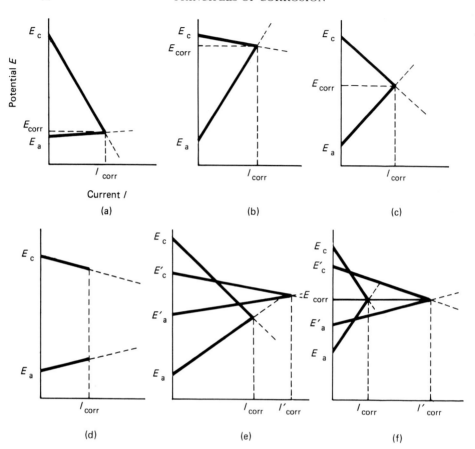

Figure 2.8 Evans diagrams (E–I) showing how the rate of corrosion reaction is controlled by electrode polarisation and by electrical resistance

Figure 2.8(a) and (*b*) show that for a given thermodynamic tendency $(E_{\mathrm{r,cell}})$ the corrosion rate of the reaction may be controlled by the dominant polarisation of either the cathode or the anode reaction (cathodic or anodic control), the other reaction being only slightly polarised; *Figure 2.8(c)* shows that both the anodic and cathodic process can contribute and *Figure 2.8(d)* shows how the resistance of the solution (R_i) and/or the resistance of films or coatings on the surface (paint films or natural calcareous deposits, $CaCO_3 + Mg(OH)_2$) can exert a controlling influence.

Figure 2.8(e) and (*f*) show that although $E_{\mathrm{r,cell}}$ for one reaction may be higher than that for another, the rate of corrosion of the former may be less than the latter because of different degrees of polarisation; it is also evident that although E_{corr} for these reactions may be the same the rates will be different showing that E_{corr} cannot be used to determine I_{corr}.

2.13 PASSIVATION

In some situations, the formation of sparingly soluble corrosion products results in a compact barrier isolating the metal from the environment.
  The conditions necessary for *passivation* to occur are:

  (i) The corrosion product must be thermodynamically stable in the environment under consideration (see zones of passivity in the E-pH equilibrium diagrams), although even when the corrosion product is not thermodynamically stable it is still possible for *metastable* oxides which pass into solution very slowly to give rise to passivation.

 (ii) The corrosion product must form on the surface of the metal as a coherent and adherent film, a condition favoured by, for example, direct anodic oxidation. If the primary anodic reaction is $M \rightarrow M^{n+}(aq)$ followed by reaction of the cation with H_2O or OH^- ions with the formation of insoluble oxide or other compound the ability of the cation to diffuse and migrate away from the surface will result in a less-adherent or non-adherent film.

(iii) The film must be mechanically stable so that during formation it is not disrupted by cracking, flaking or blistering and the metal/environment system should be such that the film is rapidly repaired if damaged.

An important property of film substances formed anodically on metals is their electrical conductivity, which must be considered briefly in relation to the passivation mechanism. Most metal oxides are semiconductors in which the conduction process ranges from almost completely electronic to almost completely ionic.
  Platinum is an example of a metal that forms a very thin electron conducting oxide film (one or two monolayers thick) on its surface when it is anodically polarised in a number of electrolyte solutions, and the conductivity of the film is comparable with that of the metal. Lead when anodically polarised at potentials $> \simeq 1.2$ V (SHE) in a number of solutions, including solutions of NaCl, will form a dark-brown film of PbO_2 (lead dioxide), which has a conductivity about 16% that of Pb. There are many other examples of electron-conducting oxides, although the conductivity may not be so high as that of PtO_2 and PbO_2.
  However, conduction of electrons through the oxide when the metal is polarised anodically will not occur until the potential is sufficiently high to oxidise a species in solution such as H_2O molecules or Cl^- ions. In the oxidation of neutral water to O_2 the reversible potential will be 0.85 V (SHE) but an overpotential of about 0.5 V is required for oxygen evolution to occur at a significant rate, whereas Cl^- ions are oxidised to Cl_2 at potentials only slightly above the reversible value of 1.3 V (SHE).

These potentials are not normally achieved unless the metal is polarised by an external source of e.m.f. or is in contact with a very powerful oxidant such as hot concentrated HNO_3 or an HNO_3 solution of CrO_3 ('chromic acid').

At the other extreme are metal oxides in which conduction is mainly by the transport of anions and cations, i.e. transport of metal cations from the metal across the metal/oxide interface and oxide to the oxide/solution interface where they combine with O^{2-} ions to form oxide, or the transport of O^{2-} ions in the reverse direction and formation of oxide at the metal/oxide interface. If the oxide is insoluble in the solution the film will thicken, providing the field (potential) across the oxide is also increased. Thus at a constant applied potential the oxide will grow to a certain thickness with decrease in the ionic current until a limiting thickness is reached at almost zero ionic current, e.g. in the case of Al anodised in a buffer solution the limiting thickness of the film is about 1.4 nm V^{-1}. If the current is maintained constant the field required to maintain transport of charge at this rate must increase, and potentials of hundreds of volts can be achieved before rupture of the dielectric film occurs by 'spark breakdown'. Metals such as Al, Ti, Ta, Nb, Zr, etc, form these dielectric oxides when anodically polarised in certain solutions, e.g. whereas Al produces a high-field dielectric film only in a buffer solution of about pH 7 owing to film solubility in acids or alkalis; on the other hand dielectric films can be formed on Ta in most acids but not in alkalis.

2.14 ANODIC $E–i$ CURVES

During a corrosion reaction the potential of the metal will be made more electropositive (anodic) by the redox potential† of the electron acceptor, but since the anodic and cathodic sites are inseparable the anodic reaction is difficult to study. For this reason the anodic behaviour of metals is usually investigated by making the metal anodic by an external source of e.m.f., and a counterelectrode, at which the cathodic reaction takes place.

These anodic $E–i$ curves may be determined in a variety of ways, and up to 20 years ago the constant current or 'galvanostatic' technique was used in which the steady-state potential was measured at increasing (or decreasing) current densities. However, with the development of the potentiostat, an electronic device for holding the potential constant, the

† Redox potential is the electrode potential of a reversible oxidation/reduction system (e.g. $Cu^{2+}(aq)/Cu$) and this will determine the equilibrium potential of the solution. Important redox systems in corrosion are $H_3O^+/\frac{1}{2}H_2$ ($E_{H^+/\frac{1}{2}H_2}$ ranging from 0.00 to -0.83 V) and $\frac{1}{2}O_2/H_2O$ ($E_{\frac{1}{4}O_2/H_2O}$ ranging from 1.23 to 0.4 V).

potentiostatic technique has become increasingly important, particularly for studying metals that exhibit passivity. There are several variations of this method and possibly the simplest is to hold the potential of the metal electrode at a potential slightly more positive than the corrosion potential and then to record the steady-state current (the current after an interval of, for example, 5 to 10 minutes, when it becomes approximately constant). The potential is then increased incrementally to the maximum value required. In the potentiokinetic variation of this method the potential range to be studied is swept through at a predetermined rate, the current being monitored by means of a pen recorder. Both the galvanostatic and potentiostatic techniques have their spheres of application, and the advantage of the latter is shown by the anodic E–log i curves for a metal that exhibits a transition from the active, or free-corroding state, to the passive state. *Figure 2.9* shows the galvanostatic E–log i curve in which the c.d. is held constant at a series of

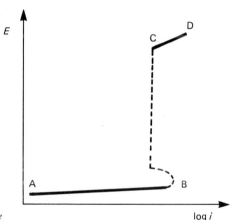

Figure 2.9 Galvanostatic E–log i curve

increasing values of i, and curve AB representing active dissolution conforms to the Tafel relationship. At B a sudden increase to C occurs due to the onset of passivation and this is followed by curve CD in which the rate again increases exponentially with increase in potential. *Figure 2.10* shows the corresponding E–log i curve ABCDEF determined potentiostatically, and in comparison with *Figure 2.9* it can be seen to have the following advantages:

(i) The magnitude of the passivation current i_{pass} is clearly defined (range between about 10^{-6}–10^{-2} A m^{-2} according to the system).
(ii) The onset of passivity is defined by the *critical c.d.* i_{crit} and by E_{pp} the *passivation potential* and its termination by the transpassive region EF

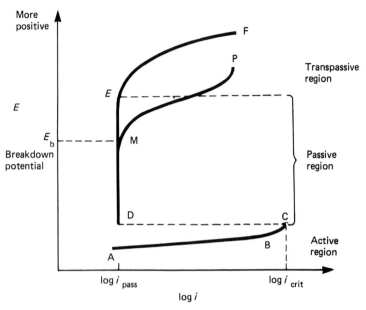

Figure 2.10 Typical E–log i curve for a metal or alloy that shows an active-passive transition determined potentiostatically

in which occur anodic reactions requiring higher potentials than those in the passive region.

It is not appropriate here to consider the detailed mechanism of passivity, but it is clear that the transition from active to passive must represent a fundamental change in the anodic process. From A–C the rate increases with potential, and it might be anticipated that divergence from the Tafel relationship at B would be due to transport overpotential. However, at potential B an oxide or hydroxide or other sparingly soluble salt becomes thermodynamically stable (or metastable) and its formation also appears to be kinetically easier (lower activation overpotential) than active dissolution at a very high rate. This results in the formation of a few monolayers of a sparingly soluble compound of relatively low electrical resistivity and a consequent sudden decrease in the current. Curve DE, the passive region, corresponds to what would be expected from an ideally polarisable electrode, i.e. there is a rapid increase in potential with an insignificant increase in current. For most metal/solution systems the upper limit of the passive region will not exceed about 1.0 V (vs SHE) and this is below the value required for the oxidation of H_2O or Cl^- ions, i.e. transport by electrons is negligible. Since the field is almost constant it means that the thickness of the oxide must be constant, and the rate of corrosion i_{pass} must be due, therefore, to slow formation of the film at the metal/film interface, transport of charge across the film and simultaneous and equal dissolution of the film at the film/solution interface.

The transpassive region EF occurs at high potentials and represents a number of possibilities of which oxidation of the metal to a soluble anion or cation of higher valence than in the active region is the most important. For example, the passivation of Cr or Cr-containing alloys is due to the formation of a film of chromic oxide Cr_2O_3 in which the valence of Cr is 3 (Cr^{3+}). At high potentials Cr is oxidised to $Cr_2O_7^{2-}$ in acid solutions and to CrO_4^{2-} in alkaline solutions and the valence is 6. This can occur quite readily through anodic polarisation by an external source of e.m.f., and by a limited number of chemical oxidants of high redox potentials such as hot concentrated HNO_3, or a solution of CrO_3 in HNO_3, and will result in the corrosion of stainless steels that are passive at more negative potentials.

The cathodic and anodic curves described could refer to the behaviour of a type 304 (18Cr-10Ni) stainless steel in H_2SO_4 in which the metal will corrode at a low rate (i_{pass}) provided the potentials *of all areas* are maintained in the passive region DE by means of an ample supply of oxygen. If Cl^- ions, or other ions conducive to breakdown of passivity are present there will be an increase in current in the passive region as shown by curve MP, which represents *localised breakdown* of passivity leading to pitting. For this to occur the cathodic curve must intersect the anodic curve at a potential more positive than E_b the *breakdown* or *pitting* potential.

2.15 FURTHER READING

DENARO, A R, *Elementary Electrochemistry*, 2nd ed, Butterworths, London (1971)
NEUFELD, P, *Elementary Aspects of Corrosion*, Portcullis Press (1975)
SCULLY, J C, *Fundamentals of Corrosion*, 2nd ed, Pergamon, Oxford (1975)
SHREIR, L L (ed), *Corrosion*, 2nd ed, Vol 1, Butterworths, London (1976)
WEST, J M, *Basic Corrosion and Oxidation*, Ellis Horwood (1980)
WEST, J M, *Basic Electrochemistry*, Van Nostrand-Reinhold (1973)

3 Marine environments

Sea water covers two-third of the Earth's surface and is an environment that ships have always had to withstand, as have the constructional materials used for the harbours between which they ply. Consequently, the marine environment has always been important, and not only the sea water itself but also the air above it which contains chloride from the sea spray. The importance of the marine environment has, however, increased considerably in the last few decades because of the exploration for natural resources in the sea and a realisation that the sea can provide energy from its wave motion and fresh water for areas of the world where this is in short supply. Marine environments are more aggressive than most inland environments and some understanding of their nature is essential if the best use is to be made of the materials exposed to them in the variety of constructions that have been and will be built either in the sea or on the sea coast. Corrosion of many alloys will often be greater on the parts that are not actually immersed in the sea. *Figure 3.1* indicates the range of different environments covered by the term 'marine environment', with an indication of the variations in the corrosion rate of steel that occur in the different zones. On land the environment near the coast is also considered to be 'marine' or 'coastal' as it is sometimes called.

The most important zone is the sea itself, because it is the chemical nature of the sea water that influences corrosion in marine atmospheres. Sea water contains chlorides which give it the high salinity, a main characteristic. There are, however, other important factors to be considered in relation to the chemical make-up of sea water.

3.1 SEA WATER

The most characteristic feature of sea water is its high salt content. The salt content of the waters of the open sea, away from inshore influences such as melting ice, fresh-water rivers and areas of high evaporation, is remarkably constant and is rarely outside the range of 33–38 parts per thousand. The common average value used for open ocean water is 35

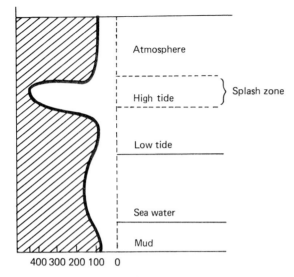

General wastage of steel ($\mu m/y$)

Figure 3.1 Marine environments: zones

parts per thousand. This is its *salinity* and is usually expressed as S‰, a convention which approximates to the weight in grams of dry salts contained in 1000 g of sea water which is obtained by titration with silver nitrate (see below).

If the salinity is known, the concentration in g/kg of the nine major ions can readily be calculated, since one of the remarkable features of sea water is that the saline composition, *regardless of the absolute concentration, has virtually constant proportions for the different major constituents.* They can readily be calculated from a table similar to that shown in *Table 3.1*. However, some of the other constituents of importance to the corrosion reaction such as the percentage of dissolved oxygen and carbon dioxide do vary and, of course, properties such as temperature, density and electrical conductivity are by no means constant.

It is difficult to determine by conventional chemical analysis the total quantity of dissolved solids in any sample of sea water. This is partly due to the fact that sea water is not a simple chemical solution and contains much organic matter, and also because of the presence of many carbonates and bicarbonates which are rarely amenable to exact and reproducible chemical analysis. Since the relative proportions of the different major constituent ions are reasonably constant, however, it is possible for any of them to be used to estimate the total amount of salts in any sample. Thus, in practice the concentration of salts in sea waters is usually expressed in terms of the chloride content. A reproducible

**Table 3.1 Major ions in solution in
an 'open' sea water at $S\% = 35.00$**

Ions	(g/kg)
Total salts	35.1
Sodium	10.77
Magnesium	1.30
Calcium	0.409
Potassium	0.338
Strontium	0.010
Chloride	19.37
Sulphate as SO_4	2.71
Bromide	0.065
Boric acid as H_3BO_3	0.026

Dossolved organic matter $= 0.001-$
0.0025 g, oxygen in equilibrium with
atmosphere at $15\,°C = 0.008$ g $=$
5.8 cm^3/l.

technique for expressing the concentration of salts in terms of the
arbitrary definitions 'chlorinity' and 'salinity' has been laid down by the
International Council for the Exploration of the Sea (ICES). Although
the figure obtained does not exactly represent the total quantity of
dissolved solids, it does represent a quantity that is very closely related—
the numerical value is slightly smaller—*and it is reproducible.* Therefore,
the use of this standard method permits an exact comparison between
titrations made by different operators at different times. The ICES
method for the determination of the salt content of a sea water is,
however, too involved for practical use, and since the halogens (predominately Cl^- ions but with small concentrations of Br^- and I^- ions),
when precipitated by a silver salt, can be estimated with precision,
sufficiently accurate measurements of the 'salinity' may be made by
determining the chlorinity $(Cl^- + Br^- + I^-)$ by titrating the chloride ions
with silver nitrate using potassium chromate as the indicator. The salinity
is then obtained by using the empirical relationship established by the
International Council that:

Salinity $= 0.03 + 1.805 \times$ chlorinity

The chlorinity of the sea is within the range of 18–20.

3.2 TEMPERATURE

The temperature of the surface waters of the oceans tends to vary directly
as the latitude, and the range is from about $-2\,°C$ at the poles to $35\,°C$

right on the equator. The temperature at any location is subject to seasonal variations, winds and currents. The surface of the sea is also affected by the weather but always to a lesser degree than any land mass. In the tropics the annual variations are smaller than those in the temperate zones where they amount to around 10 °C.

At depths below about 1000 m temperature tends to drop.

3.3 DISSOLVED GASES

Dissolved gases can be important in determining corrosion rates in sea water. The presence or absence of dissolved oxygen is, of course, a very important factor in the corrosion of metals immersed in the sea. It varies with temperature from about 8.0 ml/l for surface waters in the Arctic to around 4.5 ml/l in the tropics. Many variations of these guide figures will be found.

The concentration of the dissolved oxygen is by no means solely a function of temperature, it is also affected by the degree of movement of the water, the length of time it has been in contact with the atmosphere and very considerably by the amount of biological activity which is taking place. The photosynthesis in plants growing in the surface layers of the sea can cause a considerable increase in oxygen concentration, while the activities of some bacteria can reduce it to zero. There are diurnal variations in inshore and estuarine locations where the concentration is usually greater in the daylight hours. Outside littoral influences and below the biologically active layer the oxygen concentration diminishes. For example, in the Atlantic Ocean a minimum value of about 1–2 ml/l is found at depths of 200–1000 m. However, at a depth of 1500 m the oxygen concentration has risen again (5–6 ml/l) to near that of the surface layers and it remains fairly constant below this depth.

The movement of oxygen both outwards and inwards in sea water is much affected by evaporation, particularly in land-locked basins and docks, and by the movement of the water and the relative humidity of the overlying atmosphere.

3.4 HYDROGEN SULPHIDE AND AMMONIA

Sea water often contains hydrogen sulphide (H_2S) produced by the metabolism of the sulphate-reducing bacteria, and there are other bacteria found in the sea which are capable of producing ammonia gas and other nitrogenous compounds.

There are few estuaries, harbours, river mouths and fitting-out basins in the temperate zones which are free from patches of water contaminated with H_2S; fitting-out basins are particularly prone to heavy contamination.

The concentration of hydrogen sulphide in a sea water is difficult to estimate with accuracy in the absence of anaerobic sampling apparatus, since the sulphide oxidises to sulphate almost spontaneously in the presence of oxygen. Concentrations of the order of 30–35 p.p.m. are not uncommon, however, and there is usually a rise in concentration during the summer months. H_2S is very active in accelerating the corrosion of most ferrous and non-ferrous alloys.

The presence of ammonia in sea water has not received much attention, though it is probably present more often than is commonly thought. Not many estimations are reported in the literature of the quantity of ammonia to be found in sea water. Isolated figures for waters from the Arctic to the Pacific show that estimations vary from about 7 to 200 g/m^3. Higher amounts have been found in inshore waters and harbours.

3.5 CARBON DIOXIDE

The progress of corrosion of metals in sea water is often controlled by the presence or absence of carbonate scales. Thus, the amount of carbon dioxide (CO_2) in a sea water may have a direct influence on corrosion as well as upon the alkalinity of the sample.

In sea water the carbon dioxide can be considered to be present as bicarbonate and carbonate ions, as undissociated molecules of CO_2 and as carbonic acid (H_2CO_3) all in equilibrium with each other and the hydrogen ions present.

The free (unbound and undissociated) CO_2 consists of CO_2 and H_2CO_3 molecules which are in equilibrium with each other in the solution, roughly 1% of the CO_2 is present as H_2CO_3. The free carbon dioxide exerts a partial pressure which is related to temperature and pH. When the pH of the sea water rises at a constant temperature, carbon dioxide is released and enters the atmosphere, the free carbon dioxide in the solution falls and the amount of carbonate increases. Since the carbonate ions are divalent there is also a decrease in the bound carbon dioxide. Some of the excess base becomes associated with borate ions which are set free from undissociated boric acid. One of the results of this complicated set of reactions is that sea water is able to resist change in its pH, i.e. it is a highly buffered solution. Sea water contains more basic than acidic radicles, which means that there is an amount of 'excess base' or 'alkali reserve' which is in combination with the carbonic acid. This fraction is sometimes called the 'titratable base' or 'buffer capacity'. This excess base is equivalent to the carbonate and borate ions in the water and can be estimated by titrating the sample with a *strong* acid. Given salinity, temperature, pH and amount of excess base it is possible to calculate the total CO_2 content, its partial pressure and the concentration of bicarbonate and carbonate ions.

When an acid is added to sea water, the 'excess base' is neutralised and the CO_2 is set free, and since the evolution of the CO_2 is slow it may take as much as 24 hr for a fresh equilibrium to be established, even under conditions of continual agitation and replenishment with fresh air. It is thus difficult to change the pH of a sea water. This buffer action can affect the progress of corrosion cells, because local concentrations of acid may not be maintained in the vicinity of corroding areas for very long.

It may be pointed out that solubility data from the standard tables which are based on distilled water cannot be used for sea water. For example, the solubility product of calcium carbonate in a sea water of 19‰ chlorinity is about 530 times its solubility product in distilled water.

3.6 ELECTRICAL CONDUCTIVITY

The high conductivity of sea water is a factor in determining the amount of corrosion that occurs under immersed conditions, particularly in galvanic couples and at local situations such as crevices. The resistivity of sea water is compared with that of other waters in *Table 3.2*.

3.7 pH OF SEA WATER

Sea water is normally alkaline and the pH of the surface layers of the ocean, where the water is in equilibrium with the carbon dioxide of the

Table 3.2 Resistivity of waters (approximate values)

Type of water	Resistivity (Ω cm)
Pure water	20 000 000
Distilled water	500 000
Rain water	20 000
Tap water	1–5000
River water (brackish)	200
Sea water (coastal)	30
Open sea	20–25

Figures taken from *Cathodic Protection*, J. H. Morgan, Leonard Hill Books, London (1959).

atmosphere, lies between 8.1 and 8.3, and in the open ocean it is, again, a very regular property. However, in inshore localities and other areas where there is considerable microbiological activity these figures are not maintained. The presence of large quantities of hydrogen sulphide tends to lower the pH value (the water becomes more acid), while if there is a considerable photosynthetic activity of plants, which reduces the CO_2 content of the water, higher pH values will be found (the water will be more alkaline). The pH of sea water is altered by variations in temperature. The usual effect of a rise in temperature is to reduce the pH. On the other hand, if the heat 'boils off' the CO_2 a rise in alkalinity may be found. In the ocean depths the pH is usually below 8.0 because of the effect of pressure. The intermediate layers may sometimes become more acid due to the decomposition of organisms which sink as they die off in the upper layers.

3.8 CALCAREOUS SCALES

Although sea water contains mainly sodium chloride, it also always contains significant amounts of calcium bicarbonate and magnesium sulphate. These compounds can be of importance in the corrosion reaction where they may act as cathodic inhibitors (see Chapter 15). The cathodic reduction of oxygen or of hydrogen ions can result in an increase in pH in the vicinity of the cathode:

$$O_2 + H_2O + 2e = 2OH^-$$

and this can upset the equilibrium between dissolved calcium carbonate and CO_2:

$$CO_2 + H_2O \rightleftharpoons H_2CO_3 \rightleftharpoons H^+ + HCO_3^- \rightleftharpoons H^+ + CO_3^{2-} \qquad (3.1)$$

$$CaCO_3 + H^+ + HCO_3^- \rightleftharpoons Ca(HCO_3)_2 \qquad (3.2)$$
(insoluble)   (soluble)

Thus, an increase in pH which decreases $[H^+]$ will cause the equilibrium to proceed in the direction of the formation of insoluble $CaCO_3$. In the case of magnesium sulphate the increase in pH will result in the precipitation of $Mg(OH)_2$ according to the equation:

$$MgSO_4 + 2NaOH \rightarrow Mg(OH)_2 + Na_2SO_4$$

Under quiescent conditions, therefore, both calcium carbonate and magnesium hydroxide may precipitate at local cathodic areas. Whether they will form a dense adherent and coherent scale or a loose flocculent deposit will depend on other factors, for example, the presence of trace quantities of silicates, phosphates, etc. If the scales formed are coherent and adhere to the metal surface the rate of the cathodic process will be reduced considerably (and also the anodic reaction since the anodes and cathodes are usually in close proximity). The calcium bicarbonate and/or magnesium sulphate may thus be regarded as cathodic inhibitors.

The presence of these protective scales is particularly important in the cathodic protection of marine structures and ocean-going ships since they lower the power consumption of anode materials or decrease the current requirements in impressed systems. These scales reduce the true area of metal to be protected and in consequence a much lower current output is required to lower the potential of the structure to the value required for protection. In cathodic protection by sacrificial anodes it is possible to design anodes which give a high initial current density (c.d.) to facilitate the rapid formation of these scales, e.g. those anodes in which relatively thin fins are integrally cast with the block of magnesium. In impressed current systems a high c.d. can be used in the early stages of protection and once the scales have consolidated on the surface of the metal the current required to maintain the correct protective potential will fall (see Chapter 16).

3.9 GLOBAL VARIATIONS IN SEA WATER

A useful review of the variability in sea water at different global locations is provided by Dexter and Culberson[1]. This shows the wide variations that can be expected. The authors point out that most of the existing data is ill-suited for corrosion work, not being sufficiently detailed for any single location to permit the prediction of corrosion data with any accuracy. They also consider that it may be necessary to carry out a test programme to obtain information at sites where such data does not exist. Even data from a site 10 miles away may not be reliable.

The paper contains over 40 references to various aspects of the properties of sea water.

3.10 FOULING

There is a considerable amount of animal and vegetable life in sea water and it can have both a direct and indirect influence on corrosion. Bacteriological activity can have a direct effect on corrosion, particularly in polluted waters near the coast. Sulphate-reducing bacteria can cause steel to corrode under anaerobic conditions and this is considered in detail in Section 4.13. However, fouling is of a different nature and although it can influence corrosion, more often it leads to other problems. It causes problems on ships' hulls where the collection of marine growths increases the fuel consumption required to maintain speed. On offshore structures, the sheer weight of marine growths may cause problems and different types of organisms can collect on the internal sufaces of pipes and tubes causing a number of difficulties. It will restrict flow of sea water and may lead to secondary effects such as impingement attack because of this restriction of flow. It may also lead to deposit attack and—on passive-active alloys such as stainless steel—may cause pitting, because of the deposition of organisms on the surface. On the other hand, marine growths may, under some circumstances, be protective—on, say, sheet steel piles. Fouling varies in different parts of the world. Attempts have been made to produce maps showing overall fouling but they have limitations because of the influence of factors such as water temperature, salinity, distance from the shore, and depth.

A number of surveys of fouling have been made, in particular that by West (see 3.10.1 below). There is usually a season when fouling of a particular type is most active, although often certain types of fouling may occur throughout the year.

The general progress of fouling is as follows:

(i) Slime composed of bacteria and diatoms (single-cell plants).
(ii) Weeds grow from spores and become entrapped in the slime.
(iii) Animals, such as barnacles, tube worms and hydroids, become attached.

3.10.1 Further reading on fouling

HARRIS, J E, *Report on Anti-fouling Research 1942–1944*, *J. Iron Steel Inst.*, No 2 (1946), pp 297P–233P
Fouling of Ships' Bottoms: Identification of Marine Growths, *J. Iron Steel Inst.*, No 11 (1944), p 143P
WEST, J, *Fouling Distribution in the Major Oceans*, British Ship Research Association, Tech. Mem. No 397, 1971

3.11 CORROSION UNDER IMMERSED CONDITIONS

General data on the corrosion of ferrous metals is given in Chapter 5 and on non-ferrous metals in Chapter 6.

The corrosion of alloys is influenced by a number of factors, some of which are considered below. If these factors are not taken into account, data obtained from tests may provide misleading information for many practical situations.

3.11.1 Temperature

The effect of a rise in temperature would be expected, as with other chemical processes, to lead to an increase in corrosion. However, this does not always occur. For example, steel specimens immersed in tropical waters do not necessarily corrode at a greater rate than those in temperate climates. This situation may arise from factors other than temperature, e.g. by difference in oxygen content and fewer marine growths in colder waters.

In one series of tests carried out in sea water at different temperatures, the corrosion rate of carbon steel at a temperature of 25 °C was nearly double that at 10 °C.

Results quoted for the corrosion rates of copper and aluminium in tropical waters compared with those around the British Isles indicated a doubling of the corrosion rate for every 10 °C rise in temperature[3].

3.11.2 Depth

There is not a great deal of published data on the effects of depth on the corrosion of metals. Oxygen varies with increasing depth, tending to drop at 1000–2000 m and then to rise again. On the other hand, the temperature falls with increasing depth. Generally, the corrosion rate decreases with depth, although there appear to be exceptions. In results reported by Rowlands[3], based on tests carried out by the US Naval Civil Engineering Laboratory and Naval Research Laboratory and compared with results obtained at surface level by International Nickel, the general trend was for a reduction in corrosion rate with depth. There were, however, exceptions. The corrosion rate of zinc rose from 0.015 mm/y at the surface to 0.150 mm/y at a depth of about 2000 m. Both Cu–30Ni and Monel 400 also appeared to experience increased corrosion at depth. Mild steel corroded less with increasing depth to about 2000 m.

It is possible that the variatiohs in corrosion rates found in immersed tests near the surface of the sea may be repeated at depth, depending on the location of test.

Apart from the general corrosion rate, other factors such as pressure and mechanical effects ensuing from it may well be important in determining the types of corrosion that may occur.

3.11.3 Velocity

The velocity of sea water can influence the corrosion rate of metals in a number of ways. It can result in an increase in the amount of oxygen that reaches the metal surface, to the removal of protective films and may cause the formation of differential aeration cells. At high velocities, it can result in mechanical effects leading to cavitation.

Velocity is very important in piping systems or where movement through the sea is involved. The effect of the movement of sea water in relation to reasonably static structures and the exterior surfaces of pipelines will be influenced by the amount of abrasive material entrained with the water, the shape of the structural component and the actual alloy under consideration. Generally, carbon steel and zinc corrode at greater rates with increased velocity. The corrosion rate of steel may double where the conditions change from static to 1 m/s; for zinc the increase is even greater.

These figures are, however, based on laboratory-type tests and in practice other factors such as fouling, scaling etc may influence the rates.

Stainless steel and Monel 400 may pit at low velocities but show little corrosion at velocities over 2 m/s. On the other hand, copper alloys may exhibit increased corrosion at higher velocities.

3.12 ATMOSPHERIC MARINE ENVIRONMENTS

There are two specific types of atmospheric environment:

(i) At splash zones, where alternate wetting and drying occurs on metal surfaces with deposition of salts from the sea.
(ii) Situations where salt is detectable on metal surfaces; this may result from salt spray blown by wind. The amount of salt will be determined by factors such as the prevailing wind and general topography. This environment may occur within a few metres of some coasts and may continue over an area covered by a distance of some kilometres from the coast.

These situations can be influenced by industrial activity in the general area, and the presence of chemical pollutants—sulphur dioxide and smoke—may lead to increased corrosion rates compared with non-polluted marine environments. For many alloys, e.g. carbon steel, the corrosion rate in air is influenced by the time it remains moist and the amount of contamination, gaseous pollution and dirt in the air. Rain, condensation of moisture and relative humidity all influence the corrosion rate. Temperature has a number of effects, aminly indirect, such as the rate of evaporation of moisture from the surface and the diurnal variations in temperature. Relative humidity has a marked influence on corrosion through its effect on the time-of-wetness and can be defined as the percentage ratio of the water vapour pressure in the

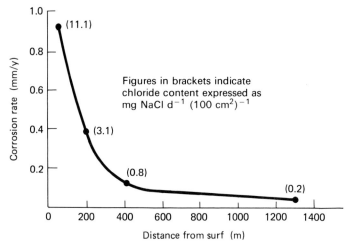

Figure 3.2 Corrosion of carbon steel at different distances from the surf

atmosphere compared with that which would saturate the atmosphere at the same temperature.

Generally, based on Vernon's work [4], it has been established that in the presence of the usual industrial pollutants such as sulphur dioxide, little corrosion of steel occurs below about 70% relative humidity (r.h.). In the presence of chlorides, however, it has been shown [5] that corrosion occurs at much lower relative humidities, e.g. about 40% r.h.

3.12.1 Spray zone (or splash zone)

This is generally the most corrosive zone to carbon steel, as demonstrated in various test programmes, mainly on steel piles (see *Figure 3.1*, p 39). It is the area that provides concern because of the high corrosion rates on offshore structures. Similar or even higher corrosion rates can occur on the coast, e.g. on tropical surf beaches. In tests carried out in Nigeria on behalf of the former British Iron and Steel Research Association, corrosion rates approximately 50 m from the surf were 950 κm/year compared with 40 κm/year some 1500 m from the surf (see *Figure 3.2*). The salt content of the air measured by exposing damp cloths to pick up salt particles and expressed as mb NaCl d^{-1} $(100 \text{ cm}^3)^{-1}$ was 11.1 at the surf and 0.2 at a distance of about 1500 m from the surf. This indicates clearly the rapid drop in salt content of the air as shown in *Figure 3.2*.

3.13 REFERENCES

(1) DEXTER, S C and CULBERSON, C, *Materials Performance*, **19**, No 9 (1980), p 16
(2) *Marine Fouling and Its Prevention*, Woods Hole Oceanographic Institution, US Naval Institute, Anapolis (1952)
(3) ROWLANDS, J C, in L L Shreir (ed), *Corrosion*, 2nd ed, Chapter 2.4, Butterworths, London (1976)
(4) VERNON, W H J, *Trans. Faraday Soc.*, **31**, 1, 668 (1935)
(5) CHANDLER, K A, *Br. Corr. I.*, July (1966).

4 Forms of corrosion

The rusting of ordinary carbon steels is the most common type of corrosion. This form of attack accounts for the major part of the costs attributable to corrosion and its control. The costs may arise from the corrosion itself, e.g. loss of steel section or perforation of sheet materials, or from the coating requirements needed to prevent or control it. This can, of course, be expensive, particularly in marine environments. There are, however, other forms of corrosion and they are often more difficult to deal with than is the rusting of steel. Steel is itself prone to forms of attack other than rusting but it is the performance of the more corrosion-resistant alloys both ferrous and non-ferrous which is particularly influenced by these other types of corrosion. Generally, stainless steels and non-ferrous alloys are far more resistant than carbon steel to marine environments but if these alloys are susceptible to localised corrosion, then they may fail prematurely, despite their general resistance to attack.

Many of these other forms of corrosion are complex and corrosion specialists do not themselves always agree on the mechanisms involved. In this chapter the detailed mechanisms will not be discussed. Chapter 2 provides a basis for the appreciation of the certain forms of corrosion and, where appropriate, reference is made to further reading for those who wish to obtain more detailed information.

General attack by corrosion is more predictable than the other forms to be considered here and much data is available for a variety of alloys in a range of environments. Furthermore, suitable control methods exist to combat general corrosion. In aggressive environments such methods may be costly but they can usually be planned. In many situations—particularly in plants—corrosion may be allowed to proceed within fairly predictable limits and replacement of components can then be made during planned maintenance periods. Some of the other forms of corrosion, however, may cause problems out of proportion to the degree of corrosion that occurs. Where practicable, some form of design audit should be established on important marine structures or plant to ensure that all reasonable precautions are taken to reduce the probability of severe and unexpected attack by these other forms of corrosion. In this chapter consideration is given to the more common forms of such corrosion that may occur in marine situations.

4.1 GENERAL ATTACK

This form of corrosion is the most common and is typified by steel rusting in air. It is sometimes called uniform attack. It is uniform in the sense that all exposed areas are attacked at more or less the same rate, but the loss of metal is rarely completely uniform over the surface. The type of attack is often as shown in *Figure 4.1*, which shows a cross-section of corroded steel.

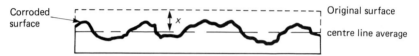

Figure 4.1 General corrosion. x indicates the average corrosion loss

Corrosion at anodic areas becomes stifled over a period of time and new anodic areas adjacent to the original ones become active. This interchange of anodic and cathodic areas leads to a reasonable uniformity of attack with time. This means that corrosion can be denoted in the form of loss of thickness of an alloy with time, e.g. μm/y. This is useful because it enables the engineer or designer to calculate the strength of the remaining thickness of alloy. It should be appreciated that corrosion rate figures are usually given for one surface only so the total loss in thickness will be the sum of the surfaces concerned.

Although considerable data exists on the corrosion rates of the common constructional alloys, engineers should treat such data with caution. Corrosion is not generally linear with time and usually—in air— it decreases (*Figure 4.2*). The data from short-term tests may not provide an accurate indication of long-term behaviour. For example, if the data obtained over a two-year period (*Figure 4.2*), had been extrapolated to eight years, the total estimate of corrosion would have been about 25% too high. Under immersed conditions, however, data from static tests may provide an optimistic view of the overall course of corrosion. Even under atmospheric conditions orientation has an important influence on corrosion, particularly of steels. In American tests[1] on specimens exposed at 30° to the horizontal, in air, over 60 ⟩ of the loss was on the underside. In other tests[2] similar variations have been obtained, e.g. specimens exposed at 45° corroded 10–20% more than vertical specimens.

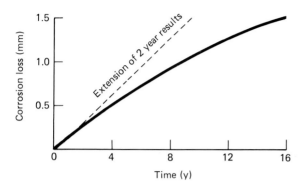

Figure 4.2 *Typical corrosion–time curve for carbon steel (Sheffield exposure)*

Sometimes corrosion rates are quoted in units other than μm/y. A common unit, particularly for laboratory tests, is milligrammes per square decimetre per day (mdd). These other units can be converted to the generally more useful μm/y provided the density of the alloy is known, e.g.:

$$mdd \times \frac{(0.00144)}{d} = inches/year \ (or \ ipy)$$

where d is the density in gcm^{-3}. Terminology such as ipy and mil/y where 1 mil = 0.001 in are still commonly used for American data. A reasonable conversion can be made from mil/y to μm/y by multiplying by 25. Factors for converting corrosion rates are given in *Table 4.1* and densities of different metals and alloys in *Table 4.2*.

Generally, data given for submerged conditions is based on static tests at ambient temperatures. The data will be different for situations where there is movement between the alloy and the solution or where temperatures are markedly lower or higher than those of the test data.

4.2 LOCALISED CORROSION

Metals corrode because of their thermodynamic instability in a particular environment and the mechanisms involve the existence of anodic and cathodic sites on the metal surface. In general corrosion, the sites change from being predominately anodic to cathodic and vice-versa. In some situations, however, this happens to a much smaller extent, or not at all, so the corrosion becomes localised at a number of anodic areas, the general surface being predominately cathodic. This type of corrosion is usually complex and unpredictable in the sense that it is not usually

Table 4.1 Corrosion-rate conversion factors

Multiply	by	To obtain
Milligrams per square decimetre per day ($mg\,dm^{-2}\,d^{-1}$ or mdd)	10	Grams per square metre per day ($g\,m^{-2}\,d^{-1}$ or gmd)
Inches per year ($in\,y^{-1}$ or ipy)	25.4	Millimetres per year ($mm\,y^{-1}$ or mpy)
Mils per year ($0.001\,in\,y^{-1}$)	25.4	Micrometres per year ($\mu m/y^{-1}$)
Milligrams per square decimetre per day (mdd)	$0.001\,44/\rho$ (ρ in $g\,cm^{-3}$)	Inches per year (ipy)
Milligrams per square decimetre per day (mdd)	$0.036\,52/\rho$ (ρ in $g\,cm^{-3}$)	Millimetres per year (mpy)
Grams per square metre per day (gmd)	$0.365\,25/\rho$ (ρ in $g\,cm^{-3}$)	Millimetres per year (mpy)
Grams per square inch per hour	372 000	Milligrams per square decimetre per day ($mg\,dm^{-2}\,d^{-1}$)
Grams per square metre per year	0.0274	Milligrams per square decimetre per day ($mg\,dm^{-2}\,d^{-1}$)
Milligrams per square decimetre	0.000 327 7	Ounces per square foot ($oz\,ft^{-2}$)
Milligrams per square decimetre per day	0.000 002 69	Grams per square inch per hour ($g\,in^{-2}\,h^{-1}$)
Milligrams per square decimetre per day	36.5	Grams per square metre per ($g\,m^{-2}\,y^{-1}$)
Milligrams per square decimetre per day	0.007 48	Pounds per square foot per year ($lb\,ft^{-2}\,y^{-1}$)
Ounces per square foot	3052	Milligrams per square decimetre ($mg\,dm^{2}$)
Pounds per square foot per year	133.8	Milligrams per square decimetre per day ($mg\,dm^{-2}\,d^{-1}$)
Grams per square metre per day	$0.365\,25/\rho$ (ρ in $g\,cm^{-3}$)	Millimetres per year ($mm\,y^{-1}$)
Grams per square metre per day	$365.25/\rho$ (ρ in $kg\,m^{-3}$)	Millimetres per year ($mm\,y^{-1}$)

Table 4.2 Densities of some alloys

Alloy	Density ($kg\,m^{-3}$)
Aluminium alloys	2.66–2.80
Copper	8.94
Copper alloys (other than aluminium bronze)	8.52–8.86
Aluminium bronze	7.78–8.16
Nickel alloys	8.27–9.22
Carbon steel	7.86
Cast irons	7.00–7.20
Stainless steels	7.65–8.06
Titanium	4.54
Magnesium	1.74

Table 4.3 Factors that lead to localised corrosion†

Heterogeneity	Area of metal that is predominantly anodic
Metal and metal surface	
Differences in metallurgical structure	Grain boundaries, more reactive phases (solid solutions, intermetallic compounds, etc)
Differences in metallurgical condition due to thermal or mechanical treatment	Cold-worked areas adjacent to annealed areas, metal subjected to external stress anodic to unstressed metal
Discontinuities in conducting oxide film or scale or discontinuities in applied metallic coatings that are cathodic to the substrate	Exposed area of substrate metal. In the case of passive metals defects in the passive film result in an active-passive cell with intense localised attack on the active area
Crevices or deposits on a metal surface or any other geometrical configuration that results in differences in the concentration of the cathodic reactant	The area of the metal in contact with the lower concentration of the cathode reactant, although there are exceptions to this rule
Dissimilar metals in contact (bimetallic corrosion)	The metal with the more negative corrosion potential in the environmental conditions prevailing (note that the standard electrode potentials are seldom applicable and the galvanic series can be misleading)
Environment	
Differences in aeration or in the concentration of other cathode reactants	Metal area in contact with the lower concentration
Differences in temperature	Metal area in contact with the higher temperature solution
Differences in velocity	Metal in contact with solutions of higher velocity
Differences in pH or salt concentration	Metal in contact with the solution of lower pH or higher salt concentration

† This table provides a general indication of the area that is likely to be anodic, but it must be emphasised that there are many situations in which the heterogeneity will have no effect or where the converse to the above may apply. From *Corrosion*, Vol 1, ed L. L. Shreir, Butterworths, London (1976)

possible to predetermine exactly where the attack will occur or even its extent. Nevertheless, the factors that are likely to cause localised attack are established and must be taken into account in the design of the structure or component (see *Table 4.3*).

4.2.1 Pitting

Pitting is a form of localised attack in which small areas of the surface are corroded with penetration of the alloy at these areas. Usually the depth of penetration into the alloy is greater than the nominal diameter of the surface corrosion. A range of pits may be formed, as shown in *Figure 4.3*. Pitting is similar in some respects to crevice corrosion. Once it has been

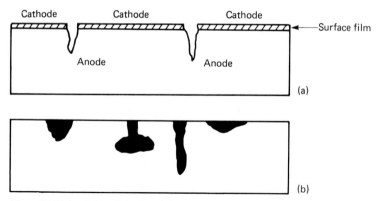

Figure 4.3 Pitting (a) small local anode/large cathode in the process of pitting (b) forms of pit

initiated its continuation is determined by reactions within the pit, which at the points of attack is anodic, with the outer surface being cathodic. Sometimes pits penetrate to a certain depth and then the downward attack stops or may continue horizontally within the metal. The effects of oxygen depletion and acid formation in the pit have to be taken into account. Sometimes intense pitting may occur over a surface so that it is similar to general attack except that some areas are virtually unattacked.

The graduation from general to pitting corrosion is sometimes vague. The term 'pit' which is difficult to define precisely is often reserved for situations where the depth of attack is greater than that of the surface cross-section. Although this is generally the case there are situations where localised shallow attack would be described as pitting (i.e. saucer-shaped pits). The total loss of metal is usually slight and in many situations pitting does not seriously affect the strength or use of an alloy. In other cases, however, it can lead to perforation of the metal, particularly where comparatively thin sections are used. This is serious if it occurs in pipes or tanks containing fluids because of the leakage that may occur.

The mechanism of pitting is complex and the likelihood and course of this type of attack is not always easily predictable. Furthermore, it is difficult on structures and plant to measure accurately pit depth or to determine the extent of attack within the metal. Even the sectioning of specimens to reveal pitting is difficult if there are only a few pits present and they have complex geometry. In many situations the cause of pitting cannot easily be established but there are certain situations where pitting can be anticipated:

(i) On carbon steel covered in millscale under immersed conditions, particularly in sea water. There is a potential difference between the

Figure 4.4 Perforation of tube with millscale

scale and steel so that small breaks in the scale lead to a small anode/ large cathode situation. The attack is concentrated on a number of small areas and this leads to pitting. Pits of 1.25 mm depth were formed in plates immersed in sea water at Gosport for six months[3]. *Figure 4.4* shows a perforation in a pipe carrying millscale.

(ii) On alloys that rely on passive films for their superior corrosion resistance, e.g. stainless steels, any break or defect in the film is likely to lead to pitting provided chloride or certain other ions are present.

(iii) From the practical standpoint, most pitting occurs in the presence of chlorides, so it is likely to occur more frequently under marine conditions.

(iv) Pitting tends to be associated with stagnant conditions and alloys that withstand pitting quite well in service, e.g. stainless steel pumps in sea water, may pit when taken out of service for prolonged periods unless they are thoroughly cleaned.

Where pitting proves to be a problem, two approaches should be considered: (i) changes in design to eliminate the situation that causes pitting and (ii) selection of alloys more resistant to pitting. In sea water the molybdenum-containing stainless steels such as Type 316 are more resistant to pitting than the chromium-nickel alloys, e.g. Type 304. Nickel alloys such as 'Hastelloy', and titanium are also resistant to pitting under most conditions. Both aluminium and copper alloys are prone to pitting under certain conditions. The alloy composition and the conditions of service have a marked effect on the probability of pitting attack. With aluminium alloys, solutions containing chlorides are particularly harmful and the situation is aggravated where local galvanic cells form on the surface, e.g. by deposition of copper from solution, which is present owing to the presence of copper or a copper alloy in the same system.

Unlike general attack, detailed data concerning pitting is difficult to obtain. Laboratory tests may provide results that do not relate well to practical conditions. Furthermore, as the measurement of pits is difficult and because of the small loss of metal the plotting of weight losses with time is not a particularly useful method of evaluating the propensity of alloys to pitting.

4.2.2 Differential aeration cells

The influence of differential aeration on a metal was demonstrated by U R Evans[4] who stated that any geometrical factor that results in a higher concentration of oxygen at one part of a metal surface compared with another will become cathodic to the area of lower oxygen concentration. In practice, a cell is often set up when the concentrations of oxygen vary over the surface. In some situations, however, the processes involved are not simple and it has been reported that less aerated zones of steel piles driven into the sea bed corroded at a lower rate than the more aerated zones[5]. In such cases the effects of the oxygen may be more complex because the high concentrations of chloride ions prevent anodic passivation of the steel, leading to a higher corrosion rate at the more highly aerated areas.

Common examples of differential aeration cells include steel partially immersed in an electrolyte solution where, initially anyway, the upper part of the plate, where oxygen is more freely available, becomes cathodic to the lower part.

Differences in temperature at different parts of steelwork can also lead to the formation of cells in a similar way. The term *concentration cell* is often used to describe the situation leading to this form of corrosion.

4.2.3 Crevice corrosion

This type of corrosion arises when a crevice is formed between two surfaces with the inside of the crevice being anodic to the external surface.

Basically crevice corrosion arises from the formation of a differential aeration cell in which the freshly exposed metal outside the crevice is cathodic to the metal inside the crevice. The large cathodic current acting on the small anodic area in the crevice results in an intensive local attack. Although this explains the increased corrosion that occurs in crevice situations, it is not now considered to be the complete explanation of the mechanism, which involves a more complex set of reactions. These are summarised below for steel in contact with a salt-containing solution:

(i) Initially the whole surface, inside and outside the crevice, is in contact with the electrolyte containing oxygen so that corrosion with oxygen reduction occurs.

(ii) Access of oxygen into the crevice occurs only by diffusion so rapidly diminishes in concentration as it is removed by the reactions occurring just outside the mouth of the crevice, i.e. by oxygen reduction and reactions producing $Fe(OH)_2$.

(iii) The cathodic reduction of oxygen on the large area outside the crevice results in anodic attack in the crevice.

(iv) The excess of positive charges (metal cations) formed within the electrolyte results in migration of OH^- and Cl^- ions from the bulk solution into the crevice.

(v) Metal chlorides then hydrolyse with the formation of metal hydroxide or oxide and H^+ ions:

$$M^+Cl^- + H_2O = MOH + H^+Cl^-$$

In the case of steel this results in the formation of $Fe(OH)_2$, which is not protective, and H^+ ions, so the pH falls to about 3.

(vi) The presence of H^+ ions and the high chloride content prevents passivation, so the anodic dissolution increases and the potential of the steel outside the crevice becomes more negative.

(vii) The increase in the rate of dissolution increases the migration of ions into the crevice, and this results in an autocatalytic reaction in which the rate increases with time.

The process is basically the same for alloys with passive films, e.g. stainless steel, where the high chloride in the crevice (it can be up to 10 times that in the bulk solution) and the hydrogen ions lead to rapid breakdown of the protective film.

In practice, irrespective of the actual mechanism involved, a situation arises where severe localised attack may occur within crevices and methods of eliminating or reducing such attack are considered in Chapter 7.

Generally the two surfaces of the crevice are metallic but one may be non-metallic, e.g. a gasket. Typical engineering situations where crevice corrosion may occur are at bolted joints, washers and flanges. Stainless steels and other corrosion resistant alloys are particularly susceptible to crevice attack under certain conditions and design plays an important role in ensuring such materials can be used in marine situations. The nature of the crevice is important because its influence is very much affected by the overall size and strength of the joint concerned. To function as a crevice, the configuration of materials must be such as to allow the entry of liquid, but sufficiently narrow to ensure stagnant conditions. The width of the gap is important and although no precise recommendations can be made, it can range from two uneven surfaces in contact to gaps of about 2 or 3 mm. With carbon steels, however, corrosion occurs in a slightly wider gap; the rust itself may lead to a situation where a crevice can form. This type of corrosion can lead to the shearing of bolts as the corrosion products form in the crevice and exert pressure on the joint, but such a situation is unlikely to occur with thick steel sections. In such a situation the strength of the steel prevents deformation by corrosion products and the corrosion is usually stifled. On thin gauge steel materials, buckling can, however, occur.

A form of crevice corrosion, sometimes called *deposit attack*, may occur when materials such as sand deposit on the metal surface causing a shielded, stagnant area. The effect under the deposit is often similar to that inside a crevice.

4.3 SELECTIVE LEACHING

This form of attack is sometimes called de-alloying, which explains its nature. Alloys are composed of different metals and, with certain of them, under specific conditions selective leaching of one metal can occur. The most common examples are the selective removal of zinc from brasses—dezincification—the selective removal of iron from cast iron, known as graphitisation, and the de-alloying of aluminium bronze in sea water. Other alloys are also susceptible to this form of attack, particularly in acids, but those noted above are most likely to be encountered in marine situations.

Dezincification occurs in two distinct forms. One is uniform with the alloy being affected in layers. The other is of a plug-type similar to pitting with the rest of the alloy being virtually unattacked.

Dezincification results in the loss of zinc leaving a weak matrix of porous copper. Most brasses can be affected in this way and the attack is encouraged by the presence of chlorides, particularly in warm environments.

Condenser tubes may be prone to this plug-type of attack leading to perforation.

The graphitisation of cast iron occurs in situations where the iron is preferentially attacked and leached out of the alloy. Graphite is cathodic to iron, so galvanic cells are set up, resulting in a porous mass of graphite and rust. Although the 'cast iron' loses its strength it may appear to be unaffected because dimensional changes do not occur. Graphitisation is usually a slow process and the degree of loss in strength of the cast iron depends upon the depth of the attack.

This form of attack occurs only on those cast irons with a suitable graphite network capable of acting to retain the product as the iron corrodes.

4.4 INTERGRANULAR CORROSION

In some alloys under certain environmental conditions grain boundaries are much more active areas than the matrix. In such cases preferential corrosion may occur at the grain boundaries, leading to intergranular or intercrystalline corrosion (see *Figure 4.5*).

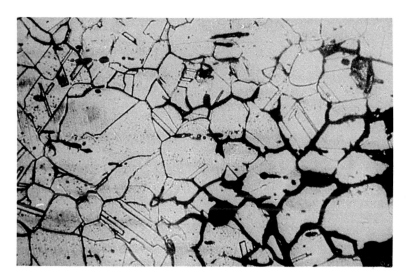

Figure 4.5 Intergranular corrosion

A common example of this form of attack occurs in austenitic stainless steels. When these alloys are heated within a certain range of temperature, they become what is termed 'sensitised' or susceptible to intergranular corrosion. This susceptibility results from the depletion of chromium near the grain boundaries. As stainless steels depend upon chromium for their corrosion resistance, a marked reduction in the amount of this element leads to a much reduced level of resistance. The chromium depletion arises from the formation of chromium carbides, which are virtually insoluble at these temperatures. The chromium is therefore removed from the grain boundaries, leaving an area prone to corrosion attack. Additionally, galvanic cells are set because the matrix is more noble than the grain boundaries.

This particular form of attack can be avoided by suitable additions of carbide-forming elements such as titanium and niobium which form carbides preferentially to chromium, or by using steels with a very low carbon content so that there is no excess for carbide formation. A particular type of this attack occurs near welds and is called *weld-decay*. A zone in the heat affected zone of the weld reaches the sensitising temperature and intergranular corrosion may occur under some service conditions.

Another manifestation of this type of corrosion is termed 'knife-line' attack, which may occur after welding titanium or niobium (colombium) stabilised austenitic stainless steels. The mechanism of this form of attack was established and reported in 1951[6]. It is similar to weld decay but knife-line attack occurs in a narrow band immediately adjacent to a weld in contrast to weld decay, which occurs at some distance from the weld. Furthermore, it occurs in steels that have been stabilised to resist weld decay.

A summary of the mechanism is as follows: Both titanium carbide and niobium carbide dissolve completely when steel is heated to just below the melting point, i.e. during welding. On cooling rapidly, a situation may arise, particularly on thin sections where neither titanium nor niobium carbide has time to precipitate. The chromium carbide also does not precipitate, but if the weld is then heated to within the critical temperature range, e.g. during stress-relieving, chromium carbide will precipitate rapidly but titanium or niobium carbide will precipitate very slowly. Consequently, the narrow band, which during welding reached a temperature sufficient to dissolve the carbides, becomes sensitised and susceptible to intergranular corrosion. This form of attack can be avoided by heating the steel, after welding, to about 1070 °C, at which temperature chromium carbide dissolves but the carbides of the stabilising elements form.

Apart from stainless steels, some aluminium alloys may be susceptible to intergranular attack. The high strength aluminium alloys depend for their strength on precipitates such as $CuAl_2$, which form along grain boundaries and this can lead to a situation where there is an attack at or near such grain boundaries.

4.5 VELOCITY EFFECTS ON CORROSION

The velocity of sea water has a marked influence on the corrosion performance of alloys (see *Table 4.4*). At lower movement speeds, the effects are mainly concerned with the rate of the corrosion process, which is affected by the ability of cathode reactants to reach the surface. This usually results in an increase in the corrosion rate but in some circumstances—particularly in the case of passive alloys—may reduce it. Velocity can also influence local corrosion effects, e.g. in crevices, by increasing the rate at which oxygen is brought to the external surface of the metal (cathode). At higher velocities a number of specific forms of corrosion can be manifested and these will be considered below.

4.5.1 Erosion-corrosion

Movement of sea water and, incidentally, other fluids, can cause mechanical action, i.e. erosion, as well as the corrosion arising from its electrochemical nature. Impingement attack and cavitation are extreme forms of this type of corrosion and will be considered separately.

Erosion-corrosion can lead to the removal of protective films from the surface of alloys and even removal of the metal itself if abrasive particles are entrained in the water. Generally, the rate of attack increases with velocity of sea water. Some indication of the effects of velocity in sea water is given in *Table 4.4* taken from tests carried out under various conditions but illustrative of the effects of velocity.

Table 4.4 Effect of velocity of sea-water on corrosion of metals $(g^{-2} d^{-1} \times 10^2)$

Metal	Velocity $(m s^{-1})$		
	0.3	1.2	8.2
Carbon steel	3.4	7.2	25.0
Cast iron	4.5	—	27.0
Copper alloys (brasses and bronzes)	0.1–0.7	0.1–2.0	17–34
Cupro-nickels	0.2–0.5	—	4–20
Monel	<0.1	<0.1	0.4
316 Stainless steel	0.1	0	<0.1
Hastelloy C	<0.1	—	0.3
Titanium	0	—	0

Data source, International Nickel Co.

The actual loss of metal will be determined by the nature of the sea water, e.g. whether it contains sand or silt, and the service conditions.

There can be other erosion or mechanical types of effect. For example, the corrosion of sheet steel piling on sea shores is affected by the abrasive action of sand, not only as the tide moves on to and recedes from the piling, but also during low-tide it may have an abrasive action arising from winds. In either case protective rust layers are removed.

The effects of erosion-corrosion may be general or localised. Often the attack may produce a grooved appearance and may have a pattern resulting from the direction of flow.

4.5.2 Impingement attack

As its name implies, this form of attack occurs when solutions impinge at a fairly high velocity on metal surfaces. This typically happens in pipe systems where an abrupt change in the direction of flow at a bend can lead to a localised failure, the rest of the pipe being virtually unaffected (*Figure 4.6*). This form of corrosion can occur in any situation where there is impingement of water usually—but not necessarily—containing entrained air bubbles at velocities in sea water as low as 1 m/s.

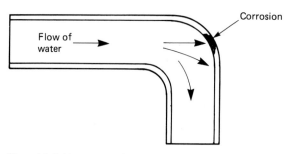

Figure 4.6 Impingement attack

Examples include half-open valves, partial obstructions and poorly mated flanges. *Figure 4.7* shows a typical attack arising from impingement of water from an inlet port. The entrance end of condenser tubes is a typical situation for impingement attack. The British Nonferrous Metals Research Association (as it was then called) developed a jet apparatus to test different types of copper alloy for their resistance to impingement. This led to the replacement of Admiralty brass by aluminium brass and cupro-nickel alloys for the condenser tubes. Although the early work at BNFMRA involved entrained air bubbles in the test, LaQue has pointed out that in tests carried out at Kure Beach, the presence or absence of air bubbles had little effect[15]. *Figure 4.8* shows a typical pattern of attack arising from impingement.

4.5.3 Cavitation damage

This is a form of attack caused by the formation and collapse of bubbles in sea water, or other liquids, at the metal surface. Conditions of high

Figure 4.7 Impingement attack on cylinder

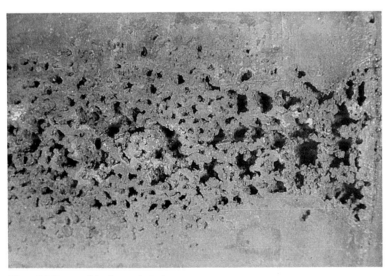

Figure 4.8 Pattern of impingement attack

velocity flow and pressure changes lead to cavitation attack. Attack is usually localised and severe and it occurs where—at areas of low pressure—water 'boils' and areas of partial vacuum (vapour cavities) are formed. When the water returns to normal pressure the cavities collapse with a release of energy of the order of 1–2 GNm^2. This leads to damage to the surface of the alloy, the extent of which depends in part on the type of alloy. This type of damage may appear to be purely mechanical but cavitation damage in sea water is greater than in distilled water, suggesting a corrosion element in the attack. The exact nature of this is not clear but it may result from the mechanical removal of protective films from the metal.

Additionally, the mechanical removal of part of the metal may leave a very active surface for a short time. Generally cobalt-base hard facing alloys, e.g. 'Stellite', are the most resistant to this form of attack. Titanium alloys, austenitic stainless steels and some nickel alloys are reasonably resistant but aluminium alloys and carbon steels are not normally used when cavitation damage is anticipated.

4.5.4 Controlling corrosion arising from effects of velocity

A number of approaches can be considered in relation to the control or prevention of corrosion arising from the effects of velocity. Clearly the choice of metals resistant to this form of attack would be the first consideration and the choice will be determined to some extent by the velocities involved. Stainless steels may pit at lower velocities, yet be excellent at rates over 3–4 m/s. On the other hand, some copper alloys may perform in the opposite way where velocity is concerned.

Design is probably the most important overall factor to be considered. The following general guidelines are worth pursuing.

  (i) Ensure smooth lines in the design and a smooth finish on the metal surfaces to produce lamellar flow.
 (ii) Consider increasing the diameter of pipes and designing out sharp changes in direction of pipework.
(iii) Eliminate protrusions into the flow.
 (iv) Where problems cannot be avoided, design to allow easy replacement of affected parts. Short tube pieces can be fitted to inlet ends to allow for replacement during shut-down. They must, of course, be designed so that they do not interfere with the smooth flow of solutions.
  (v) Special plates and baffles can be fitted to reduce attack on the main component.
 (vi) Utilise filters and screens to remove debris and abrasive particles.
(vii) Zinc blocks can be fitted to cathodically protect vulnerable parts.

The effects of velocity of sea water on corrosion are summarised in *Figure 4.9.*

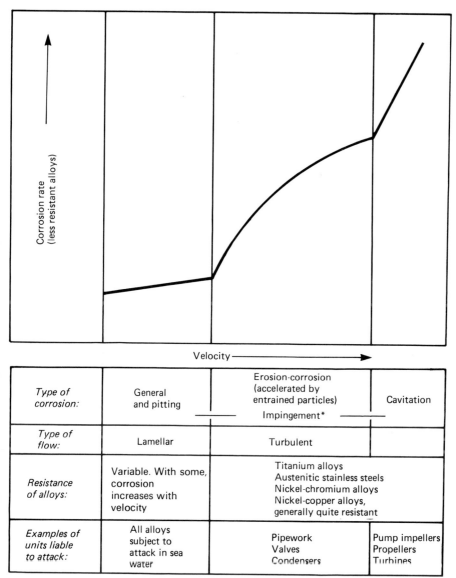

Type of corrosion:	General and pitting	Erosion-corrosion (accelerated by entrained particles) ————— Impingement* —————	Cavitation
Type of flow:	Lamellar	Turbulent	
Resistance of alloys:	Variable. With some, corrosion increases with velocity	Titanium alloys Austenitic stainless steels Nickel-chromium alloys Nickel-copper alloys, generally quite resistant	
Examples of units liable to attack:	All alloys subject to attack in sea water	Pipework Valves Condensers	Pump impellers Propellers Turbines

* Impingement attack results mainly from design features such as sharp changes of direction

Figure 4.9 Effects of velocity of sea water on corrosion

4.6 STRESS CORROSION CRACKING

The general corrosion of structural metals leads to the loss of cross-section so eventually they may fail under a load that they could originally sustain. Protective coatings are applied to steels to control the loss of metal and hence prevent mechanical failures. There are, however, situations where the simultaneous effects of a tensile stress and corrosion are greater than the sum of these factors acting separately. Their effect is described as stress corrosion cracking and can lead to rapid failure of an alloy, unlike a structural failure arising from ordinary corrosion over a long period.

There may be some confusion over terminology because 'stress corrosion' and 'stress corrosion cracking' are both used. Stress corrosion cracking is the term used to describe the type of failure arising from the joint action of corrosion and tensile stress, in which small cracks extend and eventually lead to mechanical failure. Some investigators use the same term for all cracking failures occurring in corrosive environments, including those arising from hydrogen embrittlement. Others separate the two types of failure because they are considered to be different phenomena.

Other authorities use the term 'stress corrosion' to cover all forms of corrosion where the stress accelerates a failure that would occur in its absence. An example of this type of failure is the acceleration of intergranular corrosion of stainless steels under tensile stress conditions. In this particular section only stress corrosion cracking will be considered. This is a phenomenon in which alloys are often virtually unattacked over a large area of their surface, but at certain points cracks form and follow either intergranular or transgranular paths in the microstructure. *Figure 4.10* shows a typical pattern of stress corrosion cracks. The velocity of the crack growth is usually slow initially, usually from 10^{-7} to 10^{-2} mm/s[7]. If the static load remains reasonably constant as the cracks grow, the cracks eventually reach a size that leads to rapid mechanical failure.

In marine environments a number of alloys are susceptible under certain sets of conditions, including austenitic stainless steels. Aluminium alloys of the Al/Zn/Mg, Al/Mg and Al/Cu types are also susceptible. They tend to have comparatively high strengths resulting from the compositions and heat treatment which cause grain boundary precipitates and locked-in stresses.

Other alloys may be susceptible in conditions where the sea water is contaminated with chemicals, possibly from process liquids.

The mechanism of stress corrosion cracking is complex and in practice there may well be different mechanisms for different situations. These mechanisms will not be considered here but for those wishing to study the matter in more detail a list of references is provided at the end of this section. From the engineering standpoint the following basic points should be borne in mind.

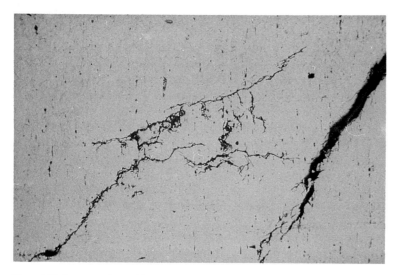

Figure 4.10 Stress corrosion cracking

(i) Stress corrosion cracking may occur at stresses well below the yield point of the alloy.

(ii) All constructional alloys are susceptible in specific environments under certain conditions, e.g. brasses crack in ammoniacal solutions but not in chloride-containing solutions, whereas austenitic stainless steels crack in chloride environments but not in ammoniacal solutions.

(iii) This form of attack occurs only under tensile stresses.

(iv) The stresses may be externally applied or induced in the alloy, e.g. by cold working.

(v) Generally, stress corrosion tends not to occur under conditions of general corrosion but in situations of local attack, e.g. pitting.

(vi) Conditions for stress corrosion to occur may be specific as regards concentration of solution and temperature and these may arise locally, e.g. at a crevice.

(vii) There is a threshold stress below which cracking either does not occur or occurs at an extremely slow rate. This stress will depend upon the environment for a particular alloy.

4.6.1 Factors involved in stress corrosion cracking

The main factors to be considered are the environmental conditions, the stress conditions and the alloy under consideration.

(i) *Environmental conditions*

For most constructional alloys, stress corrosion cracking occurs only in conditions specific to that particular alloy. As already noted, austenitic stainless steels are susceptible in chloride-containing environments. This does not mean that such steels will always crack if used under marine conditions. The concentration and temperature of the chloride-containing solutions and the stress level all determine the likelihood of such attack, so it is only under a certain set of conditions that problems may arise. Because of the combination of conditions it is not always a straightforward matter to determine the probability of attack, particularly in view of the importance of the local environment at the surface. The classic example of stress corrosion cracking of riveted mild steel boilers (caustic embrittlement) occurred even in situations where the alkaline water contained less than the concentration of sodium hydroxide considered necessary to promote cracking. This was because of the evaporation of water from the seams leading to concentration of sodium hydroxide, which led to stress corrosion cracking.

Similar situations have arisen with stainless steel back-boilers using domestic water where chlorides have concentrated. This is also an illustration of the problems that can be accentuated under heat transfer conditions.

It is not possible to provide more than a general guide to the influence of environmental factors on the susceptibility of alloys to stress-corrosion. Most constructional alloys are susceptible under some sets of circumstances, but this does not mean that problems are likely to arise in more than a few cases. Unfortunately, it is not easy or in some cases even possible to predict the exact situations where cracking will occur. Apart from the chemical composition of the solution in contact with the alloy surface, temperature and, in some cases, the presence or otherwise of oxygen will affect the process. It has been reported that in sea-water Type 316 steels, particularly where the metal surface becomes encrusted with salt, may be susceptible to stress corrosion cracking at temperatures above 140°F (60°C).

(ii) *Stress*

Stress is clearly one of the most important factors in determining the probability and course of stress-corrosion cracking, but it is not an easy matter to predict stress levels at which failure will occur. This, in part, arises from the nature of the stresses that can cause cracking of alloys. They may be applied but frequently they are internal stresses arising from fabrication procedures such as welding and cold working. In tests carried out under standardised conditions, curves similar to *Figure 4.11* are often

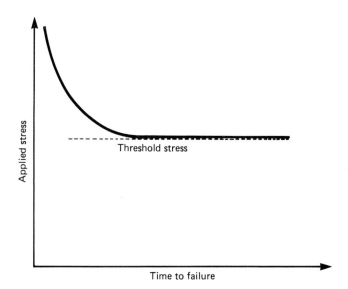

Figure 4.11 Stress corrosion: relation between stress and time to failure

obtained. This indicates that as stress levels are lowered the time to failure increases eventually reaching a limiting or threshold level. This could be taken to mean that, for design purposes, stresses below this threshold would be safe. This might well be the case for conditions that were exactly the same as those in the test but in practice account must be taken of local factors such as crevices, where acidic solutions may be present. By using a fracture mechanics approach, it is possible to determine the growth rate of stress corrosion cracking as a function of stress intensity.

Curves such as that shown in *Figure 4.12* indicate that below a particular value of the stress intensity, Kiscc, crack growth rates are negligible or zero. When Kiscc is exceeded, the rate of crack growth increases rapidly with small changes in stress intensity until it reaches a value at which fracture occurs.

It might appear that because there is a stress intensity below which cracking is negligible or does not occur, then this could be used for design purposes. Although such data is useful to specialists Kiscc is not an intrinsic property of an alloy but rather a factor obtained from the combined effects of the alloy and the environment in which it is used or tested. Consequently, such data should be used only as a general guide to performance.

(iii) *Alloys*

As the electrochemical characteristics of alloys influence their susceptibility to stress corrosion cracking, changes in alloy composition would be

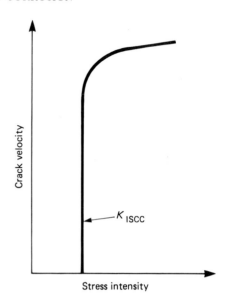

Figure 4.12 Stress corrosion: relation between rate of crack growth and stress intensity

expected to have a marked effect on thier performance in relation to stress corrosion. This is clearly demonstrated by the family of steels termed 'stainless' where increases of nickel content above about 8–10% in austenitic steels increases the resistance to stress corrosion—and with 40–45% nickel, the resistance is greatly enhanced. An interesting point is that a reduction in nickel content below 8% also produces an improvement because of the change in metallurgical structure. Heat treatments can also influence an alloy's susceptibilities in various ways, e.g. by affecting internal stresses. Further comments on susceptibility to stress corrosion cracking is provided in the sections of the book dealing with specific alloys.

(iv) *Minimising stress corrosion cracking*
Although a considerable effort has gone into research investigations of stress corrosion, no precise design data is generally available. In a complex phenomenon, such as this, it may indeed be dangerous for non-specialists to attempt to design using corrosion data in a manner similar to that used for data on yield stresses. It is, therefore, preferable to approach the situation by considering the broad factors likely to cause problems. In marine situations it is known that certain alloys may be susceptible, e.g. high-strength steels, austenitic stainless steels, and wrought magnesium/aluminium alloys. If these are involved in the design then attention should be paid to the environmental conditions, in particular the

possibility of local concentrations of salts arising through the design of the structure or plant or from heat transfer situations.

In situations where alloys are potentially susceptible, it is important to ensure sound quality control procedures in the fabrication process and where necessary and practicable, to carry out stress-relief treatments. As stress relief treatments may lead to distortion or metallurgical changes, advice should be sought before undertaking them.

Stress relief of welds, particularly on high strength carbon steels, should be considered but such treatments must be carried out correctly to avoid effects in other parts of the structure.

Shot peening (blast-cleaning) may be used to induce compressive stresses into the material. If carried out with care this may eliminate problems at, say, welds or at least delay the onset of cracking.

It is particularly difficult in most situations to control the stress in a structure or component because even where tests are carried out to determine the design data it is difficult to reproduce the exact nature of the environment. Consequently, the data so obtained may not be sufficiently precise for practical situations where variations in the local environment at the alloy surface are commonly encountered.

The fracture mechanics approach does provide a means of defining allowable initial defect sizes and, by monitoring slow cracks growth in service, enables the taking of necessary action before failure occurs. It is beyond the scope of this book to consider these approaches in detail but a short bibliography has been provided at the end of this section for those wishing to study the matter in more detail. Coatings can be used in some situations to minimise stress corrosion cracking. Organic coatings are most widely used, particularly for carbon steel and aluminium alloys. These must be applied correctly to properly prepared surfaces.

Metal coatings can also be used but in some situations problems may arise with anodic coatings such as zinc, if hydrogen is evolved. If this enters higher strength steels, they may fail by hydrogen embrittlement.

Cathodic protection can be used to lower the potential and so avoid crack propagation, but care is required to ensure that hydrogen is not produced in any situation where a possibility of hydrogen embrittlement may arise (see Section 4.6.2).

4.6.2 Stress corrosion cracking of high-strength steels

In recent years, there has been an increasing demand for steels with higher strength levels. Under most circumstances, ordinary structural steels are not subject to stress corrosion cracking. However, the higher-strength steels (over $1.4 \ GN/m^2$) tend to be more susceptible. There is not complete agreement on the mechanisms involved but it is generally accepted that hydrogen plays an important role and the terms 'hydrogen cracking' and 'hydrogen embrittlement' are commonly used when describing stress cracking of high strength steels. There have been a considerable number of investigarions and many papers on this phenomenon and for readers who wish to study the matter in more detail, some appropriate references are provided at the end of this section.

The basic cause of hydrogen embrittlement arises from the penetration of atomic hydrogen into the steel and its effect on crack growth, e.g. by accumulating near sites of dislocations or by other mechanisms of slip interference.

From the practical standpoint, any situation that is likely to lead to the transfer of hydrogen into high strength steels must be looked at carefully. Sources of hydrogen arise from pickling in acids, electroplating, cathodic protection and the sacrificial action of some protective metal coatings, e.g. zinc. Hydrogen can also be absorbed during some manufacturing and heat treatment processes. Information should be sought when designing with high strength steels regarding the advisability of phosphating, electroplating or using zinc coatings. Baking at suitable elevated temperatures is often a satisfactory way of removing hydrogen from components such as bolts, provided any protective coatings applied, e.g. Zn or Cd, are sufficiently porous to allow egress of the gas. It has been stated that some paint strippers can lead to hydrogen absorption and should be checked before use[8].

4.6.3 Summary

Stress corrosion is a complex phenomenon and continual research into its mechanism and ways of avoiding or minimising its effecis will produce new solutions to problems. At present it is not possible to provide more than broad guidelines, as discussed above. The general approach is, therefore, based on the following points and methods of control must be chosen in relation to the actual situation under consideration:

(i) Does available data indicate that the particular alloy is susceptible to stress corrosion cracking in the particular environment?
(ii) Are the stress levels likely to cause problems?
(iii) How much service data on control methods is available?
(iv) If stress corrosion cracking occurs, how serious will this be? Will it lead to a dangerous situation or high production losses, or can it be solved comparatively easily by changing a component?

The possible methods of preventing or minimising stress corrosion then fall into the following categories.

(i) Change of alloy to one not susceptible in the environment.

(ii) Treatment of alloy to remove internal stresses or to provide a structure resistant to the environment.

(iii) Examination of the design to ensure that aggressive local environments do not arise, particularly at heat transfer surfaces and crevices.

(iv) Consideration of the possibilities of changing the main environment, e.g. by de-aeration or use of inhibitors.

(v) Examination of local weather data to check whether alloy temperatures may reach critical levels.

(vi) Protection by coatings.

(vii) Cathodic protection but not where alloy is susceptible to hydrogen embrittlement.

(viii) Reduce operating stresses and/or internal stresses by, for example, annealing.

(ix) Shot peening of surface.

(x) Institution of quality control procedures to avoid stress concentrations.

(xi) Consideration of possibilities of monitoring the structure or plant.

4.6.4 Short bibliography

LOGAN, H L, *The Stress Corrosion of Metals*, John Wiley, New York (1966)

KNOTT, J F, *Fundamentals of Fracture Mechanics*, Butterworths, London (1973)

STAEHLE, R V (ed), *Proc. Conf. Fundamental Aspects of Stress Corrosion Cracking* (1969), National Association of Corrosion Engineers, Houston, Texas, USA

'Stress Corrosion', No 4 of *Guides to Practice in Corrosion Control*, UK Department of Industry, London

SHREIR, L L (ed), *Corrosion*, 2nd ed, Chap 8, Butterworths, London (1976)

PARKINS, R N, 'Stress-corrosion cracking of welded joints', *Br. Weld. J.*, **2**, November (1955), p 495–501

Conf. Stress Corrosion Cracking and Hydrogen Embrittlement of Iron Base Alloys (1973), National Association of Corrosion Engineers, Houston, Texas, USA

DE KAZINCZY, F J, 'Theories of hydrogen embrittlement', *J. Iron Steel Inst.*, **177**, 85 (1954)

Proc. Conf. Fundamental Aspects of Stress Corrosion Cracking (1969) p 398, National Association of Corrosion Engineers, Houston, Texas, USA

MORLET, J G, JOBUSARY, H H and TROIANO, A R, *J. Iron Steel Inst.*, **189**, 37 (1958)

4.7 CORROSION FATIGUE

Fatigue is the tendency for a metal or alloy to fracture under conditions of repeated cyclic stresses at loads below the ultimate tensile strength of the material. Corrosion fatigue is the term used to describe the failures of alloys by cracking under the influence of both fatigue and corrosion acting together. The importance of conjoint action should be emphasised because, as with stress-corrosion, the separate effects are not usually as severe as the joint effects. The view is sometimes expressed that fatigue is always really corrosion fatigue, except where it occurs in a vacuum, because all environments are corrosive to some extent. There may be some truth in this assertion but there is a marked difference between the behaviour of most alloys subjected to fatigue conditions in marine environments compared with similar stress cycling in laboratory air conditions.

Corrosion fatigue was first reported over 60 years ago and concerned the failure of steel cables in the sea water. The first comprehensive series of investigations of the phenomenon was carried out some 10 years later and the term 'corrosion fatigue' was coined. In recent years, the reported failures arising from corrosion fatigue has increased and it is now considered to be a major cause of in-service failures. This is certainly the case in marine situations where the environment is aggressive and structures are often subjected to cyclic stressing.

Corrosion fatigue is a complex phenomenon and there is not complete agreement on either the mechanism or the conditions under which it will occur with different alloys. A considerable amount of investigational work, particularly on mechanisms, is being carried out and much of this was reported in the proceedings of a symposium on the topic held at the University of Connecticut in 1971[9]. Although a good deal of information is available on corrosion fatigue, considerably more will be required before all the problems connected with it can be solved. In this section some of the basic points will be considered but, like many other aspects of marine corrosion, considerable strides in the understanding of the processes involved and practical data on the topic are likely to be published in the next few years.

Although there are similarities between corrosion fatigue and stress corrosion, particularly at very low stress cycle frequencies, they are essentially different. The environments for stress corrosion cracking tend to be specific tro particular alloys, but all alloys that corrode in a particular environment are susceptible to corrosion fatigue in that environment.

Generally, fatigue is assessed in laboratory tests by plotting a series of initial stresses (S) against the number of cycles (N) required to produce failure. Such plots are called S–N curves as shown in *Figure 4.13* where a similar curve for a corrosion fatigue situation is also plotted. These curves are for steel fatigue tested in air and in a 3% NaCl solution. As can be seen, in the presence of a corrosive environment—sodium chloride—the number of cycles to failure at a particular stress is less than for the steel fatigued in air. Furthermore, there is no fatigue limit.

It seems probable that there is no single mechanism covering corrosion fatigue of different alloys in a range of environments. A number of

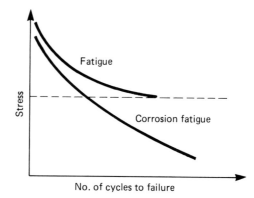

Figure 4.13 S–N curve for corrosion fatigue

different mechanisms are involved, but not necessarily the same for all situations. The original investigations considered that the following steps were involved: (i) stress concentration at a pit caused by corrosion, (ii) electrochemical attack on plastically deformed areas and at disrupted protective films, (iii) propagation of cracks.

These mechanisms presume an initial incubation period, i.e. the pitting of the surface, followed by crack propagation. Recent studies have indicated that the process is more complex and that pitting is not an essential step. It must, however, be assumed that stress concentrations are in some way set up and that electrochemical attack takes place, possibly following the rupture of protective films.

It has been demonstrated that corrosion fatigue can be stopped by the application of suitable potentials and that in one series of tests, preliminary fatigue in air had little effect on the subsequent corrosion fatigue life of wires.

It is not considered to be appropriate here to discuss the various work concerned with mechanisms. For those who wish to study the matter further some references to useful work in this area are included at the end of the chapter[9],[13]. An essential point to note is that an alloy may be susceptible to corrosion fatigue in an environment where in the non-stressed condition it is resistant. This is because the protective films that provide corrosion-resistance to some alloys, e.g. aluminium and stainless steel, may rupture during cyclic stressing, so allowing non-protected areas of the alloy to be in contact with the corrosive environment.

4.7.1 Corrosion fatigue data

Most corrosion fatigue data has been obtained from tests using standard fatigue test apparatus modified to allow for the application of some form of corrosive solution. S–N curves are produced and this provides basic data. Such data cannot always be used to evaluate the life of a component in service. Often the fabrication or assembly of a component in a structure results in a flaw or defect and it is the development of a crack from such a defect that becomes the limiting factor in the corrosion-fatigue performance.

More recent applications of linear elastic analysis to crack growth problems has led to a different method of presenting data. Using a pre-cracked specimen, the crack extension per stress cycle as a function of maximum stress intensity per cycle is measured and the data is presented on the stress intensity range per cycle, as in *Figure 4.14*. Empirical formulae can be developed to assist in the prediction of life. Information derived from curves of the type shown in *Figure 4.14* can, in the hands of

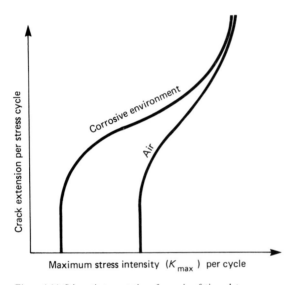

Figure 4.14 Schematic presentation of corrosion fatigue data

specialists, provide useful data in determining or—perhaps more correctly—comparing the environmental effects on fatigue crack propagation and so estimating the life of a component. It is also of assistance in determining the monitoring and inspection techniques required for any potential problem areas.

Although considerable test data is available on different alloys subjected to tests in sea water conditions, usually in sodium chloride solutions, this data cannot easily be applied to practical situations without some considerable understanding of fatigue and corrosion fatigue processes. Failures continue to occur and, as they are investigated, further data becomes available. Normally, though, engineers and designers are not in a position where they can necessarily select materials or provide designs that will eliminate the problems arising from corrosion fatigue.

Two papers presented at the UK National Corrosion Conference, 1982, demonstrated the problems of testing for corrosion fatigue. As one author states[10], 'In practice, offshore structures are not subjected to constant amplitude loading and the magnitude of each individual stress cycle will depend principally on wave height and direction'. Another author[11] found the rate of crack growth to be sensitive to cyclic frequency, electrochemical potential and sea water temperature. His concluding remarks are worth noting: 'In the absence of a significant data base from fatigue tests on tubular welded joints in sea water, with and without cathodic protection, the complex interaction of several variables makes it very difficult to come to an uncontroversial judgement on how one should cope with the influence of environment in design and inspection codes'.

4.7.2 Control of corrosion fatigue

(i) *Design*

As with most corrosion problems, the design of the component or equipment has an influence on performance. The surface condition and avoidance of sharp edges is of particular importance when considering corrosion fatigue because of its relation to flaws and stress raisers. Of course, corrosion fatigue can still occur on initially plain surfaces if the stresses are sufficient to cause even a small amount of plastic deformation in the alloy which, with cyclic stressing, will provide continual disruption of protective films and in suitable environments corrosion of the underlying metal. Nevertheless, careful attention to design can reduce the probability of attack. Areas of design that should be considered include the following:

(a) Elimination of crevices because they can lead to a concentration of salts and may lead to a highly corrosive situation.
(b) Sound fabrication procedures to ensure smooth fitting together of components and general clean lines; in particular, good welding.
(c) Allowance for proper drainage in situations where solutions can collect.

In a series of tests carried out on welded girders exposed in sea water, it was concluded that defects in the K-welds between the web and flange influence the fatigue strength more than the transverse butt welds in the flange itself[9]. Furthermore, the fillet welds loaded longitudinally, e.g. in connections between the webs and flanges of the girder, were considered to be more harmful than is generally supposed because of the occurrence of invisible notches at the roots of such welds.

Residual fabrication stresses and weld areas in particular can lead to situations where any localised corrosion could cause plastic deformation due to the enhanced level of stress and this will allow corrosion attack to be sustained. Bolt holes particularly where they are misaligned can lead to the initiation of corrosion fatigue cracks. This has been reported as a primary cause of failure in a number of investigations of aluminium alloys. The roots of gear teeth can also act as stress raisers.

(ii) *Compressive stresses*

As with components susceptible to stress corrosion, surfaces can be treated to induce compressive stresses, e.g. by shot peening. This may delay the onset of cracking as they will occur only in the presence of tensile stresses. It should, however, be appreciated that once the layer has been removed or penetrated by corrosion it will no longer be effective.

(iii) *Protective coatings*

The protection of alloys, particularly steel, by organic coatings is commonly used in marine situations and this is an effective method of controlling corrosion fatigue but not necessarily fatigue itself. Tests on steel girders, already referred to, demonstrated the value of protective coatings in sea water[9]. Sprayed metallic zinc, epoxy systems and zinc silicates all prevented corrosion in those tests. It should be noted that welds are potential problem areas when structures are protected by coating systems. As these are also areas where corrosion fatigue can be initiated, particular attention to coating procedures is required at such areas.

Electro-deposited coatings have also been used to control the corrosion fatigue behaviour of steel. Nickel is generally not considered particularly suitable, despite its excellent atmospheric corrosion resistance because of the tensile stresses in the coating which decrease the fatigue strength. This can be overcome by using a nickel coating with an internal compressive strength, which can be produced by using a stress-reducer such as saccharin or by using a nickel sulphamate bath.

Zinc plating produces a coating in compression and is resistant to sea water. It should be noted, though, that electrodeposited zinc coatings are usually only a few microns thick and will have a strictly limited life in sea water. It has been shown in some test work that electrodeposited zinc is superior to hot-dipped galvanised and diffused, sherardised coatings, but this may not be the case over long periods of exposure to sea water.

Aluminium sheet material may be clad with more corrosion-resistant alloys and organic coatings are frequently applied to aluminium alloys.

(iv) *Cathodic protection*

It has been demonstrated in various investigations that the corrosion rate is reduced and the corrosion fatigue performance improved by cathodic polarisation. In one series of experiments, on mild steel exposed to a sodium chloride solution, it was shown that cathodic protection was able to remove the effects of corrosion and an S–N curve was obtained, similar to that in air. No detrimental effects were apparent from the cathodic protection process. Caution is required with high strength steels as there may be a danger of hydrogen embrittlement arising from the application of cathodic protection. A recent paper[12] summarises the situation regarding the cathodic protection of welded steels subject to fatigue in offshore structures. In some situations polarisation is beneficial, whereas in others it may be detrimental so far as corrosion fatigue life is concerned. The author concludes that caution must be exercised with regard to overdesign of cathodic protection systems and potential distribution or uniformity. It has been postulated that one advantage of cathodic protection in arresting crack growth arises from the development of calcareous deposits within the crack with a corresponding reduction in the effective intensity range.

Work in progress will provide data on which a further assessment can be made of the advantages of cathodic protection in reducing the influence of corrosion fatigue of steels in sea water.

(v) *Materials*

The choice of materials with improved corrosion fatigue characteristics is possible in some situations. Corrosion resistant materials based on their static performance will not necessarily have improved corrosion fatigue properties, particularly if their corrosion resistance arises from protective films. The importance of corrosion fatigue as a limiting factor in design is being increasingly recognised, and Cohen[19] in a paper describing problems in the aerospace industry makes the point that the trend to develop aluminium alloys having high ultimate strength without taking into account their increased susceptibility to corrosion fatigue (and stress corrosion) is being questioned. The current demand is for increased structural reliability and high ultimate strength is no longer always the primary requirement in alloy selection. Other factors—such as fracture toughness and corrosion fatigue resistance—are considered to be very important in the selection of aluminium alloys.

4.7.3 Summary

Most constructional alloys when subjected to cyclic stressing in corrosive environments may be susceptible to corrosion fatigue depending on the service conditions. Situations that may cause local high stresses with even a small amount of plastic deformation are likely to be sites for corrosion attack and should, where practicable, be avoided. The use of techniques developed from fracture mechanics is providing data on which predictions of performance and life of components can be assessed. Methods of control such as coatings, material selection and the application of surface compressive stresses can be used to improve performance under corrosion fatigue conditions.

4.8 FRETTING CORROSION

Fretting corrosion is the damage that occurs at the interface of two closely fitting surfaces when they are subjected to slight relative movement or slip. Although the terms fretting and fretting corrosion are both used, often indiscriminately, some fretting is really mechanical and more akin to wear than to corrosion. The corrosion element is not necessarily electrochemical because fretting can occur in the absence of electrolytes and in this sense is an oxidation process. A typical example is shown in *Figure 4.15*. The mechanisms of fretting corrosion are essentially based on

Figure 4.15 Fretting corrosion

wear and oxidation. One view of the mechanism is based on the concept that reactions between the two moving surfaces leads to the rupture and fragmentation of the metal. These small fragments, partly because of the heat due to friction, are oxidised. The process continues, leading to the loss of metal and accumulation of oxide particles. Another view of the process is based on the presence of oxide films on alloys, which during the oscillation of the surfaces, are removed allowing further oxidation to occur. The final result is the same—the surface is damaged and oxides are produced. The presence of an oxide film is not, however, essential because even noble metals can be damaged by fretting and the accumulated material is not necessarily composed only of oxides.

The mechanism of fretting corrosion probably varies with different materials and conditions. The main requirements for attack to occur are (i) the materials must be touching and under load, (ii) relative motion must occur between the two surfaces at a suitable frequency.

The factors influencing the degree of fretting can be summarised as follows:

(i) *Environment*
Oxygen is not essential (fretting occurs in inert atmospheres), but generally its presence increases the damage. The effect tends to lessen with higher relative humidities.

(ii) *Frequency and amplitude of motion*
The amplitude of motion to cause damage is usually less than 25 µm and it can occur at much smaller amplitudes. For steels there is a reduction in damage as the frequency is increased to 10 Hz, but at frequencies higher than this, there is some evidence to suggest that fretting damage increases again.

(iii) *Temperature*
At temperatures of 130–200 °C, a thick oxide glaze may be formed instead of the loose oxide debris and this generally results in less metal loss than at lower ambient temperatures.

Fretting can occur with most of the constructional alloys. Spliced couplings, press fits and bolted assemblies are all vulnerable and where fretting occurs it can lead to loss of dimensional accuracy and the reduction in fatigue strength of a component (see Section 4.9). This arises because the loosening of the component permits excessive strain and the pits formed by fretting act as stress-raisers. Components that are unlikely to suffer from fretting in service can be damaged during transit, because of vibrations arising from movements in vehicles or ships. The fretting resistance of different materials will vary with circumstances but the following combinations have been listed as having poor fretting resistance:

Aluminium/cast iron
Aluminium/stainless steel
Cast iron/chromium plating
Hard tool steel/stainless steel
Chromium plating/chromium plating

(iv) *Control methods*

(a) Where surfaces are not intended to move relative to each other, attempts should be made to prevent such movement, either by eliminating the cause of vibration or by increasing the friction between the surfaces. Bearing surfaces that will be subjected to vibration during transit may be coated with lead. The lead is soon removed when the bearing is in service.
(b) The use of grease may reduce or exclude oxygen and reduce the amount of contact between the surfaces.
(c) Steel bearing surfaces can be phosphated and impregnated with oil.
(d) Where appropriate, the amplitude of vibration may be lowered.

4.9 FRETTING FATIGUE

Fretting can reduce the fatigue strength of an alloy and this has been demonstrated in tests where a fatigue curve has been obtained using specimens to which is attached a device to cause fretting damage[14]. Only a small amount of damage to the surface is necessary to cause this considerable decrease in fatigue strength of some alloys. The sensitivity to fretting is generally greatest in the low slip amplitude range and this tends to be the level of slip that occurs on many components.

Fretting may accelerate the development of fatigue cracks by super-imposing on the cyclic stresses necessary for fatigue additional high intensity stresses by the following mechanisms: (i) the development of pits, which act as stress raisers, (ii) local stress intensity caused by the fragments of oxide and other materials formed between the two surfaces. Fretting fatigue has become of increasing concern in some fields and titanium—which is generally a highly corrosion-resistant alloy—appears to be susceptible to this form of attack, as are other high strength alloys used in aircraft construction. Unlike many corrosion processes where marine conditions lead to increased attack on alloys, fretting is not generally worse in such environments as compared with others.

4.10 FILIFORM CORROSION

This form of corrosion is not particularly important in relation to marine engineering but will be briefly discussed as it is somewhat unusual. Filiform corrosion occurs under protective films and particularly under lacquers used for office furniture or 'tin cans'. It does not affect the metal to any extent, mainly affecting appearance. It progresses into a network of narrow filaments (usually less than 2 mm wide) under the lacquers. It is considered to arise at relative humidities between 65 and 90% and presumably arises initially from salts on the surface. Moisture reaches the surface of the metal because of the permeability of the thin coatings and some workers consider it progresses as a form of crevice attack.

4.11 BIMETALLIC CORROSION

The practical aspects are dealt with in some detail in Chapter 7 'Design'. It is also described as 'galvanic corrosion' and may occur whenever two metals or alloys are connected in the presence of an electrolyte. As explained in Chapter 7, there are, however, many practical factors that will determine the likelihood or extent of attack.

4.12 HIGH-TEMPERATURE OXIDATION

Although, for convenience, high-temperature oxidation has been included in this chapter, it does not fall into the same category as the other forms considered. The others, with the possible exception of fretting, are variations of aqueous corrosion, taking into account factors such as stress and local conditions, either on the alloy surface or in the environment. High-temperature oxidation is basically different because it is 'dry' corrosion and does not involve electrolytes. It is not, therefore, really a form of corrosion that is particularly relevant to marine conditions but for completeness it will be briefly discussed here.

As explained in Chapter 2, 'Principles of Corrosion', in 'dry' corrosion the corrosion product forms on the metal surface so, provided the film is continuous, further corrosion or oxidation, as it is generally called, can occur only by movement of ions through the oxice film. It follows, therefore, that oxidation usually follows a course in which the rate decreases with the increase in film thickness, i.e. it drops off markedly with time. If, however, the oxide film is not continuous or for some reason deteriorates in some way, then the corrosion-time curve will be similar to that typically obtained in aqueous corrosion.

The most important practical consideration when considering alloys exposed to elevated temperatures, i.e. over about 450 °C, is the type of oxide formed on the surface. Early work by Pilling and Bedworth published in 1923 took into account the fact that oxide films have a different volume from the metal involved in the oxidation process. However, unlike aqueous corrosion, where rust always has a greater volume of the steel from which it is produced, some oxides have a smaller volume than the steel from which it is produced, some oxides have a smaller volume than the metal from which they are formed whereas others have a greater volume. These workers argued that the oxide layer would be continuous if its volume was greater than that of the metal consumed and conversely that the layer would be porous and discontinuous if the volume was less than that of the metal involved in its formation. They based this theory on assumptions that have since been proved to have a limited validity; consequently what was at one time considered to be a useful rule, i.e. that oxidation resistance was related to the volume ratio of oxide and metal, is not now accepted as being correct. Many investigations have been carried out into oxidation processes and there are few simple rules for the engineer to follow. Generally, alloys are chosen to suit the temperature conditions. In atmospheres where the process is one that primarily involves oxygen, then the temperatures are known below which alloys may be expected to perform with only a small loss of metal due to oxidation. However, in more complex atmospheres containing SO_2, SO_2 and steam, general data is less readily available. In oxygen, the oxidation process can be represented as follows:

$$M \rightarrow M^{+2} + 2e \qquad \text{(metal–scale interface)}$$

$$\tfrac{1}{2}O_2 + 2e \rightarrow O^{2-} \qquad \text{(scale–gas interface)}$$

i.e. $M + \tfrac{1}{2}O_2 \rightarrow (M^{2+}O^{2-})$ in the oxide crystal lattice

The process continues with new sites produced in the lattice at one or other of the interfaces.

In dealing with the effects of high temperature oxidation, the aim is to provide an alloy, which at the required temperature will react to produce a protective oxide film. The alloy must also of course retain its strength, creep-resistance and other required properties at the working temperature. Generally, ordinary carbon steels are resistant at temperatures up to about 450 °C provided the environment is only mildly

Table 4.5 Alloys for high-temperature service

Alloy type	Main constituents†				Maximum scaling temperature (°C) for use under oxidising conditions
	Cr	Ni	Mo	Si	
Carbon steel					450
Ni-Cr-Mo	1	1.5	0.3		540
Cr-Mo	2.5		0.75		600
Cr-Mo	8		0.75		700
Cr-Si	8			3	800
Stainless steels					
410	12				750
430	17				840
304	18	8			900
446	27				1030
309	24	12			1090
310	25	20			1150
Co base special alloys					> 1150

† The percentages provide only a general indication of content.

oxidising and not particularly corrosive. For higher temperatures, alloy steels are required. The maximum temperature at which low alloy steels can reasonably be used is about 700 °C. Above this temperature, chromium steels are used, essentially stainless steels up to about 900 °C. Above this temperature, a number of nickel-chromium alloys will resist temperatures up to about 1100 °C. A broad indication of the scaling resistance of alloys is shown in *Table 4.5*. Only the main constituents of the alloys are given, although the carbon manganese, sulphur and phosphorus levels may be important in the low-alloy steels. In practice, either standard or proprietory alloys would be purchased.

For readers who wish to study high-temperature oxidation in more detail, the following references will prove to be useful:

KUBASCHEWSKI, O and HOPKINS, B E, *Oxidation of Metals and Alloys*, Butterworths, London
HAUFFE, K, *Oxidation of Metals*, Plenum Press, New York
SHREIR, L L (ed), Chapter 7, 'High-temperature corrosion', in *Corrosion*, 2nd ed, Vol 1, Butterworths, London (1976)

4.13 THE ROLE OF BACTERIA IN CORROSION

Bacteria can play a role, sometimes an important one, in the corrosion of metals and other materials such as concrete and wood. This form of corrosion can occur in many situations in both soils and waters. It is by no means confined to marine situations but does occur in sea water and can cause serious problems in oilfield equipment.

The nature of the corrosion depends upon the form of microbiological activity and may be quite complex. In this section only the more common types of bacterial attack will be considered. Microbiological processes can affect corrosion in a number of ways:

(i) By influencing the nature or rate of corrosion by their effect on the anodic or cathodic processes.
(ii) By creating a corrosive environment as a result of the metabolic process, e.g. by producing acid reaction products.
(iii) By producing deposits that may affect surface reactions, e.g. concentration cells.
(iv) By the production of surface films on the metal surface, e.g. sulphides.

Colonies of bacteria-producing substances may be in close contact with the alloy surface providing the necessary concentration of corroding agent. However, this is often maintained only locally and may not be detected in the surrounding environment. Bacteria causing corrosion in this way need only produce very small quantities of corrosive substances for very severe localised attack. Moreover, any bacterial colony growing on a metal surface can act as an oxygen screen which will set up a differential aeration cell. In other situations, the corroding medium may be generally infected with biological activity, which if it produces substances which are able to catalyse corrosion reactions, may render the environment corrosive even after the biological activity has ceased.

Bacteria will flourish only in the presence of certain substances such as organic food nutrients, e.g. sugars, inorganic salts, sulphates etc, necessary for their growth and multiplication. Sea water, especially estuarine water, contains many bacteria capable of causing corrosion, but in practice only certain groups cause problems. The process may be different in the presence of bacteria and the formation of products from the microbiological process may. cause problems, e.g. blockages of equipment, even though the actual loss of metal may not be increased beyond that anticipated when they are absent.

Often the different types of bacteria have more than one effect on corrosion and it is not always possible to classify them in simple categories. It is, therefore, probably more useful to consider the general effects of the different bacteria on corrosion.

Microbiological effects may arise from algae, slimes and other organisms that may not strictly be bacteria. Consequently, the terminology used may not necessarily be strictly correct in terms of microbiology. However, from a corrosion standpoint, it is the overall consequences of such organisms that are important.

(i) *Sulphate-reducing bacteria (Desulfovibrio desulfuricans)*
These are probably the most important bacteria encountered in marine situations. They derive their name from their microbiological activity in which they reduce inorganic sulphates to sulphides. There are other groups of bacteria that can reduce certain forms of organic sulphur compounds but the Desulfovibrio group are the most important in reducing inorganic sulphates.

Sulphate-reducing bacteria are able to cause corrosion of steel under anaerobic conditions, i.e. in the absence of oxygen. Generally, oxygen is an essential element in the corrosion of steel, but in the presence of these bacteria, the corrosion process differs from that experienced in normal oxygenated sea water. A good deal of investigational work has been carried out to determine the nature of the corrosion processes involved in the presence of such bacteria and an electrochemical basis was proposed by von Wolzogen Kühr nearly 50 years ago [16]. Other workers have also investigated the processes, and in broad terms, the following factors have to be taken into account.

(a) H_2S is formed by the reduction of sulphates but direct corrosion by H_2S would result in the formation of FeS (iron sulphide) as the only corrosion product.
(b) If, however, electrochemical corrosion occurs, the anodic and cathodic reactions would be:

$$4Fe \rightarrow Fe^{2+} + 8e^-$$

$$8e^- + 4H_2O + SO_4^{2-} \xrightarrow{\text{bacteria}} S^{2-} + 8OH^-$$

and combining (a) and (b) to eliminate electrons, the overall equation becomes:

$$4Fe + 4H_2O + SO_4^{2-} \rightarrow 3Fe(OH)_2 + FeS + 2OH^-$$

This equation agrees more closely with the corrosion products formed in oxygen-free neutral solutions than does the direct reaction:

$$Fe + H_2S \rightarrow FeS + H_2$$

It has been suggested that the bacteria actually depolarise the cathode and this leads to an increase in the corrosion rate. However, the removal of adsorbed hydrogen atoms from the cathode depends on the presence of the enzyme hydrogenase, which allows the bacteria which produce it to catalyse the reaction:

$$8H_{ads} + SO_4^{2-} \rightarrow 4H_2O + 2^{2-}$$

The adsorbed hydrogen atoms are supplied, under anaerobic conditions, by the cathodes of active corrosion cells.

At one time it was assumed that only bacteria producing hydrogenase caused corrosion of steel. Later work showed, however, that strains of sulphate-reducing bacteria not producing this enzyme corroded steel at the same rate as those that produced it[17]. This was true, provided the partially protective ferrous sulphide film was not allowed to form on the surface. When, however, a partially protective film was allowed to form on the surface, the corrosion rate for the bacteria producing hydrogenase was greater than for those not producing the enzyme.

Apart from the ability of the bacteria to increase the rate of electrochemical corrosion, there is a direct corrosion arising from the attack of hydrogen sulphide, generated by these bacteria, on steel with the formation of ferrous sulphide.

The presence of these bacteria is uaually associated with a distinct 'sulphide' smell. Both ferrous and non-ferrous metals generally corrode at a higher rate in the presence of hydrogen sulphide, partly because sulphide films provide effective cathodic areas.

(ii) *Detecting the presence of sulphate-reducing bacteria*

Attack by sulphate-reducing bacteria is usually associated with black corrosion products and a smell of hydrogen sulphide. The steel may be quite bright under the corrosion products. Graphitisation may well occur with cast irons. These bacteria are likely to occur in stagnant situations and under deposits or debris in low-velocity flow situations, in tanks, and in pipes buried in water-logged coastal areas. They are also commonly to be found in muds and silts in harbours.

Various methods are available to determine their presence in samples obtained for such tests. Details can be found in reference books and papers. A useful guide is provided in a NACE publication[18]. Specific methods have been devised for particular conditions. The sample taken is cultured and a count of the bacteria is made. The culture medium varies but generally it must be anaerobic and sterile and should contain a suitable lactate to allow oxidation to acetic acid. Ferrous ions should be available in the medium as the formation of black areas arising from reaction with inorganic sulphates, also in the medium, indicates reduction to sulphide. An agar gel medium may be used or a liquid, sometimes called a broth, may be preferred.

The sampling is important because the bacterial activity may be localised and samples taken some way from the steel surface may not provide a true indication of the number of vacteria present. Although, with practice, it is possible for non-specialists to test for the detection of bacteria, it is not always a simple matter and it may be advisable to have such tests carried out by laboratories specialising in such work. The mere detection of sulphate reducing bacteria does not mean that serious problems will arise, although any indication of their presence should be further investigated.

(iii) *Methods of treatment and control of sulphate-reducing bacteria*

(a) In enclosed areas sterilisation and treatment by disinfectants or biocides may be economic.

(b) The environmental conditions can sometimes be changed to make them unfavourable for bacterial growth, e.g. removal of water from oils or by blowing air into water where the activity is in progress. Environmental control is generally limited to enclosed situations.

(c) Coatings of a suitable nature will control bacterial attack in many situations provided the coatings do not have any defects and are resistant to bacterial attack themselves. This should be confirmed by reference to the coating supplier because some coating materials that are highly resistant to sea water may not be resistant to all bacteria.

(d) Cathodic protection is effective in counteracting bacterial attack (see Chapter 16).

(e) Some alloys may continue to be affected even after they have been removed from the original area of bacterial activity. For example, copper alloys may be affected by the hydrogen sulphide formed during bacterial attack, which reacts to form a less resistant surface protective film than the normal oxide film. Such films should be removed as soon as possible by thorough washing. This may occur on ships' condenser tubes where polluted sea water has been used.

(iv) *Other forms of bacteria and micro-organisms*

Although sulphate-reducing bacteria are the most important, other groups of micro-organisms may cause problems in marine situations.

(a) Slime forming bacteria can occur in various situations and may contain a number of different micro-organisms which can become active by obtaining energy from organic sources present in deposits or sea water. They may grow under anaerobic or aerobic conditions or under both sets of conditions, i.e. they are facultative. Slime-forming bacteria may cause problems by blocking equipment and their presence on steel can also lead fo the formation of concentration cells by their shielding action of the metal surface.

(b) Certain bacteria produce corrosive environments in a manner similar to the H_2S produced by sulphate-reducing bacteria. Many bacteria

Table 4.6 Summary of methods of controlling different forms of corrosion

Control method	Pitting	Crevice	Selective leaching	Intergranular corrosion	Erosion corrosion	Impingement	Cavitation	Stress corrosion cracking	Corrosion fatigue	Fretting corrosion	High temperature oxidation
More resistant alloy	●	●	●	●	●	●	●	●	○	×	●
Coatings	○	●	×	×	●	●	●	●	●	×	○
Design	●	●	×	×	●	●	●	●	●	●	×
Heat treatment	×	×	×	●	×	×	×	●	○	×	×
Fabrication control	×	●	×	●	○	○	○	●	●	●	×
Reduce stress	×	×	×	○	×	×	×	●	●	×	×
Limit cyclic movement	×	×	×	×	×	×	×	○	●	●	×
Prevent stagnant conditions	●	●	○	×	×	×	×	×	×	×	×
Alter velocity of water	●	●	○	×	●	●	●	×	×	×	×
Alter environment	○	○	×	×	○	○	○	○	○	○	○
Cathodic protection (immersed)	○	○	×	×	○	○	○	●	○	×	○

●, suitable; ○, limited suitability; ×, not generally suitable. This is a general summary. In some situations, other methods may be suitable and the methods indicated may not be effective.

found in sea water produce nitrogen compounds as by-products of their normal metabolism. The most important from the corrosion standpoint is ammonia (NH_3). Bacteria which work in the nitrogen cycle can also produce nitrates, nitrites and amino-acids, which may have effects on corrosion reactions, especially on the pitting of copper-base alloys.

(c) Other micro-organisms such as iron-oxidising bacteria, algae and yeasts will not be discussed here and the reader is referred to a NACE publication 'The Role of Bacteria in the Corrosion of Oil Field Equipment' (TPC Publication No 3) for useful information on these organisms[18].

Apart from NACE publication TPC3 a useful guide to bacterial attack is provided in 'Fundamentals of bacteria induced corrosion', by R E Tatnall, *Materials Performance*, **20**, 9, pp 32–38 (1981).

A summary of the control methods that can be applied to the different forms of corrosion is given in *Table 4.6*.

4.14 REFERENCES

(1) LARRABEE, C P, *Trans. Electrochem. Soc.*, **85**, 297 (1944)

(2) DEARDEN, J, *J. Iron Steel Inst.*, **159**, 241 (1948)

(3) HUDSON, J C, *The Corrosion of Iron and Steel*, Chapman & Hall, London (1940)

(4) EVANS, U R, *Industr. Eng. Chem.*, **17**, 363 (1925)

(5) TOMASHOV, N D, *Theory of Corrosion and Protection of Metals*, Collier-Macmillan, London (1966), p 470

(6) HOLZWORTH, M L, BECK, F H and FONTANA, M G, *Corrosion*, **7**, 441–449 (1951)

(7) *Stress Corrosion Guide to Practice in Corrosion Control*, No 4, Department of Industry, UK

(8) POOLE, P, in L L Shreir (ed), *Corrosion, Vol 1, Section 8.4, Butterworths, London (1976)*

(9) DE BACK, J, *Corrosion Fatigue: Chemistry, Mechanics and Microstructures, NACE International Conf. Series*, **2** (1972), p 529, National Association of Corrosion Engineers, Houston, Texas, USA

(10) BOOTH, G S, *Proc. UK Natural Corrosion Conf.* (1982), Inst. Corr. Sci. Tech., London

(11) SCOTT, P M, *Proc. UK Natural Corrosion Conf.* (1982), Inst. Corr. Sci. Tech., London

(12) HARTT, W H, *Materials Performance*, **20**, 11 (1981)

(13) PARKINS, R N and CHANDLER, K A, *Corrosion Control in Engineering Design*, Department of Industry (1978)

(14) WATERHOUSE, R B, *The Effect of Fretting Corrosion in Fatigue Crack Initiation, NACE International Corrosion Conf. Series*, **2** (1972), National Association of Corrosion Engineers, Houston, Texas, USA

(15) LAQUE, F L, *Materials Performance*, **21**, 4 (1982)

(16) VON WOLZOGEN KUMHR, C A H and VAN DER VLUGT, L S, *Water*, **18**, 147 (1934)

(17) BOOTH, G H, COOPER, P M and WAKERLY, D S, *Br. Corros. J.*, **1**, 9 (1966), p 345

(18) TPC Publication No 3, *The Role of Bacteria in the Corrosion of Oil Field Equipment* (1979), National Association of Corrosion Engineers, Houston, Texas, USA

(19) COHEN, B, *Corrosion Fatigue, NACE International Corrosion Conf. Series*, **2** (1972), p 65, National Association of Corrosion Engineers, Houston, Texas, USA

5 Ferrous alloys

There are four main groups of ferrous metals to be considered in relation to marine corrosion:

Ordinary carbon and mild steels
Low alloy steels
Cast iron
Stainless steels

5.1 CARBON STEEL

Carbon steel is the most widely used constructional alloy for all types of marine conditions and it is rarely used without some form of applied corrosion control, usually a coating or cathodic protection. It does not, however, follow that the corrosion characteristics of bare steel are of only academic interest. Situations do arise where steel is deliverately used bare, e.g. for some piling, cargo holds of bulk carriers and most underwater steelwork of offshore structures and, of course, in any situation where the coating fails, corrosion of the steel will occur.

Carbon and mild steels are essentially alloys of iron with small amounts of carbon and manganese, together with phosphorus, sulphur, silicon and other minor elements that arise from the manufacturing process or are added deliverately to meet certain mechanical or physical requirements of the steel. Of these alloying elements, only one has any significant influence on the corrosion of steel when added in the small amounts under consideration, i.e. copper. Additions of up to about 0.2% give a marked improvement in corrosion-resistance, in many atmospheric situations, particularly where there are industrial pollutants such as sulphur dioxide (SO_2).

5.1.1 Rusting

In very clean dry air at normal ambient temperatures, a thin protective oxide film forms on the surface of polished mild steel. Unlike the films that form on stainless steel and many non-ferrous alloys, however, it is not particularly protective and in the presence of electrolytes on the surface it provides virtually no barrier to them. Consequently, water and dissolved atmospheric oxygen react with the steel and form rust:

$$4Fe + 3O_2 + 2H_2O \rightarrow Fe_2O_3 . H_2O \qquad \text{(rust)}$$

These are the essentials of what is a complex series of reactions that occur in ordinary sea water or marine atmospheres. (For a detailed explanation, see Chapter 2.) Other reactions also occur. In situations where the supply of oxygen is restricted, Fe_3O_4 (magnetite) may be formed. The essential role of oxygen and moisture (or water) are illustrated in these reactions, but severe corrosion can also occur under anaerobic conditions in the presence of certain bacteria, e.g. sulphate-reducing bacteria (*Desulfovibrio desulfuricans*) which may be present in soils, mud, silt and water (see Section 4.13). Under alkaline conditions steel passivates so the corrosion is negligible, e.g. reinforced steel in concrete. However, the presence of certain ions, particularly chlorides, even under alkaline conditions can lead to a breakdown of passivity and corrosion of the steel.

It follows from the above that steels can corrode under most circumstances, whether in air, immersed or buried. The loss of metal, which from the engineering standpoint is of major importance, will be determined by a number of factors considered below but the corrosion product itself, i.e. rust, is of almost equal importance. Although rust is essentially $Fe_2O_3 . H_2O$, other small amounts of corrosion products, particularly ferrous salts, have a considerable influence on its properties. Rust is often protective to steel, so that corrosion in many circumstances is not linear although the degree of protection depends on the actual composition of the rust and the conditions of its formation. In situations where the rust is formed on the steel surface, this may have a stifling effect with a reduction in the rate of corrosion. On the other hand, where the rust is either continuously removed or formed away from the surface, e.g. in sea water, this will have less influence on the rate of corrosion of the steel.

Apart from its influence on corrosion, rust has a marked effect on the performance of coatings if they are applied over it (see Chapter 9). This arises mainly from the presence in rust of salts such as ferrous sulphate and ferric chloride. These are formed by reactions between the steel and pollutants and contaminants from the atmosphere or sea water. Sulphur dioxide is present in industrial atmospheres and chloride is always present in marine situations. Other reactions within the rust may also have a marked influence on its ability to act in a protective fashion. As already noted, small additions of copper, up to about 0.2%, have a significant influence on the corrosion rate of steel in air. This improvement is considered to arise from the effect copper has on the rust film, possibly the formation of basic copper sulphate[1], rather than on any improvement in the corrosion-resistance of the steel itself.

5.1.2 Corrosion in air

The main factors in determining the rate of corrosion in air are (i) the time the steel surface is moist, (ii) the extent of pollution and contamination of the atmosphere.

(i) *Moisture*

Steel is often visibly wet from rain, from dew or condensates on the surface and from sea-water spray. There are many occasions, however, when the steel is not obviously moist but nevertheless thin layers of electrolyte are present on the surface. This important fact was first demonstrated by Vernon[2] in a series of classical experiments in which he showed that in 'pure air' rusting is minimal at relative humidities below 100% but that in the presence of small concentrations of impurities such as sulphur dioxide, serious rusting can occur without visible precipitation of moisture once the relative humidity of the air rises above a critical and comparatively low value. The value depends upon the nature of the pollutants or contaminants present in the atmosphere. For sulphur dioxide it is about 65–75% but when chlorides are present, rusting may occur at a relative humidity as low as 40%[3]. As the important factor is the 'time-of-wetness' of the steel surface, corrosion rates are likely to be higher in many marine atmospheres because the steel remains wet for longer periods than at inland sites.

(ii) *Pollution and contamination*

Although relative humidity plays an important role in determining the probability that corrosion will occur, the nature of the moisture on the surface will have a determining effect on the rate of attack. In industrial atmospheres, sulphur dioxide is the main pollutant that affects corrosion, although other contaminants such as ammonium salts also have an effect. Near chemical or metallurgical plants, the effect of local pollution can, of course, be particularly severe. Such plants on coastal sites can produce very aggressive conditions.

Under marine conditions, the influence of the salt content of the air is very marked, as shown in *Figure 3.3*, which gives the results of tests carried out on a surf beach in West Africa. The rate of corrosion 50 yards from the surf—0.9 mm/y—is one of the highest ever recorded and would be influenced not only by the salt content but by the continuous wetting and drying action. Similar effects of salt concentrations with distance from the sea and its effect on corrosion rates of steel have been obtained in tests carried out at the test station at Kure Beach, North Carolina, in the United States.

(iii) *Orientation of steel*

As the orientation of the steel in a structure or a building influences the time of drying of the surface, it has some influence on the corrosion rate. These effects also influence test results obtained by exposing steel coupons because the results may differ significantly with the mode of exposure (see Chapter 21, 'Testing'). The groundward side of a horizontal surface is protected from rain but it is also shielded from the drying effects of the sun and often of the wind, so that dew tends to remain in contact with the steel for longer periods. Furthermore, particularly near the coast, salts tend to collect on the underside of steel and are not washed away. Bridges near the coast and steel in harbour installations may be affected in this way although the design and the distance vertically from the water will influence the corrosion rate. Wherever the design or orientation encourages prolonged dampness of the surface, corrosion will be increased. In American tests on specimens exposed at 30° to the horizontal, over 60% of the loss was on the underside[4]. The relative corrosion on the opposite faces of vertically exposed steelwork will depend on the prevailing wind but generally in northern latitudes, the north-facing side will remain damp for longer periods than south-facing surfaces.

(iv) *Effect of mass*

The effect of steel mass and its relation to the time that dew remains on the surface is of importance when painting steel because of the lag between the ambient and steel temperature. It has been shown[5] that this effect of mass influences the corrosion rate of steel; for example, steel plate of 55 mm thickness corroded at 38 μm over a period of 12 months compared with 30 μm for a steel of thickness 5 mm.

5.1.3 Corrosion rate of steel exposed to marine atmospheres

Many tests have been carried out to determine the corrosion rate of steel in marine atmospheres. They cannot be directly compared because the conditions are not the same for all the tests. Furthermore, as has already been discussed, the corrosion rate is markedly influenced by the distance of the test site from the sea and the nature of the wind and waves wetting the steel and providing chloride particles to the surface. Consequently, the results obtained from such tests must be treated with caution with respect to their use as corrosion data. Furthermore, much of the data for the corrosion of carbon steel has been produced from short-term tests. Often the rate of corrosion drops markedly with time. The following factors influence the rates to a marked degree. The results quoted are from ASTM Special Technical Publication 435 (1968).

(i) Distance from the sea

Cape Kennedy $\frac{1}{2}$ mile from ocean	0.18 mm/y
60 yards from sea	0.90 mm/y

(ii) Height above ground

Cape Kennedy 60 yards from sea

ground level	0.90 mm/y
30 ft	0.33 mm/y
60 ft	0.28 mm/y

(iii) Actual location, i.e. effects of topography, wind, rain, etc

Dungeness, England	0.98 mm/y
Point Reyes, California	1.00 mm/y

In what might be described as marine atmospheres but either some distance from the coast or where there is little overall salt spray, the corrosion rate is similar to, but somewhat lower than, that experienced in industrial atmospheres. This is illustrated by the following results [5]. These results are not directly comparable with those noted above.

Type of atmosphere	Site	Corrosion rate (mm/y)
Industrial	Sheffield, England	0.14
	Pittsburgh, US	0.11
Marine	Calshot, England	0.08
	Apapa, Nigeria	0.03
	Sandy Hook, US	0.08
Marine/industrial	Congella, South Africa	0.11

5.1.4 Immersed conditions

The factors that affect steel fully immersed in sea water are complex and in some cases appear to be contradictory. This is because there is an interaction between the different factors and sometimes one predominates compared with the others; for example, the oxygen concentrations drop markedly with depth to about 2000 ft although the salinity may increase. The main factors are considered below in general terms and should not be considered as data suitable for predicting the actual loss of steel in a particular situation. This can only be done when all the factors are correctly interrelated and preferably when specific test or service data have been obtained.

(i) *Sea-water properties*

The high salt content of sea water compared with other waters results in a much higher electrical conductivity and this may lead to higher corrosion rates, but more importantly it permits serious pitting in situations where small anodes are in contact with large cathodic areas. This is a particular problem when different metals are connected together but if hot-rolled steel is immersed with large areas of millscale still on the surface, pitting of the exposed steel (anodic) areas can occur. Pitting depths of 1.5 mm have been reported after six months' immersion in sea water[6].

Salinity varies in different parts of the oceans and at different depths. Furthermore, there are seasonal variations, as there are also with temperature and oxygen content. The effect of salinity or salt concentration on corrosion appears to be greatest at about 10 g/l NaCl.

Tests on steels carried out under fully immersed conditions, usually at a comparatively low depth from the surface indicate a range of corrosion rates from about 75–125 µm/y, although higher rates up to 180 µm/y have been measured in tropical waters over one year[7]. The rate may fall significantly with time depending on the circumstances, e.g. in the tests quoted above[7], over a period of four years, the rate fell from about 100 µm/y to 25 µm/y. The state of the steel surface, i.e. whether it is covered with millscale or cleaned by blast-cleaning or pickling, will influence the general corrosion rate initially and will have an even more important effect on pitting. The main factors that influence the rate of corrosion of steel in sea water are:

Velocity
Oxygen content
Temperature
Formation of calcareous deposits
Marine growths
Depth of immersion

Of these, velocity and depth of immersion have a marked effect, partly because of their influence on oxygen.

(ii) *Depth*
Tests carried out in the Pacific Ocean by the US Naval Civil Engineering
Laboratory and Naval Research Laboratory on the effect of depth on
steel corrosion are shown below. At lower depths, the temperature and
oxygen concentrations are lower than at the surface but as with surface
situations, the corrosion rate will show variations with the site of testing.
In other investigations, e.g. in the Atlantic, the oxygen concentration was
greater than in the tests quoted below.

Depth (m)	Corrosion rate (mm/y)
0	0.13
704	0.04
1600	0.02
1700	0.02
2050	0.06

(iii) *Velocity*
In one series of tests, corrosion under static conditions was 0.125 mm/y,
0.50 mm/y at 5 ft/s and 0.83 mm/y at 15 ft/s. Although in many
situations the corrosion of carbon steel would be expected to rise with
increase in velocity of sea water, the actual corrosion rate would in
practice depend on the actual conditions of service.

(iv) *Temperature*
The surface temperature of sea water varies between about -2 °C and
35 °C, the temperature rising with the lowering of latitudes. There are
seasonal variations depending on the prevailing winds, tidal streams and
elevation of the sun. These variations, as would be expected, have some
influence on the corrosion of steel but the differences in temperature also
affect conditions in other ways, e.g. by growths of marine organisms with
a reduction in oxygen diffusion to the steel surface and by oxygen
solubility. In general terms, over the range of sea water temperatures,
there is a reasonably linear relation with a doubling of corrosion rate for a
rise of about 10 °C in the water temperature. In practice, the effect may
be less marked for the reasons noted.

The use of sea water in saline distillation plants and condensers has led
to much higher temperatures than those experienced in ambient sea
water and under these conditions steel corrodes at a markedly higher
rate.

The erosion of steel tends to be more severe as the temperature
increases.

(v) *Calcareous scales on steel*

Although sea water is predominately sodium chloride, it also contains significant amounts of calcium. Scales, mainly of carbonate, may form at cathodic areas during normal corrosion and this reduces the corrosion rate of the steel, (see Chapter 3).

5.1.5 Corrosion rates of steel under immersed conditions

The corrosion rate of steel fully immersed in sea water, obtained in various tests, varies but generally it falls within the range of 50–150 µm/y, typically about 100 κm/y. The data in the table below illustrates the general corrosion rates to be anticipated in long-term tests. In a series of tests carried out on bare steel piles[8], the corrosion rate ranged from 25–50 µm/y for the entire area immersed in sea water.

Often—on static structures—fouling growth may act to shield the steel with a consequent reduction in corrosion.

Site	Testing organisation	Test duration (y)	Average corrosion rate (mm/y)	Maximum pit depth (mm)
Emsworth, England	BISRA[9]	5	0.065	—
Halifax, Nova Scotia	Institution of Civil Engineers[10]	15	0.108	1.90
Plymouth, England	Ditto	15	0.065	1.70
Colombo, Sri Lanka	Ditto	15	0.090	1.65
Auckland, New Zealand	Ditto	15	0.075	1.10

5.1.6 Corrosion at half-tide level

The level between high and low tide and the general area above it is often termed the 'splash zone'. The corrosion rates may be very high in this area because of the constant wetting with salt spray. This particular area is not always subjected to high rates of attack, for example on some retaining piles dense marine growth may provide some protection; in these situations, the corrosion rate may be similar to that under immersed conditions.

On the other hand, in exposed offshore conditions, corrosion rates of 0.4 mm/y or more may be experienced. At tropical surf beaches, even higher rates may be experienced, e.g. 0.9 mm/y (see *Figure 3.3*).

5.1.7 Stress corrosion cracking

Ordinary carbon steels are not susceptible to stress corrosion cracking in normal marine environments. High strength steels may be susceptible to hydrogen cracking (see Section 4.6.2).

5.2 LOW-ALLOY STEELS

Variations in the compositions of ordinary carbon steels have little influence on their corrosion-resistance with the exception of copper. The addition of about 0.2% of copper results in a 2–3-fold reduction in the corrosion rate in air compared with a copper-free steel. Variations in other elements in carbon steels affect the corrosion rate to only a marginal degree. Generally, the rate tends to decrease with increasing content of carbon, manganese and silicon. For example, freely exposed in the atmosphere, a steel containing 0.2% silicon rusts about 10% less rapidly than an otherwise similar steel containing 0.02% silicon.

In practice, many steels contain up to 0.1% of copper, introduced through scrap, so it is not always easy to compare the corrosion characteristics of so-called copper steels and ordinary steels. Nevertheless, there is clear evidence from many test programmes to show the advantages of a 0.2% copper addition to steel[1]. The improvement in corrosion-resistance, incidentally, does not significantly increase with additional copper over the 0.2% level. Copper-containing steels have been marketed, e.g. for steel piles, but they are not generally considered to be in the category of low-alloy steels. Low alloy steels generally contain 1–3% of alloying elements and those commonly used for atmospheric purposes contain about 0.8Cr and 0.4Cu as the main alloying elements, with higher levels of silicon and phosphorus compared with ordinary unalloyed steels. This type of steel was developed by United States Steel Corporation and is called 'COR-TEN' steel. Steel with the same proprietary name is made under licence in many other countries and there are also other proprietary steels, some containing other alloying elements such as nickel, but all have less than about 3% total alloy content.

Low-alloy steels were originally introduced to provide steels of increased strength as well as additional corrosion- resistance, particularly for purposes such as railway wagons, and are sometimes referred to as high-strength low-alloy steels (HSLA). Other low-alloy steels containing chromium and molybdenum have been developed for oxidation-resistance (see Section 4.12) but these are a different group from those being considered here.

5.2.1 Atmospheric exposure

The HSLA steels have been used, without protective coatings, for buildings and structures. They became known as 'weathering steels' and have been used in many countries. Under freely exposed atmospheric conditions, the steels corrode at a lower rate than unalloyed steels. There is little difference in the initial corrosion rate compared with ordinary steels say over the first nine months to a year, but over an extended period, the rate of corrosion drops. The mechanism of corrosion of low-alloy steels has not been clearly established but the type of rust formed plays a significant role in reducing the corrosion rate.

Although weathering steels have been used for structures and buildings in coastal situations, they are probably more suited to non-marine environments. The rusts formed on these steels at coastal sites tend to be more flaky than those formed in industrial environments. There is also more localised corrosion similar to widespread shallow pitting. Although no definite rules regarding their use near the coast have been established, it is probably reasonable to consider restricting their use at distances less than 1000 m from the sea without specialist advice to the contrary. Special low-alloy steels have been developed for piling, e.g. United States Steel Corporation's 'Mariner' and Nippon Kokan's 'NK Marine'. Teses carried out by the companies concerned indicate that these have an improved corrosion performance compared with unalloyed steels[11] but in other tests carried out by the US National Bureau of Standards[8] there was no apparent difference between the performance of 'Mariner' and bare unalloyed carbon steel. For some situations the present range of low-alloy steels may well be satisfactory but for most marine situations their use without protective coatings should not be undertaken without specialist advice. Although they corrode at a lower rate than ordinary steels, they corrode at a significant rate in most marine situations. This means the addition of some millimetres of corrosion allowance or eventual maintenance with protective coatings, which may be difficult to achieve under marine conditions. There is conflicting evidence regarding the advantage of painting low-alloy steels exposed to the atmosphere. On steel that had been allowed to rust and was then wirebrushed before painting, three oleo-resinous type paints applied to ordinary steel and low-alloy steel performed far better on the low-alloy steel[12]. On average they gave an improvement of 70% in the life of the paint coating. These tests were carried out in an industrial atmosphere and the same results would not necessarily have been achieved in a marine situation. It is not surprising that the rusts formed on low-alloy steels provide a better basis for paint than those formed on ordinary steels. The main causes of paint breakdown on rusted surfaces arise from the presence of soluble salts such as ferrous sulphate and ferrous chloride in the rust, which accelerate rusting and the flaky nature of rust, which can lead to detachment of the paint film. The rust on low-alloy steels is more adherent and some of the sulphate and chloride salts may be in a less soluble form.

Where steel is blast-cleaned and more resistant paints are used, the difference in performance of coatings on low-alloy and unalloyed steel is likely to be much less than on rusted surfaces. However, the type of rust formed on low-alloy steels during exposure of the painted steel is likely to be advantageous in reducing the breakdown of the paint at least to some extent. This appears to be the conclusions to be drawn from practical experience of painted ships' hulls made from low-alloy steel (see below, 'Immersed Conditions'). There is probably little or no economic adcantage to be gained by painting low-alloy steels for use in marine atmospheres. The steels are significantly more expensive than unalloyed steels and a considerable improvement in coating performance would be required to cover this additional initial cost.

5.2.2 Immersed conditions

Tests have shown that the types of low-alloy steels used in the atmosphere (weathering steels) do not perform any better than unalloyed steel under immersed conditions. To obtain any advantage under immersed conditions it is considered that about 3% chromium must be added [9]. This would be expected to halve the corrosion rate. Although there are problems, such steels can be produced, but the additional cost is a severe disadvantage. Nevertheless, it seems possible that in the future chromium-containing low-alloy steels will be commercially available. Steels containing about 3% chromium have been used for special purposes such as bolts for lock gates. Some success has been reported on the use of low-alloy steels for small ships and barges. In a 10-year service trial on a boat operating in a highly polluted river, a low-alloy steel of unstated composition lost approximately 0.5 mm at a position slightly above the waterline. This compared with 1 mm for carbon steel plate in a similar position. Just below the waterline the losses were 1 mm for the low-alloy steel and 2–8 mm for the carbon steel. These results do not agree with those obtained in tests on small specimens but even if they were generally experienced, a corrosion rate of 100 μm/y below the waterline would not be acceptable in most situations. There have been reports—not based on any sound test results—of advantages accruing from painting the hulls of boats and yachts made from low-alloy steels. In particular, under conditions where the paintwork is easily and regularly damaged, e.g. on harbour work boats or tugs, there appears to be less under-rusting and breakdown of coatings at damaged areas. This presumably arises from the type of rust formed on low-alloy steels.

5.2.3 Summary

(i) Low-alloy weathering steels are not generally recommended for use in marine atmospheres near the sea.
(ii) Special low-alloy steels designed for marine conditions, e.g. USSC 'Mariner' steel may be advantageous in some situations for piling, but on this data is conflicting.
(iii) Bare low-alloy steels of the 'COR-TEN' type are not recommended for immersed conditions in sea water or for burial in soils.
(iv) There is some evidence that paint coatings perform better on low-alloy steels than on unalloyed steels but this may depend on the type of paint system and the environment of exposure.
(v) Advice should be sought before using low-alloy steels in marine situations.

5.3 CAST IRONS

Cast iron is the term applied to a range of ferrous alloys which have a carbon content in excess of 1.7%, and a comparatively high silicon content. As the name implies they are used in the cast form, i.e. they are not wrought materials. Because they are cast, their mechanical and physical properties are different from the usual wrought steels. Their microstructure, because of the high carbon content and method of manufacture, is also different from steels. The most commonly used cast iron is termed 'grey iron' but there are four broad categories:

(i) *Grey iron* in which most of the carbon is in the form of flakes, contains 2–4% carbon and 1–3% silicon. These are comparatively cheap and can be cast into fairly intricate shapes. However, they have little ductility and have poor impact properties.
(ii) *White iron* has practically all the carbon in solid solution. They are hard and brittle. The matrix is pearlitic.
(iii) *Ductile (nodular)*. The carbon is present as nodules or spheroids. The cast irons of this type solidify with a pearlitic matrix which can be annealed to provide ductility in a similar manner to steels.
(iv) *Malleable iron*. This is produced by heat treatment of white irons of suitable composition. The carbon forms as nodules instead of flakes and the material is ductile or malleable.

In addition to these standard cast irons there are several high-alloy materials.

(i) *High-silicon cast iron*

The addition of about 14% silicon to grey cast iron produces a material that is corrosion-resistant to many environments. Additions of molybdenum and chromium may also be made. They are commonly marketed under proprietary names and are resistant to many acids and sea water. They have been used for pumps and valves and other equipment handling process liquors. Although resistant to sea water, they are expensive and not generally selected for this purpose alone. High-silicon cast irons, particularly those with chromium additions are used as anode materials for impressed current cathodic protection in sea water (see Section 16.31).

(ii) *High-chromium cast iron*

Provided there is sufficient excess chromium over the amount taken up by carbon to form chromium carbides, usually 25–35Cr depending on the carbon content, then these are highly corrosion-resistant materials in oxidising conditions. Generally they are not used purely for corrosion-resistance in marine environments.

(iii) *High-nickel cast irons*

The addition of about 20% nickel to cast iron produces an alloy with an austenitic structure. Consequently the group of cast alloys are sometimes called austenitic cast irons but are more commonly designated by proprietary names such as 'Ni-Resist'. The Ni-Resist cast irons are available in both flake and spheroidal form; the latter known as SG irons (spheroidal graphite) have better strength and impact properties. A range of alloys has been developed with nickel contents from about 13–35%. BS. 3468 cover some of the grades, which apart from nickel also contain 1–5% Si and up to 5.5% Cr. ASTM A-346 also covers these materials. These alloys are resistant to sea water and are used for pumps, valves and water boxes.

　　Lower-alloy contents, e.g. 2% nickel cast irons, have improved corrosion-resistance over grey cast irons, but are in a different category from the Ni-Resist materials.

　　'Ni-Hard', containing about 4% nickel and 2% chromium, is a white cast iron and has good erosion-corrosion resistance in neutral and alkaline solutions.

5.3.1 Corrosion behaviour of cast irons

Ordinary carbon steels generally corrode at a reasonably uniform rate although in some situations they may pit. Cast irons, because of their structure, tend to corrode in a somewhat different manner. They contain a high proportion of carbon or graphite and this influences their corrosion performance. Carbon does not corrode and it is cathodic to iron, so in non-alloyed cast irons—although the initial corrosion may be similar to steel—various reactions can occur over a period of time. The cast iron may become covered with a fairly impervious layer of corrosion product, which will reduce the corrosion attack. On the other hand, if the layer is porous, particularly in sea water, the cathodic action of the graphite will lead to enhanced corrosion of the iron matrix; this is referred to as *graphitic corrosion* (see Chapter 4). Generally, the corrosion rate of unalloyed cast irons tends to be similar to that of steel, although the oxide scale produced during casting may reduce corrosion in some environments. In tests carried out in sea water, a comparison of cast iron carrying scale and similar specimens where it had been machined away showed a difference of about 25% in the corrosion rate over one year. Over a period of time this difference reduced but after 12 years there was still a marked difference in the corrosion rates.

Corrosion data for cast irons in atmospheric corrosion
Because of the heavy sections normally used with cast iron, atmospheric corrosion in marine atmospheres is not usually a problem. Some data has, however, been produced in a series of tests carried out at Kure Beach, North Carolina. These showed that over a 12-year period the corrosion rates (expressed in g m^{-2} d^{-1}) of different irons all as cast, and mild steel, were as follows[14]:

Ferritic ductile iron	0.72
Pearlite ductile iron	0.37
Malleable iron	0.75
Mild steel	1.35

Tests on cast iron specimens fully immersed (15 y)
Compositions: 3.41C, 0.74Mn, 1.73Si, 0.07S, 0.48P

Site	Corrosion rate (mm/y)	Maximum depth of pitting (mm)
Halifax, Nova Scotia	0.150	1.49
Plymouth, England	0.066	5.03
Colombo, Sri Lanka	0.216	Severe graphitic corrosion

Based on data from 'Deterioration of Structures in Sea Water', 18th Report of Committee of Institution of Civil Engineers, London (1938)

5.3.2 Corrosion data for high-nickel cast irons

In atmospheric tests carried out at Kure Beach, North Carolina, over $7\frac{1}{2}$ years, the corrosion rate of ordinary cast iron was 10 μm/y compared with less than 3 μm/y for Ni-Resist material.

In tests under immersed conditions in sea water, Ni-Resist corroded at a rate of 0.020–0.058 mm/y in quiet sea water compared with 0.066–0.53 mm/y for low-alloy cast irons. This improvement was maintained in sea water moving at 1.5 m/s. Ni-Resist corroded at a rate of 0.050 mm/y compared with a rate of 1.3 mm/y for the low-alloy material.

5.4 STAINLESS STEELS

The term 'stainless steel' covers a wider range of alloys with different mechanical and corrosion-resistant properties. Any ferrous alloy containing more than about 11.5% chromium falls within this category. At the other end of the range alloys containing up to 30% chromium and 20% nickel as well as other elements such as molybdenum also fall into the category of stainless steel.

Clearly all these alloys are expensive compared with carbon steels, particularly the highly-alloyed ones. They should, however, be compared with the more expensive non-ferrous alloys because they are used only for special situations, where high corrosion-resistance is essential. Over the years many alloys have been developed for specific purposes and it is sometimes difficult for non-specialists in the field to appreciate the fields of application and properties of the different groups. In this section the discussion is concerned with wrought materials. Cast steels sometimes with similar compositions to wrought materials are produced. However, the cast materials do not necessarily have the same corrosion-resistance as wrought steels.

The stainless steels depend, primarily, on their chromium content for corrosion-resistance. Other alloying elements in conjunction with chromium improve the corrosion-resistance, e.g. molybdenum. Some alloying elements affect both the corrosion-resistance and the mechanical properties of the steels, e.g. nickel. The corrosion-resistant characteristics of stainless steels arise from the passive nature of the surface oxide film, basically chromium oxide (Cr_2O_3). This provides resistance to many environmental conditions but not to all. For example, carbon steel is more resistant than commercial stainless steels to some concentrations of sulphuric acid. In the presence of oxygen, even if the passive film is broken down locally, it is usually rapidly repaired. However, under conditions of limited oxygen access, e.g. under deposits or in crevices, the film may not be repaired and pitting may occur. This is a particular problem in sea water, with its high chloride level, which increases the tendency of the steels to pitting. The commonly used stainless steels fall into three broad categories depending upon their metallurgical structures: martensitic; ferritic; austenitic.

The metallurgy of stainless steels will not be discussed in detail but a few points are worth considering. Ordinary steels, i.e. carbon steels are basically Fe–C alloys and they can be heat treated to obtain different mechanical properties. Generally, these treatments do not have a marked influence on their corrosion performance. When a plain carbon steel is heated to a high temperature it forms a face centred cubic (f.c.c.) structure called *austenite*. When this is cooled comparatively slowly to ambient temperatures it transforms to a body centred cubic (b.c.c.) structure called *ferrite*. If, however, it is cooled very rapidly, an acicular phase called *martensite* is produced.

Ferrite is soft whereas martensite is hard, the actual hardness depending on the carbon content. By transforming the structure to martensite, higher strength steels can be produced. The addition of chromium allows the formation of martensite at relatively slow rates of cooling, so steels containing this element form martensite on cooling in air without quenching operations. Special 'tempering' treatments are used to soften these steels to provide the required properties. This involves re-heating to below the temperature at which austenite will reform. In practice there is a limit of about 14% of chromium in martensite stainless steels in the absence of other alloying elements such as nickel. The presence of certain alloying elements in stainless steels results in the retention of austenite at ambient temperatures.

5.4.1 Martensitic steels

Typical compositions of martensitic stainless steels are shown in *Table 5.1*. Such steels are reasonably corrosion-resistant with high strengths or hardness. They are used for parts of valves and for other specailised parts of plant or equipment. They are the least widely used group of stainless steels for marine applications.

5.4.2 Ferritic steels

The second group of stainless steels with a higher chromium content, the ferritic steels, are non-hardenable. They have improved corrosion-resistance compared with the martensitic steels but have limitations in mechanical properties. They have good resistance to atmospheric corrosion and are widely used for car trims. They can be used for certain

Table 5.1 Compositions of some typical ferritic and austenitic stainless steels

Type	Steel designation	Major elements (%)			
		C	Cr	Ni	Mo
Ferritic	430S15	0.10 max	16–18	—	—
	434S19	0.10 max	16–18	—	0.9–1.3
	442S19	0.10 max	18–22	—	—
Austenitic	304S12	0.03 max	17.5–19	9–12	—
	304S15	0.06 max	17.5–19	8–11	—
	304S16	0.06 max	17.5–19	9–11	—
	315S16	0.07 max	16.5–18.5	9–11	1.25–1.75
	316S12	0.03 max	16.5–18.5	11–14	2.25–3.0
	316S16	0.06 max	16.5–18.5	10–13	2.25–3.0
	317S12	0.03 max	17.5–19.5	14–17	3–4
Martensitic	410S21	0.09–0.15	11.5–13.5	—	—
	420S29	0.14–0.20	11.5–13.5	—	—

architectural and engineering applications in marine situations. Typical compositions are shown in *Table 5.1*. If the chromium content is increased to levels of about 25–29% with molybdenum content of about 4% and generally some nickel, a range of steels highly resistant to sea water are produced. These are considered in Section 5.5.

5.4.3 Austenitic steels

The third group, austenitic stainless steels, are the most widely used for marine applications. Suitable alloying with elements such as nickel suppresses the formation of martensite or ferrite, so the austenitic structure is retained at ambient temperatures. The original austenitic steel contained approximately 18% chromium and 8% nickel but many variations on this composition have been developed, particularly the addition of molybdenum to improve resistance to pitting. The addition of molybdenum has resulted in other additions to maintain the austenitic structure. Typical compositions are given in *Table 5.1*. As with the ferritic stainless steels, newer steels with higher levels of chromium and nickel have been developed and they are considered in Section 5.5. The austenitic types are the most corrosion-resistant group of the ordinary stainless steels and are widely used in marine environments. Their advantages and limitations are considered in Section 5.4.5.

5.4.4 Precipitation hardening steels

This group of steels was developed to provide higher strength levels with reasonably good corrosion-resistance and better weldability than the general martensitic steels discussed above. They are a somewhat specialised group of steels developed for the aircraft and space industries and will not be considered here in any detail. For those wishing to consider them in more depth, references are provided [16],[17]. These steels generally contain from 14–17 chromium and 4–7 nickel with smaller amounts of other alloying editions of molybdenum and copper and sometimes aluminium. There is an AISI designation 630 for cast materials—17-4 PH and 17-7 PH—but generally the steels are marketed under proprietary names. The 17-4 PH alloy has been used for propeller shafts, for boats and for shafts in cast iron casings.

5.4.5 Corrosion characteristics of stainless steel in marine environments

(i) *Air*

Generally, stainless steels are resistant to marine atmospheres in the sense that the total loss of alloy is small. The martensitic and ferritic steels do, however, pit and this may result in rust staining and, over a period of time, some overall surface corrosion. Even Type 304, an austenitic grade, pits sufficiently to produce rust staining in marine environments. The molybdenum-bearing steels, e.g. Types 315 and 316, pit only slightly and are recommended for purposes where appearance is important [18]. For general engineering purposes, any of the stainless steels are very corrosion-resistant in atmospheric marine situations provided the design of the component or equipment does not lead to the collection of chloride-containing moisture in crevices. In such situations, corrosion can occur because the localised situation is akin to immersion in sea water.

Furthermore, it is possible in tropical situations where the collected solution can reach temperatures above 60 °C that stress-corrosion of austenitic steels could occur, although this would depend on a number of design factors, the stress concentrations and whether the steel had been sensitised.

(ii) *Immersion*

The corrosion performance of stainless steels under immersed conditions in sea water will depend upon a number of circumstances, in particular the velocity of the water and the design of the component or equipment. Generally, the 13% martensitic steels are not suitable for sea water immersion. The austenitic steels are those generally used for immersion in sea water but under some conditions are subject to fairly severe pitting.

(iii) *Velocity*

Velocity has a significant influence on the corrosion performance of stainless steels. They perform best when in contact with sea water at velocities over about 1.2 m/s. Below this flow rate, stagnant conditions arise in situations where marine organisms become attached to the alloy surface or where crevice conditions arise. Although under some conditions little corrosion may occur in static sea water there is always a possibility of localised corrosion under stagnant conditions. This is illustrated by tests in which Type 316 steels were tested under both static conditions and with sea water flowing at 1.2–1.5 m/s. In the static situation the specimens became covered with marine growths, whereas there were few on those exposed to the flowing conditions. This was reflected in the depth of pitting. The average pit depths were 1.8 mm for static conditions compared with 0.05 mm in the flowing sea water. Austenitic stainless steels can be effective for pump impellers and tubes where sea water is flowing. Problems may, however, arise during periods of shut down if sea water is allowed to remain stagnant in equipment. The precipitation hardening and austenitic types are considered to have a high resistance to cavitation damage and impingement attack up to about 9 m/s in sea water.

(iv) *Welds*

Provided welds are well produced with welding rods having a corrosion-resistance at least equal to that of the parent metal and the welds are properly dressed, they should have a corrosion performance similar to that of the welded alloy. Difficulties can arise with poorly produced welds with a rough surface finish but the main problem concerns intergranular corrosion in the heat-affected zone when austenitic steels are not stabilised.

(v) *Intergranular corrosion*

When non-stabilised stainless steels are heated through a critical temperature range, a chromium-containing carbide precipitates at the grain boundary, depleting the area adjacent of chromium, so reducing its corrosion-resistance. The steels are then sensitised, i.e. susceptible to intergranular corrosion. During welding, the steel goes through this temperature range and if the welds are immersed in sea water, the heat affected zone is likely to be preferentially attacked. This is overcome by using either stabilised steels or steels containing a carbon content below about 0.03%.

Stainless steels are stabilised with additions of niobium (Columbium) or titanium. These elements have a greater affinity for carbon than does chromium. If added in sufficient amounts, about 10 × carbon content for niobium and 5 × carbon content for titanium, they combine with the carbon to prevent precipitation of chromium carbide and so eliminate this type of intergranular attack. If the carbon content is reduced to below 0.03% this generally prevents the formation of sufficient carbides to cause intergranular attack.

5.4.6 Free-cutting grades of stainless steel

To improve machinability of stainless steels, in particular for nuts and bolts, austenitic grades containing additions of sulphur and selenium are produced. Such grades may have a much lower corrosion-resistance than similar grades without these additions, particularly in sea water. Generally, they are not recommended for sea immersion.

5.4.7 Galvanic corrosion

In the passive state stainless steels are noble to all the common constructional metals except titanium. When, however, stainless steels corrode they become active and their position in the galvanic series changes to a position just above carbon steel, so that all metals except carbon steel, cast iron, aluminium alloys, cadmium, zinc and magnesium and its alloys are more noble; this includes other stainless steel components in the passive state. However, the problems that arise with stainless steel coupled to other metals depend on the relative areas of the alloys and usually the stainless steel components are protected. As discussed in Chapter 7, severe corrosion of painted carbon steel can occur when stainless steel nuts and bolts are used as fasteners for structures immersed in sea water. This arises if the coating on the carbon steel is damaged, providing a small anode compared with the larger stainless steel cathode.

5.5 SPECIAL STAINLESS STEELS FOR SEA WATER

Austenitic stainless steels, e.g. Type 304 (18Cr–8Ni), have been widely used for power plant condenser tubes where fresh water is used for cooling. Type 316 has also been used in applications using sea water as the coolant but these have not always been successful because of pitting and crevice corrosion where deposits formed on the tubes. These austenitic steels have tended to be replaced by titanium or copper alloys.

In recent years, however, sea water or saline water has been increasingly used as the cooling medium for heat exchangers. This has resulted in a growing interest in materials that will resist both the sea water and the process liquids being cooled.

A new range of stainless steels has been developed to overcome the limitations of the usual austenitic steels. The newer steels are either austenitic stainless steels with a high molybdenum content or ferritic steels containing high amounts of chromium and molybdenum. The austenitic steels contain 17–20Cr, 18–25Ni and 4–6Mo; some contain 0.5–2Cu. The ferritic steels contain 26–29Cr and 3–4Mo; some with about 0.5Ti. The carbon contents are usually 0.02–0.03%.

A range of proprietary steels is available with varying compositions. They have been developed to take advantage of newer steel melting processes, e.g. argon oxygen decarburisation (AOD), or the ability to use high chromium and molybdenum in ferritic steels because there is no need to counterbalance the ferrite formers with austenite forming alloy additions, such as nickel, which are necessary in the austenitic steels. Furthermore, recent advances in steelmaking techniques make it possible to produce highly alloyed ferritic steels with low levels of carbon and nitrogen. The approach to the development of such steels was considered at a recent symposium [19].

There is insufficient service experience to determine the corrosion performance of such steels under a variety of conditions but the results so far available indicate considerable promise. In a series of laboratory tests covering crevices, pitting and stress-corrosion cracking, the steels have shown superior resistance to the conventional austenitic steels of Types 316 and 317 [19].

In tests carried out in natural sea water, again these newer steels proved to be more resistant to localised corrosion than the standard steels. Service tests of these newer steels, particularly in condenser tubing, shows their superiority over Type 316 steel for contact with sea water.

Problems may arise with galvanic corrosion when tubes of these alloys are in contact with tube sheets of different materials, particularly those of conventional copper alloys. The problem may be overcome by using cathodic protection where the tube sheet materials cannot be changed.

Because these steels are mainly marketed under proprietary names and vary in composition, it is difficult to generalise regarding their corrosion characteristics. Nevertheless, laboratory results indicate a high level of resistance to crevice corrosion in sea water and there are an increasing number of practical applications of these steels.

Recent publications have provided indications of their use in heat exchangers, flare booms and well-head piping systems. Tests on seamless tubes on the Swedish west coast over three years at temperatures varying between 4.3 °C and 17 °C, in both above and below water line levels, have shown that while Type 316L austenitic stainless steel pitted, there was virtually no attack on the 'newer' types of stainless steel[20].

Ferritic-austenitic stainless steel, e.g. (AISI 329), containing 26Cr, 5Ni, 1.5Mo, have been used for tubing applications in oil production and high pressure lines for wet gas[21]. Other steels of this type containing copper (about 2%) are claimed to have improved resistance in polluted sea water and sour well production, both of which may contain hydrogen sulphide[22]. A number of highly alloyed stainless steels, e.g. Type 310 (25Cr 25Ni) are used for high temperature resistance. Additionally, there are many high-alloy proprietary steels that are used for specialist marine applications.

5.6 REFERENCES

(1) COPSON, H R, *Proc. ASTM,* **45**, 554 (1945)
(2) VERNON, W H J, *Trans. Faraday Soc.,* **31**, 1, 668 (1935)
(3) CHANDLER, K A, *Br. Corros. J.,* **1**, July (1966)
(4) LARRABEE, C P, *Trans. Electrochem. Soc.,* **85**, 297 (1944)
(5) SHREIR, L L (ed), *Corrosion,* 2nd ed, Chapter 3.1, Butterworths, London (1976)
(6) HUDSON, J C, *The Corrosion of Iron and Steel,* Chapman & Hall, London (1940)
(7) SOUTHWELL, C R and ALEXANDER, A L, *Materials Protection,* **9**, 4 (1970)
(8) *NBS Monograph 158,* US Department of Commerce, National Bureau of Standards (1977)
(9) HUDSON, J C and STANNERS, J F, *J. Iron Steel Inst.,* **180**, 271 (1955)
(10) FRIEND, N J, *18th Report of Institution of Civil Engineers Sea Action Committee* (1940)
(11) Nippon Kokan, *Technical Report—Overseas,* March (1970), p 36
(12) *Sixth Report of the Corrosion Committee, Special Report No 66,* Iron Steel Inst., London (1959)
(13) *Mat. and Meth.,* **24**, 117, Manual No 102, February (1954)
(14) MANNWEILER, G B, *Proc. ASTM,* **72**, Appendix 1, 42 (1972)
(15) HART, A C in L L Shreir (ed), *Corrosion,* 2nd ed, Vol 1, Chapter 3.7, Butterworths, London (1976)
(16) TRUMAN, J E, *Special Report No 86,* 84, Iron Steel Inst., London (1964)
(17) IRVINE, K J and LLEWELLYN, D T, *J. Iron Steel Inst.,* **192**, 218 (1959)
(18) CHANDLER, K A, *J. Iron Steel Inst., Publication No 117,* 128 (1969)
(19) *Advanced Stainless Steels for Sea Water Applications, Proc. Symposium University of Placenza,* Climax Molybdenum Co (1982)
(20) BERNHARDSSON, S et al, *Proc. UK National Corrosion Conf.* (1982), p 137, Inst. Corr. Sci. Tech., London
(21) GROENEWOUD, K, *Proc. UK National Corrosion Conf.* (1982), p 151, Inst. Corr. Sci. Tech., London
(22) GUHA, P and CLARK, C A, *Proc. UK National Corrosion Conf.* (1982), p 109, Inst. Corr. Sci. Tech., London

6 Non-ferrous metals and alloys

The main groups of non-ferrous metals and alloys used for marine service are based on copper, aluminium, nickel and titanium.

Some of the nickel alloys, in fact, contain more iron than nickel but as a group they fall more conveniently into the non-ferrous category. Non-ferrous alloys are widely used for marine situations, sometimes for structural purposes but also for pumps, condensers and other equipment involved with sea-water systems. Generally, their overall resistance to corrosion is considerably superior to that of carbon steels and the problems that arise are due to local corrosion and the more specialised forms considered in Chapter 4. Additional problems arise from the effects of using different non-ferrous alloys in conjunction with each other and with steels, in systems and environments that can cause bimetallic corrosion. In this chapter, the general properties of the main groups of alloys will be considered. Corrosion data is given for some of the alloys. This is based on tests carried out and is a useful way of indicating the general corrosion characteristics of the alloys. The data should, however, be treated with caution when used for design purposes. The data may be useful where it has been obtained under conditions very similar to those for which it is required in practice. In other cases, however, the test conditions may be quite different. Furthermore, unlike carbon steel, which tends to corrode reasonably uniformly, many non-ferrous alloys are subject to local corrosion such as pitting and it is difficult to provide simple corrosion rate data in such circumstances.

General corrosion data for a number of commonly used alloys immersed in sea water is given in *Table 6.1*. This is based on information in the Copper Development Association Publication No 80[14] and reproduced with their permission. The data for general and crevice corrosion was determined using samples fully immersed from a raft moored in a harbour for one year. The crevice corrosion results were obtained with samples held in a perspex jig to provide the crevice conditions.

The corrosion/erosion tests were carried out using the Brownsdon and Bannister test in which air is forced at a high velocity through a

Table 6.1 Resistance to general corrosion, crevice corrosion and impingement attack in sea water

Alloy	General corrosion rate (mm/y)	Crevice corrosion (mm/y)	Corrosion/ erosion resistance (ft/s)
Wrought alloys			
Phosphorus deoxidised copper C106 or C107	0.04	<0.025	6
Admiralty brass CZ111	0.05	<0.05	10
Aluminium brass CZ110	0.05	0.05	13
Naval brass CZ112	0.05	0.15	10
HT brass CZ115	0.18	0.75	10
90/10 copper-nickel	0.04	<0.04	12
70/30 copper-nickel	0.025	<0.025	15
5% aluminium bronze CA101	0.06	<0.06	14
8% aluminium bronze CA102	0.05	<0.05	14
9% aluminium bronze CA103	0.06	0.075	15
Nickel aluminium bronze CA104	0.075	See note (1)	
Aluminium silicon bronze DGS1044 0.06	<0.075	See note (1)	
17% Cr stainless steel 430	<0.025	5.0	>30
Austenitic stainless steel 304	<0.025	0.25	>30
Austenitic stainless steel 316	0.025	0.13	>30
Monel (nickel-copper)	0.025	0.5	>30
Cast alloys			
Gunmetal LG2	0.04	<0.04	12
Gunmetal G1	0.025	<0.025	20
High tensile brass HTB1	0.18	0.25	8
Aluminium bronze AB1	0.06	<0.06	15
Nickel aluminium bronze AB2	0.006	See note (1)	
Manganese aluminium bronzes CMA1/CMA2	0.04	3.8	14
Austenitic cast iron (AUS202)	0.075	0	>20
Austenitic stainless steel 304	<0.025	0.25	>30
Austenitic stainless steel 316	<0.025	0.125	>30
3% or 4% Si monel	0.025	0.5	>30

Note (1) The Defence Standard Data Sheets from which the figures in Table 6.1 are taken give 'up to 0.5 mm/y' as the crevice corrosion rate and 14 ft/s as the corrosion/erosion resistance limit for nickel aluminium bronze AB2 or CA104 and 8 ft/s for aluminium silicon bronze DGS1044. Ship Department Publication 18 'Design and Manufacture of Nickel-Aluminium-Bronze Sand Castings', Ministry of Defence (PE), 1979, gives the following corrosion data:

Self-corrosion rate: For general long-term use over several years a reasonable design value is 0.05 mm/y but under ideal conditions for nickel aluminium bronze in sea water a black film slowly forms which reduces the corrosion rate in accordance with an equation of the form: corrosion rate ~ (time)$^{-0.2}$.

Crevice corrosion: After the initiation period which can be about 3–15 months with negligible corrosion the crevice corrosion propagates at about 1 mm/y.

Impingement resistance: 4.3 m/s is an appropriate design value in clean flowing sea water.

(2) Table 6.1 is based on data in Copper Development Association Publication No 80 and is reproduced with their permission.

submerged jet. From the minimum air jet velocity required to produce corrosion/erosion in a 14-day test, the minimum sea water velocity required to produce a similar effect under service conditions was estimated on the basis of known service behaviour of some of the materials.

The metallurgical structure of many non-ferrous alloys has an important influence on their corrosion performance. Consequently, the effects of comparatively small compositional changes and variations of heat treatments and manufacturing processes may be more significant in relation to corrosion than is normally the case with ferrous alloys. It is, therefore, important for manufacturing processes to be properly controlled if suitable corrosion characteristics are to be obtained. Advice should be sought from manufacturers or suppliers when selecting alloys for specific situations.

In the following sections the corrosion characteristics in marine conditions of the main categories of non-ferrous alloys are considered. No attempt, however, has been made to consider the detailed metallurgy of the materials or the manufacturing techniques used to produce them.

6.1 COPPER AND COPPER-BASE ALLOYS

Copper-base alloys cover a wide range from those containing less than 1% alloying additions to those containing only about 50% Cu. Many of the alloys have been developed for marine conditions, particularly for ships, e.g. propellers, rudders, valves and pumps. There are dozens of alloys, some with only slight compositional variations. The range covers differences in mechanical and physical properties and, of course, cost. So there is often a choice for a particular situation. For example, propellers for marine conditions are usually made from copper-base alloys and they vary from high-duty aluminium bronzes, of very high corrosion-resistance to lower quality alloys, which have only a moderate resistance to sea water. They are, however, made in large numbers and are, of course, cheaper than the specially manufactured aluminium bronze propellers.

The main groups of copper alloys are as follows:

Brasses	(Cu-Zn)
Bronzes	(Cu-Sn)
Aluminium bronzes	(Cu-Al)
Gunmetal	(Cu-Sn-Zn)

Cupro-nickels   (Cu-Ni)
(High purity copper is also used)

The metals in brackets indicate the main components of the alloys. The simple classification provided above would have been valid at one time but designers are nowadays faced with a complex system of alloys because additional alloying elements have been added to provide improved properties and it is sometimes difficult to classify alloys into simple classes. The basic alloys will be considered in relation to corrosion performance, but it should be appreciated that other properties are important in determining the choice of copper-base alloys for particular applications.

A number of papers containing data on the corrosion rates of copper and its alloys in sea water have been published. These include tests carried out in tropical waters[1],[2], in waters around the United Kingdom[3] and tests in American waters[4]. These results provide a general indication of the low rate of corrosion anticipated from copper and its alloys when immersed in sea water but should be used cautiously as design data because factors such as temperature and contamination of sea water can influence the corrosion rate.

Tests carried out under atmospheric marine conditions have indicated corrosion rates from about 0.6–2.5 µm/y[6],[7]. The general corrosion rate of copper and its alloys is low in most marine situations. Attack is likely to be increased in polluted harbours and the high-tin bronzes, aluminium bronzes and cupro-nickel alloys probably provide the best resistance to such conditions. The main corrosion problems with copper alloys arise when sea water is moving at a reasonably high velocity, and from localised action. These are considered in more detail under the specific alloys.

The metallurgical structure of some copper-base alloys has an influence on their corrosion-resistance. Often, the phase produced is directly related to the composition of the alloy, but in some cases, the heat treatment or manufacturing process may lead to the formation of certain undesirable phases. For example, in aluminium-copper alloys (aluminium bronze) incorrect cooling from above about 600 °C may lead to the formation of a gamma phase which is detrimental to corrosion-resistance. The duplex nature of copper alloys can also lead to a form of corrosion where certain phases are selectively attacked. This arises from the different potentials of the phases and with some alloys corrosion may occur between the anodic phase and the surrounding cathodic phase. This occurs most commonly in some duplex brasses, Muntz metal, naval brass and manganese bronze (high tensile brass). It can also occur in some compositions of aluminium bronze, particularly under crevice conditions in marine situations.

6.1.1 Copper

There are commercial wrought coppers ranging in purity from 99.22 to 99.95. Apart from those containing beryllium or arsenic, which may have some effect, there is little difference in the corrosion performance of the different coppers under immersed sea water conditions.

Copper is widely used for piping in ships but it is not particularly suitable for use in contact with sea water flowing at more than about 1 m/s. However, it is used for services where sea water has a low velocity of flow or where flow is intermittent.

Copper is sensitive to moist hydrogen sulphide and ammonia and is therefore attacked to a greater extent in polluted sea water, for example in some harbours, than in clean sea water. It is, therefore, advisable to wash out copper supply lines carrying sea water with fresh sea water after a stay of any length in a polluted harbour. Although copper pipes are widely used for fresh water, it should be noted that water containing substantial amounts of carbon dioxide, often found in water from evaporators, may attack copper.

6.1.2 Brasses

Brasses are basically alloys of copper and zinc, and depending on their metallurgical structure, are divided into three groups:

(i) Single-phase α-brasses with up to 27% Zn.
(ii) Two-phase α-β-brasses which begin to form at about 37.5% Zn.
(iii) Single-phase β-brasses containing about 46% Zn.

Further increase in the zinc content over 50% leads to brittleness, so such alloys are not of industrial value.

(i) *Alpha-brass*
Their strength increases with content of zinc and they can be hardened by cold work and softened by annealing.

For marine purposes the most important alloys are:

70Cu 29Zn 1Sn   Admiralty brass
76Cu 22Zn 2Al   Aluminium brass

All the α-brasses can be used in clean sea water but to prevent dezincification (see Section 4.3) the addition of 0.02–0.05% arsenic is necessary. In moving aerated sea water where corrosion-erosion may occur, only aluminium brass can be used. Admiralty brass has been superseded by improved materials for modern marine condensers, but it can be used for fresh water condensers, provided that it contains arsenic to inhibit dezincification, and is considered to be economically advantageous.

Aluminium brass was developed to withstand the failures caused by impingement attack on Admiralty brass. It has the best resistance to sea water corrosion of the α-brasses and will withstand velocities of aerated sea water up to about 3 m/s. It is not, however, recommended for use in polluted sea water for prolonged periods. Aluminium brass has been used in desalination plants, particularly for the large heat-recovery condensers.

Although arsenic additions have generally overcome the dezincification of α-brasses, problems can arise when copper-zinc brazing alloys are used for joints. A capillary joint may last for a long time but it is preferable to use a more corrosion-resistant jointing material such as a silver solder.

(ii) *Alpha-beta-brasses*

In contrast to the α-brasses which are cold worked, α-β-brasses are essentially hot-working alloys.

They can be formed by most hot-working processes such as forging, hot-rolling and extrusion. Elements other than copper and zinc are frequently added, e.g. aluminium, iron and manganese, to provide 'high-tensile brasses', which may be hot worked or cast, e.g. for propellers.

Dezincification is a problem with α-β-brasses and neither Muntz metal (Cu60–Zn40) nor naval brass, which is essentially Muntz metal with an addition of 1% Sn, are generally suitable for use in sea water. Although both alloys are in fact used in sea water, e.g. in small boats such as ocean-going fishing vessels, they are rarely as economic as alternative materials, such as gunmetal and aluminium bronze. There are situations where these alloys are not seriously affected by sea water but the conditions under which this can occur are not clearly established, so except where long-term service conditions point to the good performance of these alloys, it is wiser for designers not to specify them for sea-water use. Naval brass is used for condenser tube plates—and while they suffer from dezincification—their thickness is often such as to reduce the problem to an acceptable level. Manganese bronze (high-tensile brass) does not have the same resistance to sea water as do the aluminium and silicon bronzes.

Although the basic composition of manganese-bronze is a 60-40 Cu-Zn alloy, the ratio of the two metals can vary depending upon the other alloying elements which are added in small amounts. The structure also is similarly affected by the actual composition and, while it usually falls into the α-β-class, some compositions fall within the β-range. The most frequently used alloying elements are tin, iron, manganese, nickel, iron and aluminium, all of which improve the mechanical properties but not necessarily the corrosion characteristics. For sea water use it is advisable to choose compositions that will resist dezincification. Although all high-tensile brasses are prone to dezincification, they can be used for propellers because the high-flow conditions are not conducive to this phenomenon.

(iii) *Beta-brasses*
These alloys containing about 50% each of copper and zinc suffer rapid dezincification in sea water and are not used for marine purposes and the brazing solders made from these alloys should not be used for joints that are likely to be in contact with sea water.

6.1.3 Bronzes

Originally 'bronze' was the term applied to alloys of copper and tin but the terminology is now somewhat vague, e.g. what is essentially a brass is called manganese bronze. Copper alloys containing both zinc and tin are generally termed gunmetals, although in some parts of the world they are called bronzes, and aluminium bronze contains little or no tin. Sometimes the term 'tin-bronze' is used, indicating that bronze is a broad term covering almost any copper alloy, provided that the other metal used in addition to copper is prefixed. The terminology is not, however, important, provided that designers know the composition and the corrosion characteristics of the alloys they select for a specific situation.

(i) *Phosphor-bronze*
Phosphor-bronze is widely used for marine engineering and in the shipbuilding industry. The alloy has about 8% Sn and up to 0.4% phosphorus. The higher the tin content the better the resistance to marine environments, but problems can arise if there is more than 8% Sn because a tin-rich phase is produced which impairs the corrosion-resistance.

(ii) *Aluminium bronzes*†
Aluminium bronzes contain 5–11% Al, sometimes with additions of up to 6% Fe, 3% Mn, 7% Ni or 3% Si. Other minor constituents may include Pb, Sn and Zn. This range of alloys is of increasing importance for many marine applications because of their excellent corrosion-resistance and higher strengths compared with other copper alloys. They are produced in both wrought and cast forms.

† A useful publication dealing with aluminium bronzes is available from the Aluminium Bronze Advisory Service, Copper Development Association, Orchard House, Mutton Lane, Potters Bar, Herts, England—Publication No 80, 'Aluminium Bronze Alloys: Corrosion Resistance Guide'. This lists 30 wrought and 15 cast alloys in common use with their standards designations.

Control of composition and manufacture is important in determining the corrosion-resistance of aluminium bronze alloys. The effect of metallurgical structure on corrosion performance is now well understood but materials produced some years ago did not always possess the most favourable structures and their performance was sometimes poorer than that of current alloys of similar composition. Consequently, older data may not provide a correct indication of present-day performance.

A $\gamma2$-phase may occur in Al–Cu alloys if the cooling rate from about 600 °C is too slow, or in alloys containing more than about 9.1% Al. This phase may impair the corrosion-resistance, but this can be avoided by correct attention to composition and cooling rate. This phase is not formed in alloys containing other elements such as iron or nickel; usually about 2% Fe is sufficient in sections of diameter up to about 75 mm. Manganese additions also suppress the phase but may modify the β-phase making it more susceptible to corrosion. The effect of metallurgical structure on the corrosion characteristics of aluminium bronze alloys is covered in a Copper Development Association publication[14].

Although there is little data available for atmospheric exposure, it is known that these alloys preserve their 'golden colour' for long periods under many conditions of exposure to air, and are among the most resistant of the copper alloys. They are also resistant to sulphur dioxide and high-temperature oxidation, so are particularly useful where resistance to such service conditions is required, e.g. for the inert gas fans in oil tankers. The good corrosion-resistance of these alloys arises from the ability of the aluminium to provide a thin, adherent surface oxide film, which is very protective and heals when damaged. Furthermore, it is not susceptible to localised breakdown in the presence of chlorides; consequently, these alloys are particularly useful for marine conditions and, in fact, are probably more widely used in sea water service than in any other environment.

They are less susceptible to crevice corrosion than are many other alloys (see *Table 6.1*). It has been reported that crevice corrosion of these alloys takes the form of minor selective phase de-alloying, which results in little reduction of strength and practically no impairment of surface finish[14]. These alloys are, therefore, used for pump shafts and for valve spindles where pitting in crevices makes many other alloys unsuitable.

Aluminium bronzes are subject to de-alloying, commonly called 'dealuminification', a type of attack similar to dezincification in duplex brasses in which selective dissolution of the aluminium leaves a residue of copper, which—although retaining the original shape—has little strength. A review of de-alloying of cast aluminium bronze and nickel aluminium bronze in sea water service has recently been published[8]. Generally, however, by controlling the composition and, where appropriate, the cooling rates and working temperature, metallurgical structures are produced that do not suffer de-alloying to a significant extent under most service conditions. Aluminium bronze exhibits better resistance to stress-corrosion cracking than do the brasses but is less resistant than the copper nickel alloys.

It has been used as a replacement for high-tensile brasses, which are subject to stress corrosion, for many underwater fastenings.

Nickel-aluminium bronzes containing about 4–6.5% Ni are widely used for large marine propellers, not least because of their excellent resistance to corrosion fatigue in sea water. Manganese-aluminium bronze is also used for large propellers and it has good corrosion fatigue properties, although inferior to that of the nickel-aluminium bronzes.

Aluminium bronzes are resistant to corrosion/erosion and to cavitation damage. In tests carried out in 3% NaCl solution, cavitation erosion resistance was shown to be superior to both Austenitic Stainless Steel 321 and to high-tensile brass[14].

An indication of the resistance to impingement attack and general corrosion performance of cast aluminium bronze alloys is provided in *Table 6.2*.

(iii) *Other bronzes*

A group of Cu-Al alloys closely related to the aluminium bronzes contain 11–15% Mn with 1.5–4.5% Ni and 2–4% Fe. Another alloy with 2–2.4% Si is also used for some marine applications. They are, however, subject to de-alloying in static or slow flowing water.

(iv) *Service applications*

Aluminium bronze alloys are widely used for marine service. Some of the more common applications are:

(a) Fasteners and underwater fittings.
(b) Propellers, shaft brackets and rudders.
(c) Pump impellers and bodies.
(d) High duty valves.
(e) Tube plates with titanium tubes.

Although aluminium bronze would be suitable for piping for sea water, it is not widely used for this purpose.

6.1.4 Gunmetals

The gunmetals—sometimes called 'G' bronze or 'red brass'—are basically alloys of copper, tin and zinc, sometimes with Pb additions. On

Table 6.2 Resistance of cast copper alloys to impingement attack and general corrosion in sea water

Alloy	Composition % (balance Cu)					Depth of impingement attack (mm)		General corrosion weight loss (mg cm^{-2} d^{-1})	
	Al	Fe	Ni	Mn	Zn	28-day jet impingement 20 °C	14-day Brownsdon & Bannister 20 °C	Water in slow motion	Water speed 10 m/s
Aluminium bronze	8.2	1.7	—	—	—	0.04	0.19	0.15	0.17
Nickel aluminium bronze	8.2	2.9	4.3	2.4	—	0.00	0.32	0.04	0.10
Nickel aluminium bronze	8.8	3.8	4.5	1.3	—	0.00	0.28	0.04	0.16
Manganese aluminium bronze	7.6	2.8	3.1	10.0	—	0.01	0.24	0.04	0.11
High tensile brass	0.8	0.8	0.2	0.5	37.0	0.03	0.08	0.09	0.73
	Sn	Zn	Pb						
Gunmetal	9.7	1.4	0.6			0.02	0.32	0.14	0.74
Gunmetal	5.1	5.0	4.3			0.23	0.39	0.22	1.66

The data in *Table 6.2* are taken from a paper, 'The Resistance of Copper Alloys to Different Types of Corrosion in Sea Water', by Sigmund Bog of the Ship Research Institute of Norway, presented at the 7th Scandinavian Corrosion Congress, Trondheim, 1975 and is reproduced from Copper Development Association Publication No 80.

account of their good corrosion-resistance, they are widely used for valves, pumps, water boxes, etc. Problems can arise with castings produced from these alloys, resulting in leakage under pressure. Lead is often added to improve casting characteristics and pressure tightness. The corrosion characteristics of the various gunmetal compositions are generally similar. All are resistant to corrosion-erosion but the 88Cu 10Sn 2Zn alloy (Admiralty gunmetal) is superior in sea water to the 85Cu 5Sn 5Zn 5Pb ('eighty-five-three-five'). For sea water service it is preferable to choose a gunmetal with a tin content above 5% and with a low percentage of lead. Gunmetal compositions containing nickel and lead are also produced. One composition of such an alloy is 6–7Sn, 1.5–2.5Zn, 0.1–0.5Pb, 5–5.5Ni; remainder copper. This is strong and corrosion-resistant and has been used successfully for high duty valves and components in moving sea water.

6.1.5 Copper-nickel alloys

There is a range of alloys containing nickel and copper. Some have a higher percentage of copper than nickel and these will be considered here. Another group containing more nickel than copper, e.g. 'Monel' alloys, will be considered under 'nickel' alloys (see Section 6.2). The copper-nickel alloys, sometimes described as cupro-nickel or copper-nickel iron, are single phased and a number of alloys fall within the group. The most commonly used of these alloys contain 90/10 Cu-Ni, or 70/30 Cu-Ni. Other elements that may be present include:

Manganese generally present in commercial alloys.
Iron for marine applications, up to 2% Fe is added. It is considered essential for erosion/corrosion resistance.
Chromium can be used to replace some of the iron and over 1% improves the strength of alloys.
Niobium and silicon may be used for castings.

A number of standards cover the copper-nickel alloys and there are variations in the exact compositional requirements. For marine applications the 70/30 alloys should contain 2% iron and 2% manganese (BS. designation CN108) rather than CN107 with lower amounts of these alloying elements. The limits for lead in specifications may not guarantee weldability by all techniques and it may be necessary to seek specialised advice.

The alloys are available in many forms including sheet, plate, tube, wire and castings.

The 90/10 and 70/30 alloys exhibit good corrosion-resistance to sea water and the 90/10 alloy to bio-fouling. Their corrosion-resistance arises in part from the protective surface film that forms. This may take some weeks to develop, so precautions should be taken to ensure that it forms properly. Contact with other less noble metals or with cathodic protection systems may interfere with the proper development of the film, although cathodic protection may be satisfactorily used in condenser tube ends. These alloys are used extensively for sea water cooled heat exchangers and in clean sea water at velocities up to 1 m/s general corrosion rates of 2.5–25 μm/y have been experienced. At velocities up to 3.5 m/s the alloys have shown satisfactory resistance to impingement attack. Their corrosion-resistance is lower in polluted sea water and the 70/30 alloy is preferred, but is not immune to attack. Under abrasive conditions, e.g. with entrained sand in sea water, the 90/10 alloy is considered to be generally unsuitable and the 70/30 containing 2Fe and 2Mn is superior.

The cupro-nickels are sensitive to certain conditions found in some condensers. Tubes of these alloys may be attacked if they are not maintained in a clean condition. If a heat exchanger is run at a sufficiently low speed to allow sludge to be carried into the system, any settlement will encourage overheating, which may lead to attack under the deposit. The 90/10 is superior to the 70/30 alloy, which has a tendency to pit. However, it is advisable under these conditions to flush out with clean water at the earliest opportunity. The other type of attack, known as 'hot-spot' corrosion occurs on the cooling side of a tube opposite a local high-temperature area on the steam side. 'Denickelification', i.e. de-alloying of the nickel, may occur at locally overheated hot spots and the 90/10 is more resistant than the 70/30 alloy but careful attention should be paid in situations where such attack may occur.

These alloys are reasonably tolerant to crevice corrosion and are resistant to stress-corrosion. They are used for a variety of purposes and can be soldered, brazed and welded by a number of standard methods. However, because of the fairly aggressive situations in which the alloys are generally used, care should be taken to ensure that the correct procedures are used for the specific service conditions. For example, copper-phosphorous and copper-silver-phosphorous brazing alloys should not be used, due to the possibility of intergranular penetration and consequent embrittlement. Choice of filler metals for welding can be influenced by the possibility of causing bimetallic corrosion; 70/30 alloys are slightly more noble than 90/10.

These alloys are used for a number of applications wherein their corrosion-resistance in sea water and anti-fouling properties are advantageous. These include pipelines handling sea water, condensers, heat exchangers and sea water intakes. Certain parts of offshore platforms could be clad with 90/10 cupro-nickel but as yet this is not being carried out on a large scale. This is because cathodic protection used on the submerged steel could possibly allow marine fouling on the alloys and cause pitting problems. Cladding systems in which the steel and cupro-nickel are insulated are being developed.

Experimental work and trials using copper-nickel alloys for the hulls of boats and small ships have been carried out with some success. The anti-fouling properties of the alloys are of particular value in such applications and both copper-nickel clad steel plate and sheets of the alloy adhesively bonded to steel have been used, but as yet these methods have not been widely adopted. Copper-nickel clad steel is also available in thicknesses from 6 mm upwards; the cladding is carried out by explosive or hot roll bonding and has been used to fabricate large pipes for sea water application.

The alloys are also used for the cages in which fish are reared in sea water. The anti-fouling properties and strength have proved superior to nylon mesh and the alloys are in use in many parts of the world. Furthermore, the adoption of these alloys has enabled the use of much larger enclosures with a low level of maintenance requirement.

Cupro-nickel alloys are used widely for multistage flash distillation plants[9]. A typical large plant may contain 500 tons of these alloys. The tube plates, large water boxes and elbows are fabricated from copper-nickel or 90/10 clad steel plate. In the heat rejection section, the recommended alloy is a 70/30 copper-nickel containing 2% iron and 2% manganese. In the heat recovery section 90/10 copper-nickel alloy is preferred. Cupro-nickel alloys are also used for pipelines handling sea water for offshore platforms. Seamless pipes and larger diameter seam welded materials are available for this purpose. Ninety-ten alloys are at present being tested for use, in expanded-metal form, as screens in sea water intakes.

A useful guide to the properties of copper-nickel alloys is obtainable from the Copper Development Association[15].

6.2 NICKEL ALLOYS

Three broad groups of nickel alloys are used for marine conditions:

(i) Nickel-copper, containing about 30% copper, sometimes with smaller additions (under 3%) of other metals such as Fe, Mn and Al, and with about 60–70% nickel. This group covers the 'Monel' alloys, 400 and K-500 (with 3Al).

(ii) Nickel-chromium-iron, containing about 15–20% Cr and usually up to 10% Fe, although there may be higher amounts in some alloys.

(iii) Nickel-chromium-iron-molybdenum, containing about 3–15% Mo, with up to about 20% Cr and up to 20–30% Fe.

Groups (ii) and (iii) are represented by a range of proprietary alloys such as 'Hastelloy', 'Inconel' and 'Incoloy'.

'Monel' alloys, e.g. 400 and K-500, are the most widely used of the nickel alloys for marine conditions. Molybdenum-containing alloys such as 'Hastelloy C' containing 16Cr and 16Mo are very resistant to marine environments but are generally too expensive for general use. See *Table 6.4* for the nominal compositions of some typical nickel alloys. The corrosion performance of nickel and its alloys depends on a surface film the exact nature of which is not clear, although it is probably an oxide or hydrated oxide film capable of self-repair under suitable conditions: when the film breaks down alloys may be subjected to pitting, particularly in the presence of chlorides. Nevertheless, these alloys have a high resistance to marine environments both immersed and atmospheric.

An indication of the corrosion performance of nickel alloys in sea water is provided in *Table 6.3*. These results are based on tests carried out over 2

Table 6.3 Corrosion of nickel alloys in two-year sea water tests (based on Niederberger's et al data[10])

Main alloy constituents in the nickel alloys						Weight loss and pitting behaviour in sea water			
						Quiet		Velocity 0.3–0.6 m/s	
Cr	Mo	Fe	Cu	Al	W	Wt loss (g)	Pitting†	Wt loss (g)	Pitting†
22	9	2				nil	A	0.25	A
7	16	4				0.50	A	1.45	A
16	16				4	nil	A	0.10	A
21	3	30				0.25	A	0.20	B
20	5	6				0.20	B	0.15	A
35		2				9.30	C	7.60	C
20	47					15.70	C	24.50	C
16	7					11.90	C	12.60	C
			30	3		19.50	B	24.40	B
			35			26.40	B	21.10	B
27						42.80	B	54.50	B

† Average pit depth (mm): A 0.0–0.30; B 0.30–0.8; C 0.8–4.0.

years in moving sea water at Harbor Island, North Carolina, on panels 305×76 mm in size and reported by Niederberger *et al*.[10]. They demonstrate the excellent performance of the 'Hastelloy C' type (16Cr, 16Mo) and 'Inconel 625' type (22Cr, 9Mo) of materials. The 'Monel' alloys exhibit a general corrosion-resistance in sea water that is somewhat superior to most of the copper base alloys but generally inferior to them in

Table 6.4 Some typical nickel alloy compositions

Alloy†	Cr	Fe	Cu	Mo	Other
		Composition (%) (nominal)			
Inconel 600	15	8			
Inconel	22	5		10	
Incoloy 800	20	Balance			32Ni
Hastelloy F	22	20		6.5	2Nb + Ta
Hastelloy G	22	20	2	6.5	2Nb + Ta
Incoloy 825	21	Balance	2	3	40Ni 1Ti
Hastelloy C276	15.5	5.5		16	4W
Hastelloy C4	16	1 max		15	
Hastelloy N	7	5 max		16.5	
Hastelloy B		5		28	
Monel 400		2.5 max	30		1Mn
Monel K500			30		3Al + 4Nb − Ta 1 max Co

† The names listed are trade marks of the producing companies. (Nickel-balance unless otherwise noted.)

resistance to crevice corrosion (see *Table 6.1*) and pitting in quiet sea water. The 'Hastelloy C' and 'Inconel 625' materials are very resistant to pitting even under fouling and crevice corrosion conditions. Nickel alloys generally show only slight resistance to fouling in quiet sea water and are inferior to the copper base alloys and about equal to or slightly better than titanium and stainless steels. At velocities over 2–3 m/s the nickel alloys resist corrosion and are in the same class as titanium and the stainless steels and are superior to copper-base alloys and are therefore well suited for pumps and valves.

The molybdenum-bearing nickel alloys are very resistant to cavitation damage. The 400 and K-500 alloys—although subject to some damage— are still highly resistant. All the nickel alloys are resistant to erosion-corrosion.

In marine atmospheres, nickel alloys perform well and where surfaces are washed by rain, there is little or no build up of corrosion products, although there may be some under sheltered conditions.

Apart from use in the wrought and cast form, nickel is also widely used as an electroplated coating for steel. Nickel is cathodic to steel, so at pores and damaged areas in the coating it does not protect the ferrous substrate as do anodic coatings such as zinc.

Nickel alloys are also used as cladding and overlays for steels. 'Monel 400' steel, usually about 1.02 mm in thickness, is sometimes used to protect steel in the splash zones of offshore platforms. 'Inconel 625' has been used in the form of weld overlays for particular areas where a high corrosion-resistance is required, e.g. propeller shaft applications. These overlays have also been used for offshore situations and for pump castings. Other methods of cladding with nickel include roll bonding and explosive bonding.

Wrought and cast nickel alloys are widely used for many marine situations. These include the following 'Monel' applications:

(i) Propeller shafts, often the K-500 alloy, containing 3Al is used in the age-hardened condition where high strength is required.
(ii) Pump shafts and impellers.
(iii) Valve stems.
(iv) Water boxes, where 'Monel' is used for tubes; more commonly used with cupro-nickel tubes when solder wiping is employed.
(v) Fasteners.

The higher alloyed materials are generally reserved for critical or very aggressive situations and may be considered for some heat-exchanger applications. Alloy 625 has been used for marine wire rope, springs and bellows-type expansion joints.

6.3 ALUMINIUM AND ALUMINIUM ALLOYS

Aluminium and its alloys—because of their combination of lightness, strength and corrosion-resistance—are used for many engineering and structural applications. In marine situations, their major use is in shipbuilding, although they are also employed for buildings and building components.

Pure aluminium has a number of desirable properties, including corrosion resistance, high ductility and good working properties. However, it has a low mechanical strengths, so a number of alloys have been developed to produce materials of higher strength with properties such as high temperature resistance, castability and workability. In wrought alloys, the strength levels are obtained either by cold working, sometimes in combination with annealing, or by heat treatment, depending upon the type of alloy. Aluminium alloys are produced in many forms including extrusions, sections, plate and tube, as well as in cast form.

The general corrosion-resistance of aluminium and its alloys when exposed to marine atmospheres or immersed in sea water is good, although many of the alloys are prone to some form of localised corrosion and its nature and extent usually determines the value of different alloys from the corrosion standpoint.

Aluminium and its alloys are very reactive metals with a high affinity for oxygen and it is this reactivity which is the reason for their corrosion-resistance; an aluminium oxide film forms on the metal surface which is very protective in many situations.

The wrought alloys most widely used for marine conditions are:

(i) Non-heat treatable, aluminium-magnesium alloys containing between about 2.8 and 5.5% magnesium. There is a range of alloys in this group.

(ii) Heat treatable, aluminium-magnesium-silicon alloys containing about 0.5–1.5Mg and 0.4–1.3Si with additions of manganese or chromium.

The heat treatable alloys are stronger than those in group (i) but they cannot be reheated without loss of mechanical properties unless controlled re-heating can be carried out. The corrosion-resistance of the heat-treatable alloys is also somewhat less than that of the non-heat-treatable group.

Where strength is not important, e.g. for decorative trim, other aluminium materials such as unalloyed aluminium or suitable low-strength alloys may be used.

Aluminium alloys are also widely used in the cast form and a range of such alloys is produced. The compositions vary depending on the requirements, e.g. ductility, casting properties and strength. Generally, they contain either 3–11Mg or 3–13Si, with other additions such as Mn in some of the Mg-containing alloys. Alloys containing both Mg and Si are also produced. Poor casting procedures, particularly lack of attention to the selection of suitable scrap materials, can influence corrosion performance, so careful control is required.

Of the commonly-used aluminium alloys, the age-hardening high strength alloys that contain appreciable amounts of copper (2.6%) are *not* recommended for use in marine conditions.

Corrosion-resistance and strength are not always of an equally high order in aluminium alloys and for some situations clad alloys are used. This takes advantage of high-strength alloys for the core with more corrosion-resistant material for the cladding which is metallurgically bonded as a relatively thin layer. This method has proved satisfactory for tubes and pipes carrying sea water. After the initial pitting, corrosion tends to proceed laterally, so penetration of the pipe wall generally does not occur. Furthermore, the possibility of stress-corrosion cracking of the higher-strength core material is reduced. Flame sprayed aluminium coatings can be applied to castings and extrusions which cannot be clad.

6.3.1 Corrosion characteristics of aluminium alloys in marine conditions

(i) *General*

Aluminium and its alloys are generally corrosion-resistant by virtue of the adherent protective oxide film formed on the surface when they are exposed to oxygen or oxidising solutions. Although the alloys may be attacked outside the pH range 4.5–8.5, this is not usually a problem under normal marine situations unless for some reason highly alkaline conditions are produced. The main form of attack is by pitting (see below) and for this reason average weight-loss of aluminium test specimens is often considered to be of little value. Loss of strength is more often used to determine corrosion performance. Tests have shown the corrosion rate of aluminium immersed in static sea water is under 2 μm/y, with the corrosion rate dropping with time [5]. This reduction in rate with time does not necessarily occur in moving sea water.

The corrosion rate in marine atmospheres varies with the alloy composition, but in tests carried out at Hayling Island in England, the corrosion rate of aluminium alloys was less than 0.5 μm/y [11]. Corrosion tends to fall off markedly after the first few years. The form of exposure influences corrosion. Rain washing is beneficial in atmospheric situations, while dampness and condensation in sheltered areas, e.g. on the undersides of structures and roofs, can accentuate chloride attack. This may occur with certain alloys and improved performance is likely to be obtained from cladding with pure aluminium to provide the best combination of strength and corrosion-resistance.

(ii) *Pitting*

This is the most common form of corrosion attack with the alloys generally used in marine situations. Pure aluminium is more resistant than many of its alloys and for situations where piping carries sea water, it is considered advantageous to use aluminium clad pipe. Alloy materials, although subject to pitting, may—at sufficient thickness—be satisfactory for many marine situations.

Under reasonably mild atmospheric marine conditions, pits may become dormant, their main effect being to roughen the surface. However, pits once initiated in the presence of chlorides do tend to deepen because of the acid produced at the bottom of the pit. In these circumstances the protective oxide film does not re-form and under the acidic conditions aluminium can corrode. The probability of pitting occurring and depth of pitting if it occurs will be influenced by the nature of the alloy, in particular the types of impurities or constituents in the alloys. Both aluminium-magnesium and aluminium-silicon alloys have good corrosion-resistance. Aluminium-copper alloys, although not recommended for marine environments, have shown over a 20-year test period on the coast an average depth of pitting of only 0.15 mm, i.e. about 7.5 μm/y [12].

(iii) *Intercrystalline corrosion*

Many of the aluminium alloys are susceptible to intercrystalline corrosion because of their metallurgical structures and the heat treatments that are used to produce alloys with suitable mechanical properties. The precipitation-hardenable alloys such as Al-Cu and Al-Zn-Mg and the non-heat-treatable Al-Mg alloys containing more than 3% Mg are all potentially susceptible. This arises from the electrochemical behaviour of grain boundary precipitates and the adjoining matrix. Where there is a marked potential difference then localised intergranular corrosion may occur. In practice, the heat treatments and their influence on the amount and distribution of intergranular precipitates, e.g. $CuAl_2$ in Al-Cu alloys, may well determine the likelihood of such attack. Marine environments with their high chloride content may cause severe intercrystalline attack on susceptible alloys, although often the attack will be superficial and may be prevented by the use of suitable protective coatings.

(iv) *Stress-corrosion cracking*

This form of corrosion is limited to the higher strength alloys. Alloys such as Al-Mg at high strength levels with high Mg contents may be susceptible in the wrought or cast condition as may the Al-Zn-Mg-Cu alloys. The type of heat treatment and the amount of cold working of the alloys may have a significant effect on their susceptibility. The possibilities of stress-corrosion cracking can be limited by using stress-relieved products where possible and by reducing fabrication stresses on susceptible material. Stress cracking in aluminium alloys is always intergranular and when stressed in the transverse direction they are more likely to crack than when stressed longitudinally. This is because a path for crack propagation is readily available in the transverse direction. It is, therefore, advantageous to specify material with a short transverse direction. Where designers have only a limited knowledge of aluminium alloys and the methods of heat treatment and fabrication, advice should be sought from the manufacturers on the likelihood of stress-cracking in specific situations.

(v) *Exfoliation corrosion*

Some aluminium alloys tend to exhibit a rather long, flat grain structure, unlike many alloys which re-crystallise during heat treatment after working. This elongated structure makes aluminium alloys susceptible to an unusual form of corrosion attack called exfoliation (sometimes also called 'layer' or 'lamellar') corrosion. Attack proceeds along a number of paths in the direction of the elongated grains; usually in an intergranular fashion. The corrosion product forces the layers apart and in the extreme case results in disintegration of the alloy into separate layers of metal, which may be relatively uncorroded. This type of attack is not common, but when it occurs it may be rapid. It is most likely to occur with Al-Cu alloys and can often be overcome with adjustments to the ageing treatments of the alloy.

(vi) *Bimetallic corrosion*

Because of their position in the Galvanic Series in Sea Water, aluminium and its alloys are anodic to virtually all the constructional alloys. Only magnesium and its alloys and zinc are anodic to aluminium and most of its alloys. Some aluminium alloys may be cathodic to cadmium, but generally aluminium alloys are anodic to this coating metal. The same factors and precautions are required as with other bimetallic situations but in most marine situations when coupled to other alloys, aluminium will corrode and protect the coupled alloy. Consequently, aluminium and its alloys should not be directly donnected to more noble metals, particularly under sea water immersion conditions. Even under atmospheric marine conditions, serious corrosion can occur with more noble alloys; even steel bolts can result in corrosion of aluminium structures. Magnesium is anodic to aluminium, but if these metals are joined it is possible in sea water for alkali to be formed at the cathode, i.e. aluminium, as a result of the reactions arising from the difference in potential between the metals. This can lead to attack on the aluminium.

Fasteners such as bolts and rivets should preferably be of aluminium alloys but stainless steel and cadmium-plated steel are often used and prove to be satisfactory in service, particularly if they are insulated from the aluminium.

(vii) *Design considerations*

The design factors discussed in Chapter 7 apply to aluminium alloys. However, because of its position in the galvanic series and its reactivity where the surface oxide film is not re-formed, a few points are worth noting:

(a) Wet porous materials in direct contact with aluminium, e.g. timber, can cause serious local corrosion of the metal. Suitable insulation by painting both the wood and aluminium should avoid problems. Lead-based paints should not be used. Wood preservatives containing copper salts should be avoided.
(b) Drippings containing dissolved corrosion products from copper-containing alloys on to aluminium alloys may cause corrosion.
(c) Aluminium has no anti-fouling properties so if anti-fouling compositions containing copper are used, a good barrier coating is essential. Advice should be sought before using anti-fouling coatings containing mercury.
(d) Aluminium alloys used in marine situations or ships are likely to remain damp over prolonged periods, so care must be exercised when using other materials in contact with them.

6.3.2 Anodising of aluminium

Anodising is the production of a porous oxide coating on the metal surface by anodic treatment of the metal in a suitable solution. A number of solutions, including sulphuric acid, chromic acid and oxalic acid can be used, and details are given in a number of publications. Different coloured films can be produced and this is advantageous for architectural purposes. Depending on the conditions of anodising the oxide coating may have a variable number of very small pores, and usually the anodised layer is sealed in boiling water to reduce the effect of the pores. Although anodised films are often produced for aesthetic reasons they produce a thicker oxide film on the aluminium and this adds greatly to the corrosion-resistance of the metal. In tests carried out in a synthetic marine atmosphere, the life (to an arbitrary level of appearance) increased up to four-fold on anodised aluminium. The improvement in pitting resistance is less marked. Deeper, but fewer pits occurred on some anodised material. Generally, performance was better for thicker films and improved with the purity of the aluminium.

Aluminium is also used as a coating for steels and this aspect is considered in Chapter 11, 'Metallic Coatings'.

6.4 TITANIUM

Titanium is a very reactive metal, which has a strong afinity for oxygen and forms an extremely tenacious compact oxide film when exposed to the air. This makes it one of the most corrosion-resistant of the commercially available alloys for marine situations. It is a comparatively new material for marine applications but its use is increasing for many aggressive conditions, e.g. tubes for sea water cooling, particularly with polluted water. New methods of production and high quality purification processes have resulted in ductile materials suitable for many engineering purposes. It is a light metal with a density of about 4.5 kgm^{-3} compared with about 2.7 kgm^{-3} for aluminium and 7.9 kgm^{-3} for carbon steel.

Because of its high corrosion-resistance and good mechanical properties combined with lightness, alloys have been developed for situations where high strength/weight ratios are required, e.g. the aerospace industry. Experience with titanium and its alloys in marine situations is less than with many of the older established alloys. It is clear, however, that titanium is one of the most corrosion-resistant materials for use in marine environments. However, because of its cost it is unlikely to be used as widely as many other alloys, although its use has increased considerably in the last decade.

Titanium has a high resistance to general corrosion, to pitting and to crevice attack at ambient temperatures. Its resistance to these forms of attack is much superior to either stainless steels or cupro-nickel alloys. In sea water, the corrosion of titanium is negligible.

With the ability to rapidly self-repair its protective oxide film, titanium exhibits high resistance to erosive and turbulent conditions. It has a high resistance to impingement and cavitation damage at all practical velocities. In sea water, erosion-corrosion is negligible at flow rates as high as 18 m/s. Titanium is also resistant to the severely erosive conditions arising from the presence of abrasive particles such as sand in water. At 2 m/s, the attack under such conditions corresponded to a penetration of only 0.13 mm/y [16].

Titanium alloys are resistant to both stress-corrosion cracking and hydrogen embrittlement and there is no reported evidence that commercially pure titanium is susceptible to stress corrosion in sea water. Titanium specimens under static loads of up to 80% of their tensile strength have been exposed for 5 years in a marine environment without any sign of failure. It has been reported that after 12-months' immersion in sea water, Erichsen cup-pressed specimens have shown no tendency to crack [16].

The fatigue limit of titanium in sea water, measured on a rotating beam machine, has been shown to be similar to the value obtained in air, i.e. about 50% of the tensile strength.

When titanium is coupled to other constructional alloys in sea water, it is usually the cathode of the bimetallic cell. Consequently, titanium is not usually attacked but the corrosion of the other alloy in the couple may be significantly increased. The extent of the attack will be determined by factors considered in Chapter 7, e.g. ratio of the superficial area of the two metals, polarisation effects and differences in potential. The results of a series of tests have been reported in which the galvanic behaviour of titanium in contact with other metals was studied [16]. Various anode-cathode ratios were investigated when specimens were exposed either to sea water or to 3% NaCl mist. The results are summarised below, for a 10/1 anode/cathode ratio in sea water.

(i) Severely attacked	Carbon steel
	Aluminium
	Gunmetal (88Cu 10Sn 2Zn)
(ii) Moderately attacked	Aluminium brass
	'Monel'
	Cupro-nickels
(iii) Slightly or not attacked	18/8 stainless steel
	Aluminium bronze

With a 1/10 anode/cathode ratio only carbon steel suffered any serious attack. (Zinc and magnesium were not included in the tests but would be expected to suffer severe attack.)

In air, the effect of galvanic coupling was less than in sea water. With large cathode-anode ratios, carbon steel was attacked and aluminium, 'Monel' and cupro-nickel suffered slight pitting.

Titanium is being used increasingly for tubes in sea water services and the above results are relevant to the choice of tube plate materials. The main use of titanium in marine service is as a tubing material, where its excellent resistance to corrosion makes it an economic choice in many situations [17],[18]. It is also used as an anode material for cathodic protection.

6.5 MAGNESIUM

Magnesium and its alloys are not widely used for marine situations except as an anode for cathodic protection or as components for aircraft. These alloys have poor resistance to sea water, although they can be coated with paints if required for exposure to marine situations. The addition of manganese tends to reduce attack in sea water but not to an acceptable level. Generally the addition of alloying elements increases attack. It is one of the lightest of commercial metals and is the most anodic in the galvanic series in sea water, so coupling with any other alloy will generally result in increased corrosion of the magnesium.

6.6 OTHER METALS AND MATERIALS USED IN MARINE SITUATIONS

Other non-ferrous metals are not used for structural parts or for components to be used in marine environments. Zinc is used as an anode material for cathodic protection and both zinc and cadmium are used for coatings (see Chapter 11). Other metals may be used for specific purposes, e.g. platinum, in some anodes for cathodic protection. Others such as arsenic, niobium and manganese are important as alloying elements. However, the properties of the metals themselves are not of direct interest.

6.6.1 Cobalt

Although not used for constructional purposes, cobalt is an essential element for the 'hard-facing' alloys produced commercially under the name 'Stellite'. There is a range of materials, which may contain 25–35Cr, 5–20W, up to 6Mo, 0.2–2.5C, possibly iron, manganese, silicon and nickel additions, the remainder being cobalt.

These materials are very resistant to corrosion by sea water and to impingement and cavitation. They are also resistant to high temperatures, particularly the high chromium alloys.

6.6.2 Graphite

Although not a metal, graphite conducts current and is used as an anode in some cathodic protection systems. It is also used as a packing material and in greases. Attention should be paid to the cathodic nature of carbon in contact with other metals, particularly aluminium, where moist conditions are likely to prevail.

6.7 REFERENCES

(1) SOUTHWELL, C R, HUMMER, C W and ALEXANDER, A L, *Materials Protection*, **4**, 30 (1965)
(2) SOUTHWELL, C R, ALEXANDER, A L and HUMMER, C W, *Materials Protection*, **7**, 41 (1968)
(3) FRIEND, J N, *J. Inst. Metals*, **39**, No 1, 111 (1928)
(4) BULOW, C L, *Trans. Electrochem. Soc.*, **87**, 127 (1945)
(5) ROWLANDS, J C in L L Shreir (ed), *Corrosion*, 2nd ed, Vol 1, Chapter 2.4, Butterworths, London (1976)
(6) MATTSON, E and HOLM, R, *Special Technical Publication 435*, ASTM (1968), p 187
(7) SCHOLES, I R and JACOB, W R, *J. Inst. Metals*, **98**, 272 (1970)
(8) FERRARA, R J and CATON, T E, *Materials Performance*, February, 30–34 (1982)
(9) GLOVER, T J and MORETON, B B, *Proc. UK National Corrosion Conf.* (1982), pp 105–108
(10) NIEDERBERGER, R B, FERRARA, R J and PLUMMER, F A,, *Materials Performance*, **9**, 18 (1970)
(11) *Special Technical Publication 435*, ASTM (1967), p 151

(12) *Special Technical Publication 175*, ASTM (1956)
(13) PETERSON, M H, BROWN, B, NEWBEGIN, R L and GROOVER, R E, *Corrosion*, **23**, 5 (1967)
(14) *Aluminium Bronze Alloys Corrosion Resistance Guide, Publication No 80*, Aluminium Bronze Advisory Service, Copper Development Association, Potters Bar, Herts, UK
(15) *Copper Nickel 90/10 and 70/30 Alloys Technical Data*, *TN31*, Copper Development Association, Potters Bar, Herts, UK
(16) *Corrosion Resistance of Titanium*, IMI, Birmingham
(17) MCMASTER, J A, *Materials Performance*, **18**, No 4, 26 (1979)
(18) NEILL, W J, *Materials Performance*, **19**, No 9, 57 (1980)

7 Design

The design of plant, equipment and structures can have an important influence on the corrosion of alloys and the performance of protective coatings. Some guidelines can be listed, but there is no simple way of ensuring that the design is such as to provide the best performance for alloys and coatings. Designers must develop an 'awareness' of corrosion so that some of the more obvious problems can be avoided. Designs can be audited by specialists and this may well be the most satisfactory way of ensuring that corrosion problems are reduced to a minimum. Throughout this book reference has been made to the importance of design in avoiding situations such as localised corrosion that occurs in crevices, erosion-corrosion in pipes and stress-related problems. In this chapter some of the more common situations relating to the design of plant and structures will be considered but it should be appreciated that good design is not just the avoidance of features that promote corrosion. It is the best use of materials, control methods and coatings to provide the most economic solution to a specific problem.

Corrosion is, of course, only one of many aspects concerned with design and it is neither desirable nor profitable to consider corrosion without relating it to other requirements. For example, the strength requirements of materials do not always correlate with corrosion-resistance but the choice of materials must take into account corrosion performance even though this may not eventually prove to be the main priority. In short, design requires deliberation before choice and an understanding of basic corrosion processes. From the corrosion standpoint, design can be considered in broad terms as the attempt to provide environmental conditions that will best suit the alloy or coating chosen to provide corrosion-resistance. Additionally, it can be considered in relation to the choice of the most suitable materials and coatings to resist the environment. The environment, in this context, is the local one at the metal surface, and this can be significantly altered by various means. The overall environment can be changed, e.g. by using desiccants in a box girder or inhibitors in a water system, local changes can be made, however, by alterations in design. A typical example is the drilling of drainage holes to remove water from a channel. By doing this the environment is changed from either immersed or constantly damp conditions to an atmospheric situation where the steel and any coatings can be dried by the sun and the wind so they are moist for only short periods. Again, by reducing bends in pipes to provide more gentle changes in direction, the environment in the pipe can be changed from one causing erosion effects to a less aggressive one, resulting in a lower rate of corrosion. Sometimes it is not practicable to alter the environment to provide less corrosive conditions. In such cases suitable materials or coatings should be chosen to resist the situations that arise. Often, this can be done by using high quality materials in local areas, provided this does not cause other problems, e.g. bimetallic effects, or by designing so that materials or component in areas likely to be badly attacked can be removed and replaced reasonably easily.

Although, in practice, design is necessarily a compromise between many competing factors, unless corrosion is taken fully into account at the beginning of a project, problems of a less or greater nature are bound to arise. In particular, where different parts of an overall design are considered separately and pumps, components and parts are brought in from manufacturers, some serious problems can occur if the complete design, materials and coating requirements are not considered at an early stage.

Another important consideration in design is the necessity for maintenance. In the case of structures this usually involves re-painting and in sea water systems and process plant, the replacement of the more vulnerable parts, e.g. valves. Proper access for such maintenance work is essential if costs are to be kept to a low level. The basic requirements of design can be considered under the following headings:

(i) Features that promote corrosion.
(ii) Access for maintenance.
(iii) Bimetallic connections.
(iv) Selection of materials.

7.1 FEATURES THAT PROMOTE CORROSION

A number of design features promote corrosion because they influence the immediate local environment and also because they affect some other part of the structure or system. Some such features are mainly applicable to structures, tanks, etc, and they can be placed in a number of fairly broad groups which will be considered below:

(i) Entrapment of moisture and salts.

(ii) Shape.
(iii) Crevices.
(iv) Corrosion at ground level.

(i) *Entrapment of moisture*

Moisture, salts and dirt can collect in many parts of a structure. Typical examples are shown in *Figure 7.1*, where open channels and angles can collect water which, as it evaporates, may leave concentrated salt solutions in contact with the protective coatings. The local environment

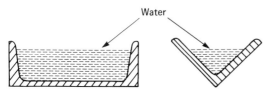

Figure 7.1 Entrapment of water

may be more corrosive than that for which the protective system was chosen to withstand, so premature coating failure occurs. There are a number of ways of dealing with the situation and often the sections can be designed so that water does not collect in them (*Figure 7.2*). Other alternatives are to weld plates over them (*Figure 7.3*) or to provide a drainage hole (*Figure 7.4*). The use of rolled hollow sections or tubes may be advantageous (*Figure 7.5*). The provision of suitable drainage holes is a comparatively simple and effective way of removing water in many situations. Generally, however, it is less efficient than preventing the water collecting in the first place. Even good drainage is likely to leave

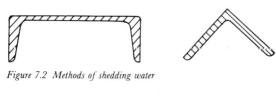

Figure 7.2 Methods of shedding water

Figure 7.3 Sealing to avoid entrapment of water and salts

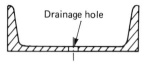

Figure 7.4 Drainage hole to remove water

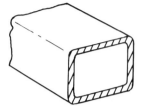

Figure 7.5 Avoid entrapment of water with rolled hollow sections

some pools of water on the steelwork. Furthermore, there is always a tendency to leave dirt and salts on the surface. In marine atmospheres, these are likely to leave damp patches for most of the time because of the hygroscopic nature of sodium chloride. There are some points that should be borne in mind concerning drainage holes:

(a) They should be at the lowest point to ensure complete drainage. This can be determined comparatively easily on existing structures but sometimes more than one drainage hole is required.
(b) They must be large enough to ensure that they do not become blocked with debris.
(c) The drainage from the hole must be away from other parts of the structure.

If necessary, drain pipes should be fitted. Their use is often useful to run water from expansion joints and similar features away from the main structural elements.

Drainage from tanks is particularly important and this is considered later under 'Tanks'.

Moisture can be a problem where absorbant materials such as wood or lagging meterial is in contact with metals, particularly steel or aluminium alloys. This can lead to serious localised corrosion in marine situations unless suitable attention is paid to the problem.

A typical example is shown in *Figure 7.7*, where a wooden deck is in contact with steel. This is, however, generally less of a problem than where soft woods are used, e.g. for roof trusses in contact with steel sheet in a humid situation. Both steel and wood should be insulated by use of a suitable barrier coating, e.g. bitumen.

(ii) *Ground level corrosion*

Problems frequently occur at ground level where steel columns enter or rest on the ground. The situation arises from the collection of water that runs from the building or structure and often there is a build-up of debris,

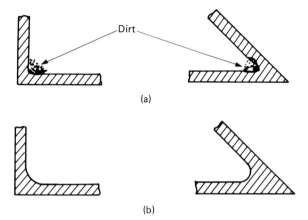

(a)

(b)

Figure 7.6 (a) Features where dirt and moisture collect are often found difficult to clean and paint (b) Rounded surfaces are less prone to corrosion and easier to paint

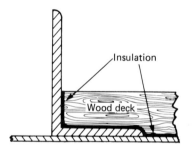

Figure 7.7 Insulate wood from steel with coatings or mastics

which acts as a poultice at the foot of the column (see *Figure 7.8*). Furthermore, there is usually an element of splashing.

There are various ways of dealing with the problem (see *Figure 7.9*):

(a) Use of concrete plinths; they must be designed to allow water to run off and away from the steelwork.
(b) Increasing the thickness of coatings at and near ground level.
(c) In less aggressive atmospheres—particularly in buildings—the addition of a bituminous coating may suffice.

(iii) *Crevices*

Crevices are formed at joints where small gaps remain between the components of the joint. Examples are given in *Figure 7.10*. On the whole, welded joints, provided they are properly designed and made, are preferable to bolted joints so far as crevices are concerned. Stitch welding can, however, cause a series of crevices and poor welding practice can

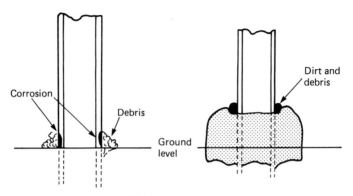

Figure 7.8 Corrosion at ground level

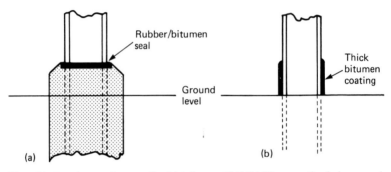

Figure 7.9 Corrosion control at ground level (a) Concrete plinth (b) Bitumen coating for less aggressive situations

lead to enhanced corrosion problems apart from crevices (see below, 'Welds').

Alloys that depend on passive films for their corrosion-resistance, e.g. stainless steel and aluminium, are prone to crevice attack because the passive film may not repair itself under these conditions. Any design detail that results in small gaps between members should be examined very carefully at the drawing board stage. In the worst situations structural damage can occur on comparatively thin steel sections. On heavier sections this is unlikely to happen but it will lead to breakdown of coatings and difficulties with maintenance. The problems arise from the nature of rusting. Rust, $Fe_2O_3 . H_2O$, has a considerably greater volume than the steel from which it is produced. Consequently, in an enclosed space either the volume of rust will lead to the buckling of the steel and the shearing of bolts or corrosion will be stifled and eventually stop. There appears to be no firm data on the relation between the forces exerted by the rust and the ability of steel to resist this force, although the formation of corrosion products may exert a pressure as high as $40 \, \text{N/mm}^2$. In

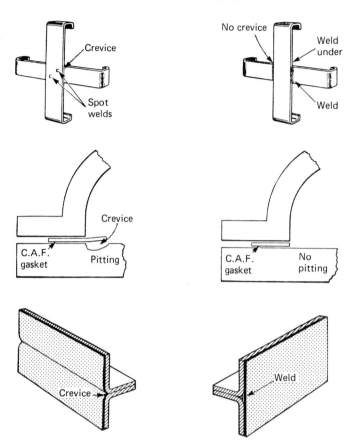

Figure 7.10 Typical crevice situations and methods of control

qualitative terms thin gauge sheet and small angles will succumb to the pressure, whereas heavy plates and sections resist this form of attack. The effect of rusting in confined spaces becomes particularly acute where steelwork is in contact with materials of low elasticity and low-tensile strengths such as brickwork, masonry and concrete. These materials tend to crack and disintegrate under these conditions (see 7.4.3, 'Buildings').

It is not always practicable to avoid crevices but in many situations comparatively slight changes in design will eliminate them or reduce their effect. For example, it may be preferable to deliberately increase a gap to avoid a crevice situation. The following methods of control can be considered:

(a) On new equipment or structures use welded butt joints in preference to bolted joints.
(b) Use continuous welding to close gaps where they exist.

(c) Where welding is impracticable, appropriate fillers and mastics can be used, but they should be examined at regular intervals to ensure that there is no shrinkage or cracks which could exacerbate the situation.

(d) On bolted joints the application of thick organic coatings to both mating surfaces and then closing the joint before the coatings dry may be an effective method of avoiding small gaps. (This method cannot be used with high strength friction grip bolts.)

(iv) *Shape*

The shape of structural members has a bearing on coating performance. Rounded sections are easier to coat and are less prone to damage than edges on rectangular sections. Although the use of rounded contours, e.g. tubes, or rolled hollow section (RHS), is not always practicable, where a choice exists they are to be preferred to other sections, e.g. I-beams. It is easier to coat tubular members and a great deal easier to wrap them. Although there are often requirements for the rounding of sharp edges on other types of section, this is not always done and paint coatings tend to be thinner at these edges and so fail prematurely unless additional stripe coats are applied.

7.2 ACCESS FOR MAINTENANCE

Once a structure or building has been erected, problems with maintenance painting may arise if access to the steelwork is difficult. In some situations access may be virtually impossible, in which case, severe corrosion problems are likely to arise. The access in this context is not that of physically reaching the steelwork with scaffolding, although this may be a quite separate difficulty, but rather being unable to examine or paint the steel because of the type of design. There are a number of obvious examples as shown in *Figure 7.11*. Back-to-back angles should always be avoided because not only are they difficult to repaint but the rust formation in the gap between the angles may cause problems, particularly under marine conditions. If, for some reason, there are parts of a structure that will be inaccessible, then these should, wherever practicable, be enclosed so that they do not come into direct contact with the environment. Again, if the spacing between members will not allow for painting, the use of mastics or—preferably—welding on of suitable plates should be considered. Even in situations where access is possible it may prove to be economic to enclose complex areas to avoid time-consuming repainting. It is cheaper and quicker to paint large flat areas rather than a

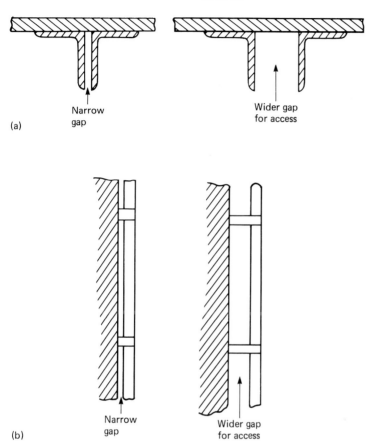

Figure 7.11 Access for maintenance (a) Structural sections (b) Pipes

number of complex bolted areas. It is interesting to note that in the
United Kingdom, trials are being carried out on large enclosures for
bridges to reduce the costs of repainting, particularly those involved in
gaining access. By enclosing a structure in a 'box' made of strong plastic
sheets the environment can be controlled to reduce corrosion and extend
coating life.

Consideration should be given to local environmental situations on
structures and where appropriate 'box-in' with steel sheets or plate, areas
that are likely to be particularly prone to corrosion or prove difficult to
maintain. Decisions on such approaches to corrosion control will be
determined by economic considerations but with the total costs of
maintenance continually rising, the environmental approach may have
merit in specific situations where enclosures of this type can be achieved
fairly easily because of the simple geometry of the structure.

7.3 ENVIRONMENT CONTROL IN STRUCTURES

Little can be done to change the environments where structures are exposed in the atmosphere. In enclosed systems or spaces it is possible to alter the environment by means of air-conditioning, use of inert gases and inhibitors. It is possible on a limited scale to choose the environment where plant and structures are to be sited and this can be a profitable exercise worth exploring to a greater extent than is generally recognised. It is a common practice to consider the route of a pipeline but the siting of plant in relation to the fumes emanating from other parts of an industrial complex and, taking into account the prevailing wind, can sometimes reduce maintenance costs to a significant extent. Again, the actual site chosen for plant on the coast may have an important influence on the treatment, e.g. filtering of water used in heat-exchangers. There are variations of environment within a land-based structure or building. In northern latitudes, the north facing surface remains damp for longer than those facing south, and undersides of beams on large structures may also be damp for prolonged periods. Generally, these differences in performances of coating systems can be taken into account by using thicker films or more resistant systems. A particular problem that arises in enclosed spaces is condensation, which can lead to serious corrosion under marine conditions because any chloride salts that collect on surfaces tend to remain almost permanently damp. In extreme cases, air conditioning may be necessary, but careful design of ventilation systems will assist in the alleviation of the problem. Generally, enclosed spaces should be either hermetically sealed or deliberately ventilated. The control of moisture in enclosed spaces by means of desiccants such as silica gel is a cheap and usually successful way of dealing with many problems where the enclosed space is reasonably well sealed and manholes are kept closed. 250 g of silica gel for each m^3 of void should be effective for about 2 years[1]. The desiccant is then renewed. It is advisable to have a means of checking when the silica gel is saturated with moisture; this is accompanied by a colour change.

7.4 SPECIAL SITUATIONS

7.4.1 Tanks

Tanks are worth specific consideration because design plays a particularly important role in ensuring that they provide long-term service (see *Figure 7.12*). The main points to be considered are:

(i) The drainage should be at the lowest point.

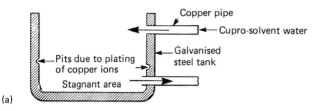

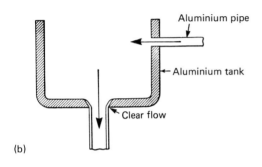

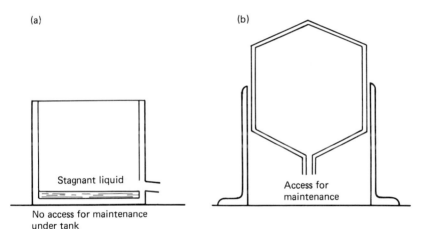

Figure 7.12 Design of tanks (a) Poor (b) Improved

(ii) Stagnant areas at outlets should be avoided.
(iii) There should be adequate access for protecting the exterior of the tanks and there should be free flow of air round them.
(iv) Tops should be designed to ensure that water runs off.
(v) Pipes and outlets materials should be compatible with those of the tank material.
(vi) On steel tanks the interior lining must be capable of withstanding the liquids stored in the tank and must be pore-free.

(vii) Internal struts or other components, e.g. heaters, inside the tank should be insulated from non-compatible material and preferably should not be of more noble material than the tank. If they are, consideration should be given to coating them.

7.4.2 Welds and fasteners

Both the common methods of joining alloys can cause problems. Welds— by their nature—tend to produce discontinuities in the surface of the alloy and these can lead to metallurgical variation and potential sites for coating breakdown. In some alloys, the inhomogeneity of the weld areas can lead to localised corrosion but there are usually ways of avoiding this. Generally, in carbon steels such problems do not arise because potential differences between the parent metal and the weld are not usually sufficient to cause serious corrosion. However, when immersed in sea water, problems may arise. There have been reports of corrosion rates of up to 10 mm/y at welds on ice-breakers[2]. This severe corrosion was attributed to galvanic effects between the weld metal and the steelplate. The use of a more noble electrode for welding was reported to have overcome the problem. Welds are often points of stress concentration which may affect alloys in relation to corrosion-fatigue and stress-corrosion performance and sound welding procedures are essential. However, welds that are perfectly satisfactory from the standpoint of strength and jointing efficiency may still cause general corrosion problems. Spot, skip or stitch welds can lead to a series of crevices and continuous welding is to be preferred. Butt welds are generally less likely than lap welds to lead to problems with moisture and dirt collection. The main problem with welds, though, arises from their effects on coatings. Splatter and irregularities can result in thin coatings, in a similar way to high profiles from blast-cleaning, and premature coating failure may occur. A number of problems may arise with bolted components. Nuts and bolts are difficult to paint because they cause an irregularity in the surface of the steel but they can also cause bimetallic problems and crevices. Although insulation is advantageous to control the possibility of galvanic corrosion, the insulators themselves can cause problems on stainless steel and aluminium alloys if they lead to crevice formation, particularly under immersed conditions.

7.4.3 Buildings

Two main design-related problems can occur with buildings in any environment but they are often accentuated in marine situations. On reinforced concrete buildings, there can be spalling of the concrete if there is rusting of the steel reinforcement (see Chapter 20) but similar problems can also occur with steel-framed buildings that are clad with concrete, masonry or similar materials. Even brick-built buildings are not immune from this form of attack. The cause is similar to that discussed under 'Crevices', where—if rust forms in a small gap, because of the increased volume compared with the steel that is corroding—pressures are built up on the two materials forming the gap or crevice. Heavy steel sections can withstand this pressure and corrosion is stifled; lighter steel sections, e.g. sheet, may buckle. However, where the gap is produced between heavy steel sections and masonry, e.g. in a building, or where the steel is completely surrounded, e.g. in reinforced concrete, then the material that is weak in tension will crack. The situation is illustrated in *Figure 7.13*. The problem is usually one of design in that moisture or water reaches the steel through gaps, leaks or other features that should not be present. It also occurs if there is insufficient cover of concrete. Generally,

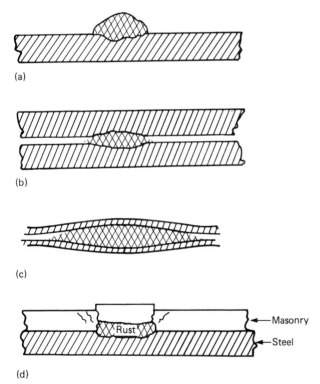

(a)

(b)

(c)

(d)

Figure 7.13 Rust formation and its effects with steel and building materials (a) Rust of larger volume than the corroded steel (b) Heavy sections of steel with rust formation stifled (c) Light sections of steel with the steel buckling because of the force exerted by the rust (d) Close contact steel and masonry showing cracking of the masonry

this type of attack on structural steel develops over a number of years and tends to be a potential problem with some older buildings. A similar form of attack may, however, occur more quickly on components and fixings used to hold heavy cladding blocks for a building. If water reaches such components and is trapped, then it is possible that corrosion could be sufficiently intense to cause loss of section in the fixing bars with eventual failure. In practice this is not generally a problem if the components are correctly chosen and properly protected.

The other design-related problem that may occur with buildings again involves attack on the steel stanchions or columns. However, it is not the effect of rust but the actual attack on the steel that may cause problems. Where steel columns are encased as shown in *Figure 7.14(a)* and water enters the space between the steel and the encasement, e.g. from a roof,

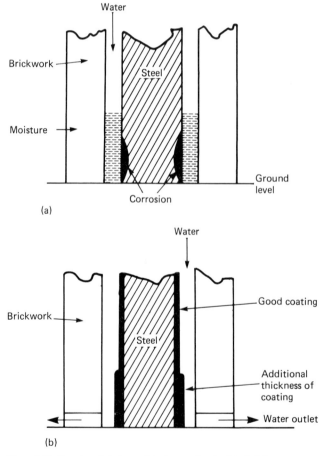

Figure 7.14 Corrosion of steel caused by ingress of water into buildings

then the water will tend to collect at the lowest level. This can produce attack on the steel. The situation can be more serious if the space is filled with old mortar, broken bricks, etc, because the whole of the steelwork may remain in contact with the damp material for prolonged periods. This type of attack is usually avoided by providing a sound protective coating on the steelwork and by allowing any water that collects to escape at the bottom of the building (see *Figure 7.14(b)*).

7.5 BIMETALLIC CORROSION

Bimetallic corrosion can cause problems out of all proportion to the actual amount of corrosion that occurs. It tends to be localised, leading to pitting, and in pipe systems can lead to rapid perforation and failure. It is basically a design problem because so many features of plant, pipe systems and structures involve the use of more than one material.

It is, of course, inevitable that different alloys will be used in complex designs and the problems are more acute in marine situations than in any other. It should, however, be appreciated that the use of different alloys in a system does not necessarily lead to serious corrosion problems. It is the way in which they are used that may cause difficulties. This is well illustrated by the classic example of steel rivets in copper and copper rivets in steel, both immersed in sea water. In the latter case, no serious problems arise, whereas in the former, the steel rivets may corrode to an extent where they literally fall out of the rivet holes. The reason for the different performance of the two assemblies arises from relative areas of anodes and cathodes in an electrolytic cell. A small anode (steel rivet) and large cathode (copper) will result in severe attack on the anodic metal, whereas a large anode (steel) and a small cathode (copper rivet) leads to far less intensity of corrosion.

The basic requirements necessary for bimetallic corrosion are (i) the presence of an electrolyte, (ii) electrical connection between the metals, and (iii) a difference in potential between the two metals. Even if these three requirements are met, marked corrosion will not occur unless the cathodic reaction is sustained in some way to prevent excessive polarisation, e.g. by the ample supply of dissolved oxygen.

7.5.1 Electrical connection

The two alloys concerned in bimetallic corrosion must be electrically connected for the cell to be operative. It is not always appreciated, however, that the connection may be outside the electrolyte. If two different alloys are immersed in a solution and joined together or are in electrical contact at some other part of the structure, bimetallic corrosion can occur. It is possible in some situations for corrosion between two alloys to occur even if they are not electrically connected. The corrosion of steel and aluminium and zinc coatings by copper depositing from cupro-solvent water is the best example. This is, however, a special case of reactions arising within the electrolyte and will be considered below.

7.5.2 Electrolyte

The electrolyte does not necessarily have to be a bulk solution. Bimetallic corrosion occurs in the atmosphere where the electrolyte is a condensed film arising from rainwater, dew, etc. Salt deposits on the surface, which are often hygroscopic, can also act as an effective electrolyte, as can soils of suitable conductivity. In fact any damp material in contact with metallic components may provide suitable conditions for bimetallic corrosion, e.g. damp lagging.

The electrical conductivity of the electrolyte is important and generally the effects of bimetallic corrosion are likely to be more severe under marine conditions because sea water has a conductivity of about 50 times that of ordinary supply water. Strongly conductive solutions such as sea water provide conditions where bimetallic corrosion may occur over a greater area of the alloy. In low conductivity solutions, the attack tends to be confined to an area near the junction of the alloys.

The pH of the electrolyte also influences the amount of corrosion, which increases as the pH is decreased, i.e. the solution becomes more acid.

Corrosion products may deposit on the more noble metal with secondary effects. This can occur in a carbon steel-stainless steel couple where the corrosion products from the carbon steel may deposit on and induce localised corrosion of the stainless steel. In enclosed systems inhibitors may in some situations be effective in reducing or eliminating bimetallic corrosion.

7.5.3 Potential difference

For any corrosion cell to operate there must be a potential difference between the electrodes. The broad differences in potential to be anticipated when joining different alloys are given in various galvanic series, which are the actual potentials measured in an actual environment such as sea water (see *Table 7.1*). This provides a good general indication of the likely intensity of corrosion which is governed by the distance between the alloys in the series.

Table 7.1 Galvanic series of alloys and metals commonly used in sea water

Cathodic	Titanium
↓	Austenitic stainless steels (passive)
	Chromium steels (11–30% Cr) (passive)
	Nickel and high nickel alloys (passive)
	Silver solders
	Monel, bronze, gunmetals, 70/30 cupro nickels, copper-nickel iron alloys
	Copper, admiralty, 70/30 brasses
	Nickel and high nickel alloys (active)
	Naval brass, 60/40 brass, high-tensile brass
	Lead, tin
	Austenitic stainless steel (active)
	High nickel cast iron
	Chromium stainless steel (active)
	Carbon steel, cast iron
	Aluminium alloys
	Cadmium
	Aluminium
↑	Zinc
Anodic	Magnesium, magnesium alloys

Temperature can influence the relative positions of alloys in the series, e.g. in sea water at temperatures near the boiling point, the positions of steel and aluminium may be reversed. Other effects such as velocity and aeration can also affect the potential difference in a couple. Stainless steel is reasonably noble by virtue of the passive oxide film formed on the surface. In situations where the passive film breaks down and is not re-formed, stainless steel adopts a potential very close to ordinary carbon steel. This can occur locally in sea water, particularly at crevices.

The flow rate of the solution also affects the behaviour of a galvanic cell. In part this arises from the increased oxygen reaching the alloy surface at higher flow rates. This influences oxide films which tend to become more protective leading to passivity in the couple. On the other hand, the velocity may be such, particularly if abrasives are entrained in the water, to disrupt protective films leading to a change in the potential of the alloy. Copper alloys tend to shift to a less noble potential with increasing flow rate. Under erosion-corrosion conditions where passive films are removed from alloys such as stainless steel, galvanic effects may be quite severe on such alloys.

In practice the environmental conditions may have an important influence on bimetallic corrosion. It is, therefore, important when considering data on bimetallic couples to consider the practical situation and design of the plant or structure with some care.

7.5.4 Anode-cathode ratios

The ratio of the anodic area to cathodic area is particularly important in sea water because of its high conductivity. The example of copper rivets in steel as opposed to steel rivets in copper has already been mentioned. Much less obvious situations can occur where small anodes are in contact with large cathodes leading to severe localised pitting.

Millscale, which is produced during the manufacture of steel, can cause serious pitting of steel if it is not removed. This is not strictly bimetallic corrosion because millscale is an oxide, but the mechanism is the same. There is a potential difference of about 0.3 V between carbon steel and millscale left on its surface and the scale is more noble. Consequently, if steel is covered with about 90% of scale, the anode to cathode ratio is 1:9, so if immersed in sea water, severe pitting can occur.

Similar situations can occur with fasteners and at welds. A particular problem related to area ratios can develop if a more noble metal is in contact with painted steel. If, for example, a coated steel tank corrodes and the carbon steel plate is replaced by welding in a plate of stainless steel, serious pitting of the carbon steel can occur if the coating is damaged (even if the damaged area is not immediately adjacent to the stainless steel).

This problem of applying coatings to anodic and cathodic areas is not always fully appreciated. Although by coating the complete couple bimetallic corrosion can be controlled, this is not necessarily so if only the anodic part of the couple is painted. Any small area of damage leads to a large cathode to anode ratio and severe pitting may occur. It is, in fact, preferable to protect the cathodic area rather than the anodic area, but much better to paint the whole of the couple. The problem really concerns the nature of many such couples. If a highly resistant material such as Monel or stainless steel is used for the nuts and bolts it may seem pointless to paint such fasteners. Nevertheless, it may be advisable to do so for the reasons given. Proper insulation is, however, the preferred method of control (*Figure 7.15*).

Under immersed sea water conditions, bimetallic corrosion is likely to be a problem. It is clearly not always practicable to avoid using more than one alloy but attempts should be made to limit the number and type of different alloys in contact. Wherever possible the more noble alloy should be smaller than the less noble ones and critical parts should— wherever practicable—be more noble.

7.5.5 Rate of corrosion

Although it is a reasonably straightforward matter to determine the anode alloy in a bimetallic couple and to decide whether corrosion is

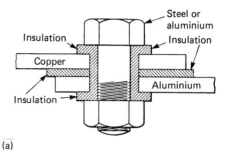

(a)

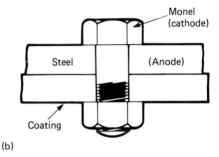

(b)

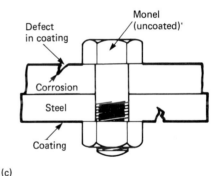

Figure 7.15 Bolted joints (a) Insulation of the joint (b) Coating of whole assembly (c) Coating of anodic areas only

(c)

likely to occur, it is more difficult to predict the course of the corrosion. This depends upon factors such as the environment, velocity of solutions and aeration but mainly it is determined by polarisation effects and these are not always easy to quantify (see Chapter 2 for an explanation of this phenomenon). With sufficient data reasonable predictions can be made but in practice such data is not usually available. Tests can, of course, be carried out but these are not as straightforward as they might at first appear to be. Simple tests in beakers may not reproduce the conditions and may provide misleading information. Consequently, generally reliance is placed on published test and service data and experience. This information is usually supplied in the form of a chart or diagram, which serves as a reasonable guide to designers provided they appreciate the points discussed in this section. Information is provided in BS. PD6484:1979.

7.5.6 Minimising bimetallic corrosion

There are a number of ways in which bimetallic corrosion can be minimised or even prevented. Wherever practicable the following guidelines should be followed:

   (i) Construct equipment and structures for marine service from one material.
  (ii) Where (i) is not possible, select combinations of alloys that are close together in the galvanic series.
 (iii) Avoid small anode/large cathode situations and ensure that critical components are more noble.
  (iv) If necessary add extra thickness to the less noble component and increase the anodic area.
   (v) Insulate dissimilar alloys. When insulating bolted joints ensure that the shank of the bolt is insulated by a non-conducting sleeve (*Figure 7.15*). Check that the components are effectively insulated.
  (vi) Consider the use of inhibitors in enclosed systems.
 (vvi) Protect the bimetallic couple with suitably resistant coatings. The whole assembly should be painted but do not paint the anodic area only because, if the coating is damaged, severe localised corrosion may occur.
(viii) Design so that anodic components can be replaced without difficulty.
  (ix) Monitor structures and plant during their service life to ensure that measures to combat bimetallic corrosion remain effective. Modifications should be checked for possible bimetallic effects. It should be appreciated that the coupling of different alloys may be outside the solution where attack will occur.
   (x) When monitoring, check the current flow in preference to the potentials of the alloys concerned.
  (xi) Where appropriate cathodic protection can be used.
 (xii) Ensure that absorbant materials are prevented from coming into contact with the alloys in the structure.

7.5.7 Compatibility of alloys with other materials

Graphite acts in a manner similar to a metal and is included in some galvanic series. It is very noble and so can cause accelerated corrosion of most alloys. Graphite may be used in certain types of heat exchange and is a component of some lubricants.

7.6 FLOWING WATER SYSTEMS

In structures exposed to the atmosphere or immersed in sea water, the constructional alloy is usually steel. Most of the corrosion design requirements are concerned with ensuring that the coatings do not fail prematurely and that they can be adequately maintained. Many of the points already considered are relevant to plant and equipment, particularly the external surfaces. However, a different set of problems arises on the internal surfaces of such equipment. Bimetallic corrosion is generally a more serious problem under immersed conditions and flowing fluids add to the difficulties that may arise. The problems related to flowing water have been considered in Chapter 4 and it is of some importance to ensure that the design is such that the various effects are minimised. Often, this is a matter of material selection, choosing alloys that will resist impingement, etc, but many of the problems can be alleviated by suitable design measures.

In pipework any feature that affects smooth flow is a potential source of turbulence or impingement, with resultant corrosion. It is, therefore, important to maintain a pattern of smooth flow as shown in *Figure 7.16*. *Figure 7.16(a)* indicates an improvement on *Figure 7.16(b)*, which is preferable to *Figure 6.16(c)*, but for many situations any right-angled change in flow may cause impingement problems, particularly with higher velocities.

Inlets of condenser tubes cause many problems and the use of plastic inserts or extending the pipe at the tube plate have proved to be effective ways of avoiding corrosion.

The use of replaceable parts, such as plastic inserts at the end of tubes, is a sound way of increasing the lives of tubes but care must be paid to the design so that smooth flow is not affected. If the inserts are not designed in this way, then the problems may be removed from the inlet to a position further along the tube.

Generally, turbulence in piping, at valves, in water boxes, etc, causes problems and the designer should make every effort to avoid such conditions. Problems are often introduced during the working life of the system by replacing and refitting components of a different size from that originally scheduled. This may change the flow pattern and increase corrosive effects. Poor workmanship, e.g. allowing gaskets or washers to be 'proud' in a water stream, can prove to be particularly troublesome (*Figure 7.16(d)*).

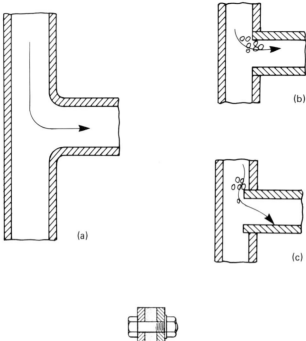

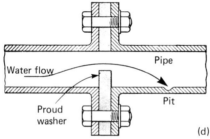

Figure 7.16 Flow of water in pipes should be smooth

The selection of suitable valves so that they are not operated in half-open conditions will reduce turbulence, which has two effects in promoting corrosion. It disrupts the 'boundary layer' which is formed on the inside surface of a pipe carrying flowing water and, by entraining air into the field, it can increase the rate of oxygen reaching the surface of the pipe, which will lead to increased corrosion of some alloys, e.g. carbon steel.

Debris, sand and other abrasive particles in the water stream can lead to erosion-corrosion effects, and also to deposit attack and oxygen concentration cells. Suitable filtering and screening will assist in reducing this problem but regular cleaning to remove deposits and fouling is a maintenance requirement.

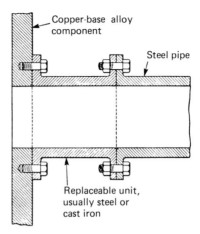

Figure 7.17 Easily replaceable parts

Other methods of approach include the following:

(i) An increase in pipe diameter will assist in reducing velocity and improving flow. This may reduce erosion-corrosion effects in some situations.

(ii) Sometimes impingement effects can be ameliorated by the use of plates or baffles that will be attached but can be readily removed.

(iii) At critical areas there are clear advantages in having easily replaceable units (see *Figure 7.17*).

(iv) At the planning stage, pipework should be designed to provide maximum smooth flow. Although they cannot be completely eliminated, elements such as bends, valves, flanges and other fittings should be carefully sited and kept to a minimum within the practical requirements.

7.7 REFERENCES

(1) *BS. 5493:1977, Protective Coating of Iron and Steel Structures Against Corrosion*, p 84

(2) UUSITALO, *Proc. 2nd Int. Congr. Met. Corr. NACE* (1963), National Association of Corrosion Engineers, Houston, Texas, USA

8 The selection of materials for marine environments

The basic aim when selecting materials is to achieve an economic solution for the particular situation but this is not necessarily an easy matter. Although the comparative costs of various materials can be obtained, it is far more difficult to predict their performance in what is often a complex situation. Corrosion performance is not always the determining factor when selecting materials even for aggressive marine conditions. The large range of alloys available has been developed as much for mechanical and physical properties as for corrosion characteristics. Furthermore, structures and plant have to be fabricated and materials may be selected for weldability and forming characteristics as well as for corrosion properties. Corrosion is nevertheless an important requirement, particularly in marine situations, and must be considered as a high priority in relation to other factors. There is rarely a single solution to a problem concerned with corrosion. This will be evident from the range of tubing materials used for condensers and the various approaches to the protection of the splash zone on offshore platforms.

Sometimes decisions are based on subjective conclusions drawn from experience or even on limited knowledge of the alternatives. Experience obviously cannot be discounted but there is always a possibility that it will not take a proper account of developments in a particular field. Again, many items of plant are purchased as a unit and it is often difficult to change items within the unit.

The specific factors that have to be considered will depend upon the type of the structure or plant but there are general points, which are considered below.

8.1 FACTORS INVOLVED IN THE SELECTION OF MATERIALS

These include the following, related to the various aspects involved in the use of the materials.

8.1.1 General considerations

(i) Cost in relation to life and maintenance or replacement.
(ii) Mechanical and physical requirements.
(iii) Availability in the form required, e.g. as a casting or in tube-form.
(iv) Ease of forming and fabrication.
(v) Weldability and/or ease of soldering or brazing.
(vi) Ease of installation, repair and replacement.

In some situations the following additional factors must be considered:

(i) Heat transfer characteristics.
(ii) Vibration effects.
(iii) Thermal expansion.
(iv) Biofouling characteristics.
(v) Strength/weight ratio.

These lists are not exhaustive but indicate the requirements that must be satisfied apart from the corrosion characteristics. The corrosion requirements will be determined by the operating conditions which may, of course, be influenced by the design of the plant or structure. These conditions must be considered in detail because a particular factor, e.g. velocity of sea water, may be a determining factor. Some of the more important corrosion related considerations include the following:

8.1.2 The service environment

(i) Will the alloy be used immersed, in air, or in a splash-zone situation? The corrosion-resistance of alloys varies in these three basic areas as do the methods of controlling corrosion. High-cost, resistant alloys may be considered as an economic solution in the splash-zone, whereas protected carbon steel may be quite suitable for other situations and cathodic protection for immersed parts.
(ii) Under immersed conditions what will be the velocity of the sea water? The effect of velocity on the performance of alloys varies but even more important the possibilities of other effects such as cavitation or erosion corrosion must be considered.
(iii) What will be the temperature of the sea water? Temperature has an effect on the general corrosion rate of alloys but perhaps more important is its influence on other forms of corrosion such as stress-corrosion, which is often related to the temperature.
(iv) Are acid dew point conditions likely to be experienced? Materials that resist gases will not necessarily resist the acid condensates that form when the alloy is cooled to the dew point. This is typified in stack liners where the design and operation may be modified to

alter the position of the acid dew point in the stack rather than to use more resistant materials.

(v) Are high pressures involved?

(vi) Is fouling likely to be a problem?

(vii) Is bacterial corrosion anticipated?

(viii) Are there likely to be problems with general contamination or pollution of the sea water or atmosphere? Typically, in sour gas or crude wells this may produce limitations on the hardness or strength of steels used.

(ix) Are the environmental conditions likely to cause excessive vibration or cyclic stressing? Such conditions may lead to corrosion-fatigue problems.

8.1.3 Design and operating considerations

(i) Is the design such that a range of alloys will be used in contact with each other? Will this have a major influence on the choice of materials for components and equipment from the bimetallic corrosion standpoint?

(ii) Can the design be changed?

(iii) Will the plant or equipment be operating continuously? Some materials may be less suitable during non-operating conditions particularly with sea water, e.g. alloys relying on maintenance of a passive film may be subject to crevice or pitting attack.

(iv) Will welding be required? Apart from the problems of welding some alloys, welds themselves may be a potential source of problems. This arises in part from variations in alloy content across the weld and partly from the different mechanical conditions in such areas which can lead to corrosion-fatigue and stress-corrosion problems.

(v) Is the structural design such that stress concentrations can be set up? Residual stresses set up during manufacture can, under certain conditions, lead to corrosion-fatigue and stress-corrosion cracking of alloys susceptible to this form of attack. Severe deformation or bending of alloys can lead to similar failures.

(vi) Are suitable fasteners available for the particular design?

(vii) Will corrosion products affect plant reliability?

(viii) Will the design or operating conditions lead to specific types of corrosion, e.g. impingement, cavitation, pitting, crevices, etc?

In practice, experienced designers do not usually consider a list such as this in detail because one or two factors predominate, e.g. good heat transfer characteristics on heat-exchangers is an essential requirement.

Nevertheless, any of these factors may, of itself, lead to problems. Furthermore, the type of structure or equipment will have an important bearing on the form and extent of corrosion that can be accepted. In some situations, e.g. tubes, a few pits—possibly even one—may result in a failure and the loss of plant time, whereas a reasonable amount of general corrosion may be acceptable.

Selection of materials is part of the overall design requirement and detailed design will influence the choice of materials. Where appropriate the design should be changed to suit the material, e.g. by altering the velocity of the sea water. In other situations, it may be more economic to choose the material to suit the design.

The operating conditions are equally important and a guide to operational requirements should be provided for plant operators. Even where alloys have suitable properties, there are few certainties in corrosion, and it is always advisable to reduce any possible risks to a minimum. This can be done by auditing the design carefully to ensure that situations which might promote corrosion are so far as is possible eliminated. In the case of heat-exchangers, for example, this would involve matters such as (i) avoidance of crevices in contact with sea water, (ii) cleaning of all welds, and (iii) incorporation of filtration to reduce deposits and fouling. Additionally the operating conditions should be specified so as to minimise corrosion problems. This might include regular cleaning, re-circulation of water during shut down periods and if necessary monitoring of the performance of the materials. The final choice of materials will usually be a compromise between conflicting requirements and costs will, of course, play a predominant role in this choice. The costs are not necessarily those of the materials. Delays in obtaining the required materials or additional time required to install special plant result in extra costs and must be taken into account. Nevertheless, in many situations the attitude to costs proves to be a major cause of ensuing problems. The selection of materials, fabrication processes, etc, will not of itself ensure sound performance, so some form of quality control is essential for most situations. This is considered below.

8.2 COSTS AND RELIABILITY

The overall cost of any structure or plant is really a primary factor in design considerations. While it can often be shown that overall, more expensive materials will—over the design life—prove to be more economic than cheaper, more corrosion-prone materials, this is not always the only point to be taken into account. If the maintenance is to be a responsibility of some party other than the manufacturer, it may be that the cheaper materials provide, initially, a more competitive product. Again, the capital budget will be limited and it may be necessary, although technically undesirable, to choose materials of a lower standard to meet the financial requirements. Of course, designers should indicate the best technical solution but their views will not necessarily be those that are finally adopted. Furthermore, the corrosion requirements and others such as strength or weldability may result in a compromise, with materials of lower resistance than would have been chosen purely on corrosion considerations.

Costs fall into many categories but in general terms the main six elements are:

(i) Materials.
(ii) Fabrication and transport to site.
(iii) Installation and erection.
(iv) Quality control and testing.
(v) Monitoring.
(vi) Maintenance and repair.

The percentage of the total costs for each of the elements will vary with the particular piece of plant and generally there is no overall ratio between the different costs of items. Sometimes an expensive material may cost more to fabricate but this is not necessarily so. Each organisation has its own method of determining choice and it is rarely done on a straightforward Net Present Value (NPV) calculation, in part because of the imponderables in the equation, not least predicting the cost of repairs and maintenance. A useful guide to methods of economic appraisals of corrosion control measures has been published by NACE[11].

Most of the decisions are not made on purely technical grounds. There are a number of reasons for this. Generally, the plant or equipment is made from a number of units and these are 'bought-in'. Although the best unit available may be purchased it is not necessarily of the design that would technically be the most suitable. It is unlikely to be economically advantageous to specially design, say, a pump or a valve for a single item of plant—although there may be situations where this is done—so a particular item is used because it is available and comparatively cheap.

Design is essentially a series of such compromises, so it is worth considering design audit and quality control as effective methods of achieving cost-effective solutions.

8.3 QUALITY CONTROL

Quality control is generally a small percentage of the total cost but this may still amount to a sizable total amount. When attempts are being made to reduce the overall costs of a project this is an area that may well be cut-back or eliminated. There is likely to be no immediate effect so this often appears to be a reasonable decision in times of financial stringency.

In broad terms, quality control is a method of improving both reliability and the probability of sound performance. The higher the probability aimed for, the more it will cost initially, but the lower will be the maintenance costs. On the other hand—with no quality control—maintenance costs may well rise steeply with time. A comparatively small amount of quality control will provide a great improvement in the probability of satisfactory performance. There is somewhere an overall minimum total cost of maintenance and quality control. This will vary with different situations because the quality control or inspection will be more beneficial in some areas than others. Some parts of the total plant are more critical than others because any breakdown or failure affects the operation of other elements of the process.

Consequently, where quality control is to be limited it should be applied to those parts of the equipment or structure that are most critical in relation to their overall effects.

Some forms of quality control, using the term in its broadest sense, are particularly cheap. For example, checking the manufacturer's specification of materials used for equipment. Often a comparatively small change can be made, e.g. replacing cadmium plated bolts by, say, stainless steel bolts, assuming the change does not involve secondary galvanic effects. This may avoid closing down vital parts of equipment for repair work. Again, the equipment may be standard for ordinary water and operate satisfactorily but less suitable for sea water.

Equipment suitable in the United Kingdom may not be suitable for tropical climates for a number of reasons and often slight changes will improve efficiency to a considerable extent. If, for instance, it will prove difficult to repaint immersed areas, the addition of sacrificial anodes, i.e. cathodic protection, may prove to be beneficial for the particular use of the equipment even though the manufacturers had not considered this to be necessary.

The other approach to ensure suitable performance standards is monitoring. This is discussed in Chapter 21. Again, monitoring adds to the overall cost and generally can be carried out only over a limited part of the plant or equipment. However, where necessary selection of critical parts for monitoring will provide the most economic solution.

8.4 CORROSION DATA

One essential requirement for the materials engineer or designer is to have data on the performance of the alloys to be selected for a project. For many situations considerable data is readily available from past experience with the use of particular alloys in plants or structures.

Of course, the data may be specific to the operating conditions of the plant and may not be directly applicable to another situation where the conditions may be slightly but significantly different. In such cases further tests may be required.

There is also a large amount of test data published in many papers and books and this is obviously useful to the designer. Assuming that the investigators who carried out the work were competent, then the data is accurate for those particular conditions. It should, incidentally, be pointed out that even apparently straightforward data obtained from tests on specimens immersed in the sea or exposed in the atmosphere may vary with different investigators. This may arise from errors, e.g. incorrect weighing of specimens, but this is rare. More often the problems arise from the techniques and methods used. The method of exposure in air influences the rate of corrosion; the removal of corrosion products is sometimes a problem and often the exposure conditions for the test are not the same in different investigations.

Data obtained under laboratory conditions may be misleading when applied to practical situations. Salt spray tests carried out on metal coatings may, in fact, provide a completely different order of performance compared with tests at a marine site. Problems may, therefore, arise in translating such data to practical conditions, particularly where it has been obtained in laboratory tests. Special laboratory rigs have been developed for certain requirements and the data collected in such tests is likely to be more realistic than that collected in standard tests. It follows therefore that the best data is that collected by the designer because, assuming the expertise is available in the project team, it can be interpreted within the requirements for the project.

Other data is obviously useful and essential, but should always be considered critically and, whenever possible, the original papers should be studied to determine the conditions under which it was obtained.

Generally, tables containing data on specific alloys should be treated with caution. A table of corrosion rates of different alloys will indicate a general order of resistance but clearly a good deal more information would be required before such information could be used for design data. In most cases only a broad order of performance is required but there are situations where the actual corrosion rate is important if reasonable economic assessments of alternative materials are to be made. Data on localised corrosion is often more difficult to interpret or translate to a specific design requirement. The conditions under which the data was obtained should be studied critically. For example, the depth of pitting with time may be more important than the number of pits produced.

Stress-corrosion and corrosion-fatigue data is particularly difficult to relate to practical conditions. It may be necessary to assess such data in qualitative terms and to either choose an alloy that appears to exhibit no corrosion in a particular situation or to use a form of redundancy or safety allowance.

8.5 MATERIALS FOR SALTWATER SYSTEMS

For many situations, alloys have been used satisfactorily over a number of years. The choice is often based on cost and for many pieces of equipment, e.g. pumps, there is a range available to satisfy many different situations. There is, however, an increasing requirement for materials to withstand more aggressive conditions and as the demand grows either new alloys are produced or materials though to be almost exotic a few years previously become relatively cheaper as greater use is made of them. This is typified in heat-exchangers, where titanium and the highly-alloy stainless steels are being increasingly used.

It is not practicable to consider the materials for all the plant, equipment and components likely to be used under marine conditions. Some of the requirements have been considered in other chapters. In many situations, basic ferrous alloys such as carbon steel or cast iron are used for the main parts of a piece of equipment, e.g. sluice gates, and problems may arise with the various other alloys utilised for pinions, fasteners, etc. The protective coatings used for items of equipment and some elements such as fasteners are not always adequate for marine conditions. Careful scrutiny of the manufacturer's specifications is advisable. Some of the components, equipment and services widely used with sea water in ships, offshore platforms and process plant constructed in coastal situations will be considered below.

Sea water is increasingly used for cooling purposes and it has to be conveyed in pipes, moved by pumps, controlled by valves and used in condensers or heat-exchangers. Materials commonly employed for the various components in the system will be considered below.

8.5.1 Piping systems

A variety of piping systems is used to carry sea water for different purposes. The design of the system is particularly important (see Section 7.6). Avoidance of sharp bends, situations that could cause undue turbulence and ill-fitting flanges will all lead to improved performance from all materials used for pipes. Other problems arise from the replacement of sections of pipe with materials different from those in the main system. This can, of course, lead to bimetallic corrosion, as can the use of valves and pumps made from incompatible alloys. Pipes must be of suitable thickness for sea water service.

Depending on the system, pipes may have to resist general corrosion, pitting, crevice attack, galvanic corrosion, impingement, erosion-corrosion effects and fouling. Fouling may cause additional corrosion if it is not removed, because the organisms may decompose in the pipework; it may also affect velocity by partially blocking the pipe.

(i) Steel

Ferrous alloys are widely used for salt water systems where long-term durability is not a main objective; both carbon steel and ductile iron are used. Steel pipes may be used without coatings but are more generally hot-dip galvanised or lined internally with cement mortar, glass reinforced plastic, epoxy and various plastics such as polyvinyl chloride (pvc). The coatings are determined by factors such as size of pipe and type of system. If bare steel pipe is used it must be thoroughly cleaned of all internal scale before use, otherwise severe pitting may occur because of the difference in potential between steel and millscale. Equally scale must be removed before coatings are applied, to ensure sound performance. Severe attack may occur on internally coated steel pipe if the coating is damaged and the pipe is connected to more noble alloys in the system, e.g. at valves or pumps. The possibility of a large cathode/small anode relationship will exist, leading to pitting of the steel pipe.

Very small diameter bare or galvanised steel pipes tend to become clogged with corrosion products so that pipes below about 40 mm internal diameter may lead to problems of blockage. The corrosion of steel increases with higher velocities and temperatures. On the other hand, steels do not tend to pit to the same extent as some non-ferrous metals. Generally, steel is well suited to large diameter pipes where weight is not of great importance. Special corrosion pieces or 'wasters' are often fitted in parts of the system where corrosion is anticipated, e.g. at joins between steel piping and non-ferrous parts. These should be designed for easy removal and should be of larger diameter than the steel pipe. Corrosion pieces are expected to fail and should not be replaced by more resistant alloys as this will merely transfer corrosion to another less accessible part of the system.

(ii) Non-ferrous alloys

Copper was at one time commonly used for pipes of small diameter but it is tending to be replaced by the more corrosion-resistant cupro-nickels and aluminium brass. At velocities of the order of $1\frac{1}{2}$–$2\frac{1}{2}$ m/s, 90/10 cupro-nickel will be several times more corrosion-resistant than copper, although the diameter of the pipe will influence the effect of velocity. At higher speeds 70/30 cupro-nickel would probably be preferred. Both cupro-nickels and aluminium brass are more resistant to turbulent conditions than copper. Cupro-nickels are also resistant to biofouling.

Generally, corrosion pieces are not used with non-ferrous pipes in the same way as with steel but it may well be useful to use short, easily replaceable lengths of pipe at positions where erosion-corrosion or impingement is likely to occur.

(iii) *High duty piping*

For some situations where the conditions are particularly aggressive because of high temperatures of saline water and contamination with, for example, hydrogen sulphide, more resistant materials than those discussed above may be considered. The increasing development of geothermal resources and highly corrosive situations in desalination plants may make consideration of comparatively expensive materials worthwhile. These include titanium alloys and various proprietary alloys containing high nickel-copper additions and the high alloy 'advanced stainless steels'. For example, a high chromium austenitic steel (20–25Cr) has been used for the piping of a cooling water system for a blast-furnace using sea water in Italy. This replaced brass pipes that failed by pitting in 6–12 months [8].

8.5.2 Pumps and valves

In sea water service pumps and valves have to withstand a variety of operating conditions, in particular water flowing at different velocities, often with entrained particles such as sand. The design of a pump is particularly important in determining its ability to withstand the various corrosion effects it may encounter. These include impingement, cavitation, erosion-corrosion, general corrosion, crevice and pitting attack. Additionally, because of the ways in which pumps and valves are constructed bimetallic corrosion is often a problem. Generally, both valves and pumps are purchased from a range of standard products although they may be specially designed for specific situations. Cost is, of course, important and often the choice is between expensive, high duty equipment with long service lives and cheaper less durable units that can be repaired or replaced. Although corrosion is an important element in the design of pumps, the main design requirement is to withstand mechanical effects arising from pressure, containment of sea water and suitable pumping capacity. Similarly, valves are designed to restrict and control flow and it may be necessary to consider the material chosen for these purposes somewhat critically where corrosion performance is a high priority.

Because of their intricate design, pumps are usually cast and such alloys do not always have the same corrosion characteristics as wrought materials of the same nominal composition.

There are many different designs for pumps and valves and only a few essential points will be considered. A number of references providing more detailed information are given at the end of this chapter[1],[2],[3].

(i) *Pumps*

There are several different types of pump such as rotary, reciprocal and centrifugal. Within each type there is a range of different designs but from the corrosion standpoint the important elements are (i) body or casing, (ii) impeller or vane, and (iii) spindle.

Design is of considerable importance in determining the performance of pumps. Any features that cause turbulence are likely to intensify erosion-corrosion and cavitation effects. Situations such as flanges that protrude into the water stream should be eliminated. Apart from the design, the operation conditions may also influence corrosion performance; for example, pumps should be selected to operate near to their designed capacity as this will reduce the likelihood of cavitation. Materials of construction are, of course, important but it is often more economic to improve the design or to change the operating conditions rather than to select unnecessarily expensive materials. Often the non-operating conditions may be important. Stainless steel may be satis-factory during the operation of sea water pumps because there is a considerable degree of aeration. When the pump is stationary, the presence of sea water may lead to crevice attack and in such situations washing out and filling with fresh water may be advantageous for long shut-down periods.

The choice of materials is influenced by a number of factors:

  (i) Cost.
 (ii) Corrosion performance.
(iii) Strength.
 (iv) Suitability for casting.
  (v) Repairability.
 (vi) Galvanic effects on other components.

The least critical part of the pump with respect to bimetallic corrosion is usually the body because it is reasonably thick and can often be repaired locally without too much difficulty. The other parts of the pump are smaller in relation to the casing and, if less noble, may corrode at an unacceptable rate. Grey cast iron should not be used for sea water pumps without some form of cathodic protection or a very good coating. Where, however, it is used it may exhibit graphitic corrosion. The graphite will be more noble than the iron and may cause galvanic problems with other components.

Ductile cast irons (SG) and low-alloy (2% nickel) cast irons or cast steel are used for casings but for long-term service they should be coated with suitable materials including epoxies, rubber, etc. Austenitic cast irons, copper alloys and stainless steels may also be used for high duty service.

The potential of stainless steel will be less noble if it is 'active', i.e. corroding, so this should also be considered in relation to the operation of the pump.

Impellers and valves must resist the flow of the water and at high velocities this may lead to cavitation effects. Generally, stainless steels, some nickel alloys and titanium are all resistant to such effects and are widely used. Nickel aluminium bronze and some copper base alloys are also used.

There may be restrictions on the compositions for alloys used for impellers and specifiers should check that the selected materials will suitably resist the operating conditions.

General bimetallic effects between alloys used for the body and the trim can be summed up as follows:

If the body is cast iron or carbon steel then it will 'protect' trim of stainless steel and non-ferrous metals. If copper alloys are used for the body, then stainless steel trim will be protected provided it remains 'passive' and similarly the potential of brass or bronze trim may be influenced by the velocity of the sea water. The area of the body in relation to that of the components must be taken into account in determining the actual likelihood of problems arising.

(ii) *Valves*

Materials for valves are determined both by the type of piping and the type of valve. The avoidance of bi-metallic corrosion is important but equally, materials suitable for proper operation of the valve are required. Cast iron bodies and bronze or stainless steel trim are often used for valves with ferrous pipes, and gunmetal or bronze bodies with trim made from Monel, nickel, aluminium bronze and other copper alloys are used for non-ferrous pipes. The design of the valve influences the degree of corrosion attack that is likely to occur. It is often preferable to use a small, fully opened or shut valve rather than a partially closed, larger one, which may set up turbulence in the system.

Valves usually have to resist fairly aggressive conditions and comparatively small amounts of corrosion, which would be acceptable for many items of equipment, may cause problems such as leakage with valves.

Valves used in offshore oil and gas production processes may have to resist not only the normal problems that arise in marine situations but additionally the effects of hydrogen sulphide (H_2S) and carbon dioxide (CO_2). A recent review of valve service in oil and gas production indicates the nature of some of these problems which have arisen in part from inadequacies in valve design, materials and quality control. The view expressed is that valve failures are no longer a problem in Middle East fields but in the North Sea more experience is required before the performance can be confirmed as completely satisfactory.

Summary
Pipe systems are chosen mainly on economic grounds. Ferrous for low initial cost and for larger diameter systems; non-ferrous for long-term requirements. The valves and pumps are usually purchased as standard items and should be chosen to ensure adequate performance for the service requirements and compatibility with the pipe system.

Although valves are usually cast, large ones in special systems may be fabricated from wrought materials.

8.5.3 Condensers and heat exchangers

Heat exchangers of various types are used with sea water as the coolant but the most common are those employing tubes with tube plates to hold them in position. Heat exchangers of this type are used on ships and in many other situations where sea water is readily available and is used in preference to non-saline water—generally for economic reasons. The tubes have to withstand a range of conditions and problems have arisen with many of the alloys that have been employed for this application. Apart from the sea water, the tubes have to withstand the process liquors being cooled. Although copper base alloys have been successfully used for condenser applications with steam, they are not necessarily suitable for other situations.

Tube materials require a range of properties if they are to be used successfully. These include:

(i) Suitable heat transfer properties.
(ii) Strength characteristics.
(iii) Ease of installation.
(iv) Acceptable cost.

And resistance to:

(v) Erosion-corrosion and impingement.
(vi) Contaminants in sea water.
(vii) Biofouling.
(viii) Deposit attack.
(ix) General corrosion.
(x) Pitting and crevice corrosion.
(xi) Stress-corrosion cracking.

A number of different materials have been used for the tubes, including cupro-nickel, aluminium brass, aluminium bronze and stainless steels (Type 316).

Cupro-nickel, particularly the 70/30 variety containing 2% Fe, and aluminium brass inhibited for dezincification by arsenic, have been successfully used for ships' condensers where the problems have generally arisen from water velocities through the tubes, contamination of the water, turbulence and deposits in the tube. However, 70/30 cupro-nickels have also been prone to what is termed 'hot spot' corrosion. Both aluminium brass and the cupro-nickel alloys appear to exhibit improved performance if there is some iron in the cooling water. This iron was usually present from the corrosion of ferrous water boxes but the use of non-ferrous materials for these boxes and the use of protective coatings and cathodic protection for those made from steel has resulted in a depletion of iron compounds. Sometimes iron plates or blocks fitted to the tube plate are used to provide iron—and incidentally some cathodic protection—but more commonly iron salts such as ferrous sulphate may be added to improve the protective surface films on the tubes.

Where sea water is used in coastal situations for condensers, a 70/30 cupro-nickel alloy containing 2% Fe and 2% Mn has been successfully used.

Tube plates for ships' condensers have commonly been produced from Naval brass or—for smaller units—high tin-phosphor bronze. Other materials such as aluminium bronze, silicon bronze and cupro-nickel have also been successfully used. On more expensive installations using 70/30 cupro-nickel, tube plates have been made from 90/10 cupro-nickel alloys. Water boxes and condenser doors are commonly made from ferrous materials, either cast iron or mild steel, suitably coated, often with sacrificial anodes. Non-ferrous metals, such as gunmetal and aluminium bronze, are also used but they tend to be heavy and are often difficult to manufacture. Monel water boxes have been used to save weight but bimetallic problems have been experienced in some cases. Ninety-ten cupro-nickel has been used as a cladding for steel water boxes and this appears to be satisfactory.

Non-ferrous alloys are still used for ships' systems, but these materials may not be satisfactory for many large heat exchangers erected in coastal situations using sea water as the cooling medium. Austenitic stainless steel tubes have been widely used but suffer from pitting and crevice attack. Two groups of alloys are being increasingly used: titanium and the higher-alloyed ferritic and austenitic stainless steels (see Section 5.4).

Titanium is a very corrosion-resistant material which until recently was considered to be too expensive for tubes in heat exchangers. However, improvements in manufacturing techniques and an increasing demand for titanium tubes has led to a relative decrease in their cost.

The decision to use titanium has proved to be successful but the first heat-exchangers were constructed in about 1968 with little service experience. Various tube plates have been used, e.g. Monel clad as a weld overlay on carbon steel. In many situations titanium has replaced the non-ferrous metals discussed above, although attention must be paid to the possibility of bimetallic corrosion of tube plates. Published experience indicates excellent performance [5],[6]. In higher temperature sea water, anodising has sometimes been used.

The higher-alloyed stainless steels have much improved resistance to pitting and crevice attack compared with Type 316 and a considerable amount of laboratory work has been carried out to investigate performance. Austenitic materials of this type have been used for a number of sea water applications [7],[8]. A monoethanolamine (MEA) system in Saudi Arabia with sea water at a maximum service temperature of 55 °C on the tube side uses these materials as replacement for 70/30 cupronickel tubes that failed after six months due to process side corrosion.

In Brazil, coolers for an ammonia plant have been in operation for seven years with sea water at a maximum temperature of 32 °C and it is claimed that they have provided good performance.

Ferritic steels in this group of stainless steels have also been used in a number of applications, again with apparent success.

These stainless steels noted above are covered by a number of proprietary materials and each company has carried out its own development and testing work. There is general agreement on the improvements in localised corrosion attack compared with conventional austenitic stainless steels. As with other materials, care must be exercised in the design and operation of the plant to obtain the highest performance.

These steels appear to have a role to play for heat-exchanger tubes and as more service experience is gained, increased applications are to be expected.

8.5.4 Fouling and cleaning of tubes

The methods of cleaning include the following:

   (i) Re-circulating abrasive slurries.
  (ii) The MAN system using nylon brushes connected to plastic baskets which are attached to each tube end. The operator automatically cleans the tube by reversing the water flow.
 (iii) Re-circulating chemicals.
 (iv) Intermittent chlorination.
  (v) Continuous chlorination.
 (vi) Cleaning with 'Amertap' sponge rubber balls.
(vii) Chlorination combined with (vi).
(viii) Mechanical methods carried out during shut-down. Generally this involves circulating plastic or rubber plugs (not metallic) through the tubes.

An investigation of the influence of some of these measure on the corrosion of heat-exchanger materials used with sea water has been reported [9]. Tests on a range of the above methods were carried out on tubes made from aluminium, cupro-nickel and titanium. The results can be summed up as follows:

  (i) The corrosion performance of titanium was not affected by aggressive cleaning. Daily intermittent chlorination of 25 mg/l for 24 min/day was an effective method of controlling microbiological fouling.
 (ii) Ninety-ten cupro-nickel has inherent biofouling resistance but unnecessary cleaning with balls disrupted the protective film forming process and increased corrosion. Optimised cleaning with 'Amertap' balls or chlorination plus the balls resulted in no significant increase in corrosion.
(iii) The tests on aluminium confirmed the effect of temperature on corrosion and fouling; both decreased with decreasing temperature below 20 °C. The protective surface film is easily damaged by mechanical cleaning; the optimised cleaning method was daily intermittent dosing with low-level chlorination plus sponge balls.

8.5.5 Propellers for ships

There are a number of established companies specialising in the manufacture of propellers and generally these are purchased either 'off-the-shelf' for small ships or boats or specially designed and manufactured for larger ships. Propellers are available in a range of sizes and many different materials are used for their manufacture. These include cast iron, carbon steel, stainless steel, various bronzes and plastics.

Such a large choice of materials indicates the range of propellers used in ships. The selection of materials is determined not only by corrosion, which can be severe under the conditions in which propellers have to operate, but also the requirements for blade contour to provide maximum propulsion at minimum power, the difficulties of manufacture of some designs and, of course, cost. From the corrosion standpoint the materials may have to withstand some or all of the following forms of attack:

(i) General corrosion.
(ii) Pitting.
(iii) Impingement.
(iv) Erosion.
(v) Cavitation.
(vi) Corrosion-fatigue.
(vii) Stress-corrosion.

There are also mechanical requirements that have to be met and the materials chosen must have suitable casting properties.

Propellers become damaged and corrode in service, so another material requirement is that they should be capable of reasonably straightforward repair. This usually involves local grinding, welding and straightening. Such repairs must be carried out carefully by competent personnel to avoid introducing notches and areas of stress concentration that could become sites for corrosion-fatigue or stress-corrosion failure. With some materials it is necessary to carry out local stress relief and this again should be done by personnel who appreciate the requirements.

The classification societies, e.g. Lloyds, lay down rules and recommendations for repairs and these should be consulted before undertaking such work.

Materials commonly used for propellers include the following:

(i) *Nickel-aluminium-bronze*
The composition may vary according to the mechanical properties required. These alloys are sold under various proprietary names. They are probably the most satisfactory materials for resistance to cavitation and erosion effects.

(ii) *Manganese-nickel-aluminium bronze*
These alloys have a range of mechanical properties and are widely used. They again generally provide satisfactory service.

(iii) *Manganese-bronze*
Before the advent of the two materials noted above, this was the standard propeller alloy. Higher manganese contents are generally used for propeller alloys to provide harder materials with more resistance to corrosion and erosion. The additon of nickel to the standard high-duty manganese bronzes leads to improved performance but they are not really in the same class as alloys (i) and (ii) above.

(iv) *Stainless steels*
Stainless steels perform reasonably well under the well-aerated conditions experienced by propellers. They are, however, subject to crevice attack when used for controlled pitch propellers. Generally stainless steel compositions tend not to be 316 or 304 but rather 16Cr 5Ni 1Mo, which is probably chosen for its cavitation resistance. Stainless steel propellers are also possibly chosen because they tend to be easier to repair than those made from non-ferrous alloys.

(v) *Low-alloy steel and cast iron*
These are not generally considered as suitable for high-duty propellers; they are, however, cheaper materials than those discussed above so they are used where the requirements are less stringent.

8.5.6 Shafts for ships' propellers

Steel, often coated with high-duty coatings such as neoprene and other plastic materials, is often widely used for shafts because of the mechanical requirements. Plated carbon steels have also been used. Whereas lead and tin may be satisfactory, nickel may lead to failures as a result of pitting. There is, of course, a potential problem with cathodic coatings because if there are defects or they become damaged, then severe localised corrosion of the steel may occur. Steel shafts are often covered with sleeves where they pass through the stern tube bearings. These sleeves have been made from various copper and copper nickel alloys and stainless steel.

A recent development has been the use of Inconel 625 weld overlays. This is applied only to the bearing area and since the deposited metal has approximately the same strength as the commonly used shafting materials, the overlay is applied locally and machined to size. It has been used for other shaft applications such as the exposed portion of the flange shaft of ships fitted with controllable-pitch propellers.

Other materials used for propeller shafts, particularly for smaller boats, include various copper alloys and Monel.

Table 8.1 General corrosion data on alloys in sea water

Resistance to form of attack	Titanium	Austenitic stainless steels	Nickel	Nickel alloys	Copper	Brasses	Bronzes	Cupro-nickel	Aluminium alloys	Cast iron	Carbon steel
Quiet sea water immersion	A[1]	A	A	B	B	B	B	B	Depends on alloy	C	C
Pitting	A	C	C	B	B	A	B	B-90/10 most resistant	B	B	B
Crevice	A	B/C	B/C	B	A	A	A	A/B	B	B	A
De-alloying	A	A	A	Some alloys susceptible	A	Some alloys susceptible	A	Some susceptibility	Some alloys susceptible	Some types susceptible	A
Fouling	C	C	B	B/C	A	B	B/C	A-90/10 B-70/30	C	C	C
Cavitation damage	A	A	B	B	C	C	C Al bronze (B)	C	C	C Alloy Cl-B	C
Velocity (> 2 m/s)	A	C pitting	—	C pitting	A/B	A/B	A/B	A	—	C	C
Velocity (2–15 m/s)	A	A	A	A	B	B	B	B	—	C	C
Intercrystalline	A	Susceptible	May be susceptible	May be susceptible	A	A	A	A	Some alloys susceptible	A	A
Galvanic[2]	1	3	3	1–5 depending on alloy	6	7	5	4	9	8	8

Note (1) Qualitative assessments of resistance: A—high, B—average, C—poor.
(2) Indicates position in Galvanic Series in sea water. If two alloys are connected, the higher number in the couple is more likely to corrode at enhanced rate but many factors are to be taken into account.

8.6 GENERAL SUMMARY

A number of the more commonly-used materials in sea water systems has been considered in Section 8.5. There are, of course, many other situations where materials have to be selected for systems involving sea water, e.g. desalination plants. Considerable knowledge of the properties and corrosion characteristics of materials is necessary before plants of this nature can be designed. The conditions in parts of the plant are similar to other heat-exchangers but the high temperatures at which the sea water is treated involve problems of a particular nature. An example concerns the chlorination of sea water to prevent marine fouling in MSF desalination plants; now a common practice. However, the processes involved in chlorinating sea water are not so well established, as they are with fresh water. Consequently, problems are likely to arise for which data is not available.

A recent paper described problems in this area [10]. Bromine is produced at the decarbonator and de-aerator units and the associated vents systems in a number of these plants. This has led to attack on stainless steel. As few materials are resistant to wet bromine, it is not practicable to change the materials, consequently an investigation is required to determine conditions under which bromine is formed so that this can be avoided.

Although general data of the type summarised in *Table 8.1* is readily available, there will often be special problems for which specialist advice or special investigations are necessary. A good deal of useful data is available from the manufacturers and supplies of alloys and full use should be made of such services. Considerable data is available in a number of books [12],[13],[14] but this must be appraised critically for the particular situation under consideration. In particular, the influences of design, bimetallic corrosion and localised effects must be taken into account.

8.7 REFERENCES

(1) NEERKEN, R F, 'Selecting the right pump', *Chem. Eng. Deskbook*, April, 87 (1978)

(2) BOOTH, M L, 'Pumps for corrosion media', *Chart. Mech. Eng.*, January, 72 (1977)

(3) CLARKE, R A and GEDDES, G, 'Which pump?', *Engineering*, November, 1089 (1972)

(4) KING, J A and BADELEK, S C, *Proc. UK National Corrosion Conf.* (1982), p 145, Inst. Corr. Sci. Tech., London

(5) MCMASTER, J A, *Materials Performance*, **18**, No 4, 28 (1979)

(6) NEILL, W J, *Materials Performance*, **19**, No 9, 57 (1980)

(7) *Advanced Stainless Steels for Sea Water Applications*, Associazione Italiana di Metallurgica and Climax Molybdenum Co (1982)

(8) BERNHARDSSON, S et al., *Proc. UK National Corrosion Conf.* (1982), Inst. Corr. Sci. Tech., London

(9) LEWIS, R O, *Materials Protection*, **21**, No 9, 31 (1982)

(10) TODD, B and OLDFIELD, J W, *Proc. UK National Corrosion Conf.* (1982), Inst. Corr. Sci. Tech., London

(11) *Direct Calculation of Economic Appraisals of Corrosion Control Measures*, NACE Standard RP-OZ-72, National Association of Corrosion Engineers, Houston, Texas, USA

Useful text books containing corrosion data on metals and alloys

(12) SHREIR, L L (ed), *Corrosion*, 2nd ed, Vol 1, Butterworths, London (1976)

(13) UHLIG, H H (ed), *Corrosion Handbook*

(14) *Corrosion Data Survey*, National Association of Corrosion Engineers, Houston, Texas, USA

(15) *DECHEMA Werkstoff-Torbelle*, DECHEMA, Frankfurt

(16) LAQUE, F L, *Marine Corrosion*, John Wiley, New York (1975)

Protective coatings for steelwork—surface preparation of steel before painting

9

Steel is the alloy most commonly used for marine constructions and to withstand the aggressive environments it is usually coated, generally with paint although other coatings such as zinc and plastic materials are also used. As paint coatings and techniques for their application have been developed to meet the requirements of such hostile environments as the North Sea and the Arabian Gulf, the surface protection of steelwork has evolved into a technology, the understanding of which is essential for reasonable economic success in protecting steelwork. The differences between the painting of simple structures on land in mildly aggressive conditions and the coating of offshore structures and marine installations in tropical coastal areas is considerable. For many land-based structures, painting is largely for cosmetic purposes and while rusted steelwork is generally considered to be unpleasant in appearance and a sign of poor management, some corrosion over most of the structure can be accepted even if proper maintenance of coatings is delayed for some years. Delaying maintenance in such circumstances is unwise from the economic standpoint but not necessarily serious so far as structural stability is concerned. In severe marine environments, the corrosion rate is much higher so there is a much greater requirement for high quality protection. This is not, however, achieved just by specifying better coatings. To obtain the standard of protection required involves high standards of preparation of the steel surface and application of the coating to provide the highest quality of dry protective film. This in turn means that specifications must be properly prepared, workmanship must be first class and proper quality control measures are required.

The importance of achieving high standards is always important but in aggressive marine conditions it may mean an increase of one or two years between maintenance re-coating and this is worth very large sums of money even by the standards of oil companies.

This part of the book is divided into a number of chapters covering the main elements involved in the protection of steelwork.

(i) Surface preparation of steelwork prior
to coating (Chapter 9)
(ii) Paint coatings (Chapter 10)
(iii) Metallic coatings (Chapter 11)
(iv) Specifications and quality control (Chapter 12)
(v) The selection of coating systems for marine
conditions (Chapter 13)
(vii) Maintenance procedures (Chapter 14)

The general principles of steel protection for marine situations are considered later but there are continuing developments in materials and techniques aimed at improving coating performance. It is essential, therefore, for those involved in the protection of steelwork to keep abreast of such developments and trends to ensure that they obtain the maximum level of performance from protective coatings.

Modern paint coatings, e.g. epoxies, differ radically from many of the more common drying oil-type paints and—while they can give out-standing performance under the right conditions—they are less tolerant and more affected by poor surface preparation of the steelwork and less than adequate application techniques for the coating.

There is not always a clear demarcation line between what are generally considered as paint materials and those called plastics. Certain materials which might fall into either category and others that are basically rubbers can be considered by virtue of their protective qualities and specialist application requirements to fall into a special class, which has been called in this book 'high-duty coatings'. This term is also used, particularly by paint manufacturers, for coatings that are considered here as paints.

Apart from the organic (and some inorganic) types of coatings (paints and plastics), the second largest group of coatings used for steelwork in marine environments is metallic coatings, mainly zinc and aluminium. Additionally there are materials used for special situations such as wrapping tapes and 'grease paints'.

Although not widely used on structural members, a series of temporary protectives are used for components and smaller items. These are generally oils and greases and are temporary in the sense that the coatings can be easily removed by simple solvents such as white spirit. Phosphate coatings are widely used for sheet steel products, e.g. cars, but are not employed to any extent for structural steelwork.

Many factors influence the performance of coatings (*Figure 9.1*) but it is generally agreed that the cleanliness of the steel surface before coating

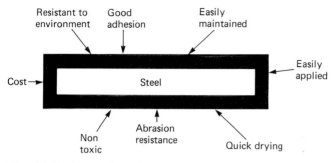

Figure 9.1 Requirements of a coating

application is probably the single most important one. As it is applicable to all coating systems, it will be discussed before considering the various types of protective systems used.

Surface preparation of steelwork before coating
The efficiency of all coatings is influenced by the nature of the surface to which they are applied. For many coatings, thorough cleaning of the surface is carried out as part of the process because without this the coating would not adhere properly. For example, pickling to remove rust and scale is always carried out before hot-dip galvanising or electroplating. Again, plastic and hot sprayed metallic coatings are also applied to surfaces that have been properly cleaned. Many paints will adhere to steel surfaces that have not been thoroughly cleaned and for some situations this may be acceptable. For most marine conditions, however, thorough cleaning of steel before painting is essential if the high potential protective value of many coatings is to be achieved. This section is, therefore, mainly concerned with paint because it is in this field that most problems arise. It should, however, be emphasised that the thorough cleaning of steel is required for most coatings.

9.1 THE NATURE OF THE STEEL SURFACE

Grease, dust and dirt must be removed before painting to ensure sound adhesion of the coating. No coating will protect the material to which it is applied unless it firmly adheres to it. Steel is no different from other materials in this respect. There are, however, additional problems that arise with steel because of the way it is manufactured and the nature of the corrosion product formed on its surface.

9.1.1 Millscale

Heavy steel sections and plates are produced to the required size and shape by hot-rolling. During this rolling operation the hot steel reacts with oxygen to form oxides:

$$xFe + yO_2 \rightarrow Fe_xO_y$$

The oxides are formed in layers, typically Fe_2O, Fe_3O_4 and FeO. This oxide layer which varies in thickness and composition depending upon the rolling temperature and size of the steel is called 'millscale'.

During rolling and handling of the steel, the millscale tends to crack and parts of it become detached from the steel surface. It is, therefore, rare for the millscale to be present as a firmly adherent, unbroken coating on the steel. Firmly adherent scale might well prove to be a satisfactory base for further paint coatings and attempts have been made to treat the surface during rolling to produce such scales. These have not, however, proved to be satisfactory in practice.

The cracking and flaking of millscale during handling is further aggravated by the rusting of the steel that usually occurs before it is finally painted.

The problems arising with millscale can be summarised as follows:

(i) Hot-rolled steel is removed from the mill. At this stage it will have some intact millscale and bare steel areas where it has flaked or cracked.

(ii) The steel is moved to the stockyard, generally open to the weather and the areas where the millscale has cracked or flaked begin to rust. Further rusting occurs at cracks. Thus rusting tends to undermine the remaining scale without necessarily leading to further flaking.

(iii) If the steel is then cleaned by manual methods, such as wirebrushing and chipping, loose scale and rust are removed but intact scale and adherent rust are merely burnished.

(iv) If the steel is then painted, the performance will depend on a number of factors:

    (a) The exact state of the surface before painting.
    (b) The environment of exposure.
    (c) The paint system.

These are discussed below.

The state of the surface is important because it may vary from almost completely intact scale to severely flaked scale with rust on the steel. Paint applied to intact scale may perform very well but generally this situation occurs only in tests where steel is painted at the mill before any handling or storage has occurred. Starting with intact millscale a range of surface conditions can occur extending to a situation where virtually all the millscale has weathered away leaving rusty steel. Furthermore, the nature of the surface will be influenced by the environment where the weathering takes place. In aggressive marine environments the scale tends to be removed more rapidly than in milder conditions.

The worst situation regarding paint performance is where some of the millscale has flaked off and some is still apparently adhering but has become under-rusted during exposure. Where surfaces of this type are manually cleaned, e.g. scraped and wire-brushed, and then painted, rapid failures may occur particularly where the steelwork is exposed in an aggressive marine atmosphere. In tests carried out in such an atmosphere paint flaked more or less completely over steel weathered for three months (*Figure 9.2*). On steel where all the scale had been weathered away before painting, i.e. a rusted surface, a similar paint system lost adhesion after about two years[1].

Figure 9.2 Failure of paint coatings applied to millscale in a marine atmosphere

It is very difficult to judge the conditions of weathered steel when the surface is partly rusted and partly scaled and whilst severe flaking does not necessarily occur, it is always a possibility. Furthermore, painting on such a surface will almost certainly lead to a reduced life for the paint system. So manual methods of cleaning are not recommended where painted steelwork is to be exposed in a marine or other aggressive environment (*Figure 9.3*).

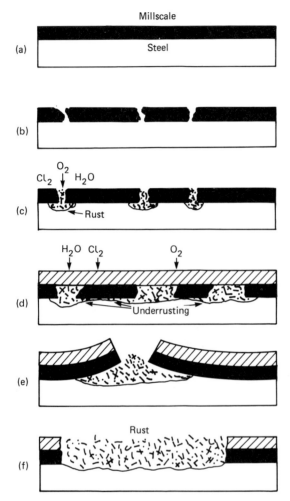

Figure 9.3 Typical flaking of paint applied over millscale in a marine atmosphere (a) Hot-rolled steel with millscale (b) Millscale cracks during cooling and handling (c) Steel corrodes in storage (d) Loose scale rust is removed and paint applied (e) Ingress of moisture and chloride under rust and scale which begins to flake (f) Scale and paint flake off leaving part paint and part rust (as in Figure 9.1)

Where all the scale has been removed by weathering so that the steel is covered with rust, severe flaking does not generally occur. Situations can, however, arise where rust itself flakes off in large areas carrying the paint with it. More frequently, the failure of paint applied to rust occurs from local deterioration, e.g. blistering. The type and thickness of paint coating will, to a large extent, determine its life. In particular, inhibitive priming coats, e.g. red lead in oil, provide improved performance on rusted surfaces. The performance of paint coatings on rusted steel is, however, always inferior to that of paint applied to a clean surface.

Table 9.1 Effects of surface preparation on coating life (125 µm paint system)

	Life (y)	
Surface preparation	Mild atmosphere	Aggressive atmosphere
Wire-brushing	6¼	2¼
Blast-cleaning	16½	11

As can be seen from *Table 9.1* based on different investigations carried out by The British Iron and Steel Research Association, a reasonable life may be obtained in mild environments where there is little pollution or contamination even though paint coatings provide a much longer life on blast-cleaned steel. However, under aggressive conditions, the life of the steel applied to manually cleaned steel is very short. Millscale is an oxide, so it does not corrode under ordinary environmental conditions but it can behave like a metal when in contact with steel in the presence of an electrolyte. It is cathodic to steel so, particularly under immersed sea water conditions, it can lead to an increase in the corrosion of steel (see Section 7.5). If the steel is largely covered with millscale (large cathode) this can result in quite severe pitting of the steel exposed at discontinuities (*Figure 9.4*). There often appears to be a misunderstanding of the role of rust as a base for painting so this will be considered below.

Rust is the product produced when ferrous metals corrode. The theoretical composition is $Fe_2O_3 . H_2O$, which is sometimes expressed in the form of α-FeOOH or β-FeOOH. Other forms of corrosion product such as magnetite, Fe_3O_4 can also be formed under certain conditions. Rusts of this form and composition are rarely formed as the only corrosion product on steel.

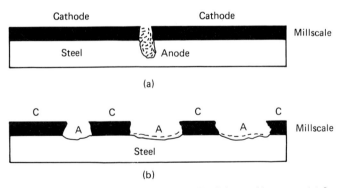

Figure 9.4 Pitting of steel partially covered with millscale immersed in seawater (a) Large cathode/anode ratio—pitting (b) Equal areas cathode and anode—general corrosion

Usually other substances are also formed, in particular soluble salts such as ferrous chloride, in marine environments and ferrous sulphate in industrial environments. These result from the reaction of steel with pollutants and contaminants in the atmosphere or from reaction with chlorides in sea water. These salts influence the performance of paint applied over them as summarised below.

9.1.2 Influence of salts on paint performance

When steel rusts in polluted or contaminated air, salts such as ferrous or ferric chloride and ferrous sulphate form in pits at the rust/steel interface. Consequently, they are virtually impossible to remove by manual cleaning methods even though in some cases the burnished appearance of the rust is mistaken for a rust-free steel surface. These salts react with the moisture that permeates through paint systems and this leads to further corrosion of the steelwork with the formation of more rust. As rust has greater volume than the steel from which it is formed, the paint coating has to provide sufficient elasticity to take account of the increased volume. As this process continues the paint film eventually cracks and flakes at local areas (*Figure 9.5*). These salts can also be a problem even when steel has been blast-cleaned, as discussed later.

9.2 CLEANING OF STEELWORK

Where steelwork is to be painted, it must be cleaned of grease, dirt, oil, etc. Large quantities of oil or grease should be physically removed by scraping with the residue being removed by emulsion cleaners followed by water rinsing. The practice of wiping the surface with solvent is not recommended because it tends to spread a thin film of oil over a wide area. Cleaning by solvent vapour decreasing can be very effective. It is also usual to remove all loose rust and scale. Cleaning at this level is carried out by a variety of methods using tools that are operated manually, mechanically or pneumatically. The more common of these include wire-brushes, scrapers and hammers. Other mechanical equipment includes needle guns, grinders and abrasive discs.

These methods do not generally remove the adherent rust from pits in the steel. Grinding can do so by removing steel to the depth of the pits but except possibly in localised areas this is not economic.

Manual cleaning is suitable for situations where steel is to be enclosed in concrete or brickwork or for the interior steelwork in buildings that are maintained in a warm and dry condition. Paint applied to properly prepared steel will, however, always provide longer-term protection.

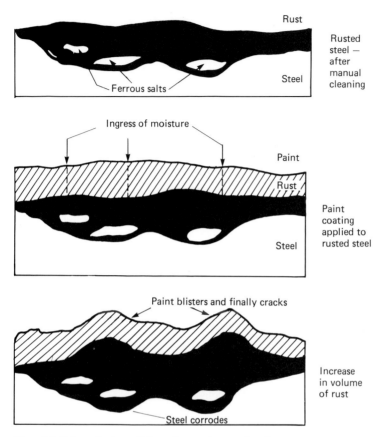

Figure 9.5 Failure of paint applied over ferrous salts on steel surface

Manual cleaning may sometimes be used to remove loose heavy deposits of rust and scale prior to the employment of more effective methods such as blast-cleaning.

Standards for manually cleaned steelwork are in Swedish Standard SIS 05 5900:1967 and The Steel Structures Painting Council Painting Manual, Vol 2 (1982), Specs. SP2 and SP3. Two grades ST1 and ST2 depicted by photographs are the basis of both Standards. In practice it is often difficult to determine the standard of manual cleaning from these photographs.

The only practical methods of removing all rust and scale from the steel surface is by pickling in acid or by blast-cleaning. Pickling is used for steel that is to be hot-dip galvanised and for continuous cleaning of strip. Generally, however, blast-cleaning is the method used to prepare structural steel for aggressive situations or to ensure the best performance from coatings.

As already noted, the advantages to be gained by blast-cleaning are illustrated in *Table 9.1*.

9.2.1 Dry blast-cleaning

Dry abrasive cleaning is the most widely used mechanical method of surface preparation for structural steelwork. Processes fall into two broad categories.

(i) The abrasive is carried by a jet of compressed air through a gun on to the metal surface. Generally the abrasive is a non-metallic type and is not circulated for re-use. This method may be used in an enclosed cabinet or on site.

(ii) The abrasive is thrown centrifugally on to the metal surface from rotating impellers in a fixed plant. Complex plants with a series of impellers at different angles may be used for large steel constructions. Generally, round iron or steel shot is used as the abrasive and after a screening to remove small particles it is recirculated. This method is widely used for plate and other simple geometrical sections. On more complex shapes it may be necessary to hand-blast areas where the abrasive has not provided adequate cleaning. A detailed description of this method has recently been published[2].

A method using a vacuum at the head so that the abrasive can be collected and recirculated is also used. This is a comparatively slow method but has the advantage that abrasive is not spread over the site. It is particularly useful for local repairs and for maintenance work.

Another portable method of cleaning, which at the same time avoids the spread of abrasive over the site has been developed. A hand-operated machine is wheeled over the surface and the design is such that abrasive is propelled on to the surface at an angle which causes it to ricochet back into the machine where it is collected. No vacuum is used in this method which is used primarily for large flat horizontal surfaces.

Dry blasting must be carried out in good ambient conditions, and care should be taken to ensure that the surface which is being cleaned, or has been cleaned, does not become wet. Similarly during compressed-air blasting the relative humidity should not be so high as to cause condensation at the nozzle, while the air used should also be dry and free from oil.

The blast-cleaning process produces a large amount of dust and debris and this must be removed from the cleaned surface, preferably by vacuum cleaning, before it is painted.

Abrasives used for dry blast-cleaning
A wide range of abrasives is available for blast-cleaning. These include sand, mineral types such as silicon carbide, aluminium, oxide, iron slag, copper slag and metallic types such as steel grit and shot, chilled iron and shot and chopped steel wire.

The size of the abrasive particles influences the speed of cleaning, e.g. coarse abrasives tend to be more effective on heavy rust deposits. On the other hand, coarse particles produce a deeper profile. Fine particles remove salt deposits more effectively because they can reach the bottom surface is important and the size of abrasive is the main factor in determining its shape and peak-to-trough height. Other factors such as the pressure at the nozzle of the gun, type of steel and impact angle of the abrasive striking the surface also affect the speed and efficiency of cleaning.

The particle shape affects the speed of removal of scale and probably its completeness. Grit tends to clean more rapidly and gives a better key for paint than does shot, but tends to be destructive of the blast-cleaning plant itself. For this reason, shot is often used in automatic plant, while grit is more commonly used in air-blasting systems.

Coated abrasives
A particular problem arises when blast-cleaning marine structures for maintenance purposes, particularly offshore. After blast-cleaning the surfaces rapidly re-rust because of the salt spray and the generally damp conditions. Consequently difficulties arise with the protection of the steelwork because of the rapid re-rusting. To overcome these problems coated abrasives have been developed. The purpose of such abrasives is to act in the ordinary way to clean the surface but additionally at the same time to apply a thin coating to the steel surface. A number of different abrasives are available for this purpose; some are coated with both a polymer and a metallic zinc powder. During blast-cleaning the coating on the abrasive is transferred to the steel surface and becomes bonded to it (*Figure 9.6*). The mechanically deposited layer is only a few microns thick and does not cover the whole surface. Tests have illustrated the beneficial effects of coated abrasives but as yet there is not sufficient practical experience to judge their merits under a variety of conditions.

9.2.2 Wet blast-cleaning

High-pressure water jetting at pressures in excess of 700 bar is a very effective method of cleaning deposits from many types of surface. Because of the high pressures it is, however, a potentially dangerous process and more effective methods have been developed for the cleaning of rust, scale

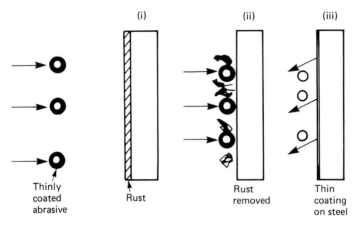

Figure 9.6 Coated abrasives. Diagrammatic representation of action on steel

and paint from steel surfaces. Lower-pressure water jets, some as low as 2–10 bar, are used and these are added to the air stream containing an abrasive, usually sand.

Over the last few years the techniques and equipment have improved with reductions in both the volumes and pressures of water used. This has made it more attractive as a method for maintenance and particularly for cleaning offshore and marine structures. The main advantages of this method compared with dry blast-cleaning are considered to be:

  (i) More effective in the removal of soluble corrosion products, such as ferrous chloride, from pitted surfaces.
 (ii) It is possible for experienced operators to selectively remove individual layers of coatings which is particularly useful for maintenance work.
(iii) Useful for cleaning local areas with minimum damage to surrounding paintwork.
 (iv) Relatively dust free.
  (v) The spark hazard is less than with dry blast-cleaning.

  The main disadvantages are considered to be:

  (i) Leaves the surface wet.
 (ii) Quantities of clean, fresh water are required. This is a particular problem offshore and in many overseas situations.
(iii) A sludge of abrasive in water has to be removed and this may prove to be difficult and time-consuming.
 (iv) The rate of removal of complete coating systems or millscale is slower than with dry blast-cleaning.

The main disadvantage is undoubtedly that the surface remains wet after the removal of surface deposits and quickly rusts. If the cleaning has been carried out efficiently and all the soluble salts have been removed, the rust that forms, provided the surface has dried before paint is applied, may not be too detrimental to the performance of the coating. Nevertheless, having cleaned the surface it is preferable not to have any rust present. Consequently, various methods have been considered to overcome this problem. The two main approaches have been:

(i) The use of water tolerant inhibitive priming paints applied before the steel has dried.
(ii) The inclusion of an inhibitor as part of the water-blasting process to prevent rusting for several hours so that the primer can be applied to clean steel.

A two-part cleaning system has been considered for offshore and marine structures where the initial cleaning is carried out using sea water followed by a further cleaning with fresh water. The advantage of such a method would, of course, be to conserve limited supplies of fresh water.

The addition of inhibitors to the water used for wet blast-cleaning is under investigation and problems can arise, particularly if too high concentrations are used. Polyphosphates at concentrations of 0.25–0.75% have been used successfully to provide inhibition of rusting of the surface for some hours with no apparent detrimental effect on the paint coating. The optimum concentrations of inhibitor and the general effectiveness of the method are still under investigation.

NACE Standard RP-01-72 is a recommended practice for water blasting prior to coating or re-coating.

9.3 ACID PICKLING

In this method, steel is immersed in an acid, usually inhibited, to remove scale and rust and, after thorough rinsing, is coated. Pickling is widely used to clean thin gauge steel particularly in continuous strip processes. It is generally only used for heavy steel products if they are to be hot-dip galvanised. It is also used for components such as nuts prior to hot-dip galvanising or electroplating. Both sulphuric and hydrochloric acids are used for pickling.

Although phosphoric acid can be used, it is seldom employed as a means of removing scale and rust because of the expense. Steel may be initially descaled in sulphuric acid and then, after rinsing, immersed in 2% phosphoric acid containing about 0.5% iron at 85 °C for a few minutes. This method is rarely used nowadays but before the advent of blast-cleaning was used to prevent surface rusting and provide a good surface for painting. This method should not be confused with the phosphating process used for thin-gauge steel, e.g. car bodies.

Acid washes and gels are not recommended for application on site except under the most careful supervision because of the difficulty of neutralising the acids and the possibility that acids will reach crevices, etc. Furthermore, the operation can be a hazard to workmen in the vicinity of the operation.

9.4 FLAME CLEANING

In this method an oxy-acetylene or oxy-propane flame is passed across the steel. The sudden heating causes millscale and other rust scales to flake off as a result of the differential expansion between the scale and the metal. Immediately after the passage of the flame, any loose millscale and rust that remains is removed by wire brushing. This generally leaves a powdery layer which must also be removed by dusting down.

The level of cleanliness obtained is generally considered to lie between that obtained by abrasive blast-cleaning and that resulting from manual cleaning.

The advantages of this process are:

(i) The relative mobility of the equipment enables it to be used at any stage of fabrication or erection.
(ii) It can be used under relatively wet and damp conditions and helps to dry the surface.
(iii) If priming paint is applied whilst the surface is still warm this ensures that there is no condensation, and it also slightly speeds up the drying of the primer.

The disadvantages are:

(i) If the flame traverses the surface too slowly, thin sections of the steel may be warped.
(ii) It is a fire hazard.
(iii) If not carefully controlled it can affect the metallurgical properties of the steel and should never be used near high-strength friction grip bolts.

9.5 STANDARDS FOR SURFACE CLEANLINESS†

Most standards are concerned with the degree of removal of rust and scale by blast-cleaning and do not take into account the influence of soluble salts present on the surface.

† The term 'cleanliness' is generally used in relation to the degree of cleaning of steel surfaces, although 'cleanness' might be considered as more suitable.

There are a number of national standards. Most of them are based on Swedish Standard SIS 05 59 00:1967 which provides a series of photographic standards and written descriptions of various stages of visual cleanness of steel surfaces after blast-cleaning. It also categorises the degree of rusting of the steel before blast-cleaning from A to D, the four grades ranging from completely scaled steel to a rusted and pitted surface. This aspect of the standard is widely ignored but is important because it is more difficult to achieve high standards of cleanliness on steel that is badly pitted by rusting.

The British Standard BS 4232:1972 'Surface Finish of Blast-Cleaned Steel for Painting' does not use photographs but provides a series of percentages of rust and scale that have to be removed, e.g. first quality is complete removal of all scale and rust.

The American Steel Structures Paints Council (SSPC) specifications employ the Swedish grades of rust A to D, but is similar to the British Standard in specifying percentages of rust removal.

A sample of steel blast-cleaned to an agreed standard which can then be used to provide a visual assessment of surface cleanness is commonly used in practice.

At present the three most widely used standards and the broad correlation between them is shown below:

United Kingdom	United States	Sweden
BS 4232:1967	Steel Structures Painting Council (SSPC)	SIS 05 59 00:1967
First quality	White metal	Sa 3
Second quality	Near white	Sa $2\frac{1}{2}$
Third quality	Commercial	Sa 2

Many countries other than those noted above have published standards for the surface preparation of steelwork. Most of them are based on the Swedish Standard SIS 05 59 00:1967 although there are some additions and variations. The German Standard DIN 55928 (1978) contains an additional section on the cleaning of welds and the Shipbuilding Research Association of Japan's Standard of 1975 includes the preparation of steel that has weathered after application of a blast primer.

Paint manufacturers' data sheets should give the minimum standard of preparation which they consider to be necessary for their materials. However, it should be remembered that this standard only refers to the visual appearance. The amount of chemical contamination left will depend on the state of the steel before preparation and on the type of cleaning process used. Higher grades of preparation than those recommended by the paint manufacturer may well be economically justified if the surface before preparation is very corroded and the subsequent exposure will be to wet conditions.

None of the standards for surface cleanliness can be considered to be completely satisfactory. Although the Swedish Standard is widely used, the photographs depicting the standard of cleanliness, e.g. Sa $2\frac{1}{2}$, do not always in practice match the appearance of blast-cleaned steel. The British Standard 4232 provides percentage figures for the amount of rust and scale acceptable at each level, e.g. second quality, but in practice it is difficult to actually measure such areas.

9.6 SURFACE PROFILE

When steel is cleaned with abrasives to remove scale and rust there is an inevitable roughening of the surface. The degree of roughness is important because this will affect the coverage of the highest points or peaks by paint. Ideally surface roughness should be considered in three-dimensional terms but there are obvious difficulties in applying such a concept, and in practice the profile is considered as two-dimensional. The surface roughness for machined surfaces is usually defined on a numerical basis. This is satisfactory where the changes in surface profile are reasonably uniform as on many machined surfaces, and the Centre Line Average (CLA) and Root Mean Square (RMS) are both used. In the United Kingdom the most commonly used method is CLA and the readings can be obtained directly with suitably calibrated instruments or graphically from a profile trace.

Other methods are also used to characterise the roughness but all are based on an assumption that the profiles are reasonably smooth. Where large deviations in smoothness occur as in blast-cleaning, these methods are of less practical value.

Various methods have been used to determine the surface profile of blast-cleaned steel. Probably the most accurate is the stylus method in which a sapphire or diamond contact point is drawn over the steel surface and a trace is obtained on a chart. The height is usually magnified by a factor of 100 or more and the length by a smaller factor. Although this method has limitations, e.g. the problem with re-entrant parts of the profile, it is the most accurate method available. Unfortunately the instruments are delicate and not suitable for field work. Replicas of the surface of blast-cleaned steelwork can be taken, and returned to the laboratory for examination and this may be useful in cases of dispute but is clearly not suitable as a field method of control because of the inevitable delays that occur before the results are available.

As a reference method, microscopic measurement under clearly defined conditions provides a good method of recording profile over a small area. The microscope is focussed on troughs, then peaks and the differences are measured on the vernier scale.

A number of other methods have been devised to measure surface profile including a pneumatic instrument. The principle of this device is that if an air orifice is pressed against a surface and air under pressure is passed through it, then there is a relationship between the loss of air through the orifice and the surface roughness. In practice this type of instrument has not proved to be particularly satisfactory.

The most commonly used instrument for measuring surface profile of blast-cleaned steel is a dial gauge, usually fitted to a frame, with a fine pointed stylus. After calibration, such instruments are used to take direct readings on the blast-cleaned surface. Measurements are made on both peaks, i.e. highest points, and troughs, i.e. invert points. This type of instrument clearly has limitations, not least the size of the stylus point and its ability to reach the bottom of narrow troughs. Although such instruments can be used directly in the field, it is generally more convenient to make a replica of the surface to facilitate measurements and to retain the sample as a record. Special tape materials are available for making a replica of the surface.

A method widely used in the United States is the technique of comparing blast-cleaned surfaces with specially prepared standards of known profile height. Such standards are available, prepared with a range of different types of abrasive and abrasive sizes. The technique is sometimes described as the 'comparator method'. Provided the comparators are carefully manufactured and checked to produce accurate surfaces for comparison the method has much to recommend it.

9.6.1 The importance of profile measurement

The measurement of surface profile is important because the profile influences the effective thickness of paint that covers and protects the steel. Furthermore, there are always some peaks that are particularly high, sometimes called 'rogue peaks' and it is a matter of chance when measuring surface profile whether these will be measured (*Figure 9.7*). Consequently, an empirical relationship between measured profile and paint film thickness has been adopted.

Dry paint film thickness = 3 × surface profile (all measurements being in the same units, e.g. μm).

Schwarb and Drisko consider that there may be a preferred profile height or a preferred abrasive for different coatings[3].

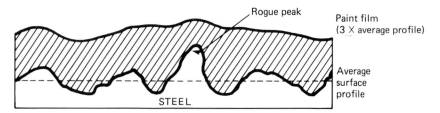

Figure 9.7 Profile after blast-cleaning

9.7 SOLUBLE CORROSION PRODUCTS

A problem that has been highlighted in recent years because of the increasing amount of steelwork that has been used for marine situations is the influence of soluble iron salts, e.g. ferrous chloride, on the performance of paint coatings. As noted previously, these salts cause the breakdown of paint coatings applied over manually cleaned surfaces but they can also be a problem when steel has been blast-cleaned, particularly where the steel had originally, before cleaning, been exposed to an aggressive industrial or marine atmosphere resulting in the formation of these salts in pits. It may prove to be difficult to remove such salts by dry blast-cleaning, although wet blast-cleaning is generally more effective in this respect.

There is no sound data regarding the concentration of soluble salts that could be considered as acceptable for different situations, although investigations to provide such data are in progress. Generally such salts are more likely to cause problems with painted steel where it is immersed or exposed to condensation conditions.

Various methods for determining the presence of such salts are available but they all have limitations. Two such methods are based on dipyridyl test papers and on potassium ferricyanide, the latter is described in Appendix G of BS 5493:1977.

9.7.1 Protection after blast-cleaning

It is advisable to protect steelwork as quickly as is practicable after blast-cleaning. If the blast-cleaning is carried out after fabrication, then the first coat of the final protective system can be applied. Where, however, blast-cleaning is carried out prior to the completion of fabrication or where the protective coating cannot be applied immediately, some form of protection must be applied. Sometimes specifications call for the application of a thin primer, variously called a 'holding primer', 'blast-primer' and 'pre-fabrication primer' within three or four hours of blast-cleaning. This is reasonable but the time between cleaning and protection will be determined by the actual environmental situation. Sometimes, under dry, clean conditions, longer periods will be satisfactory, but in damp, condensating conditions, three to four hours may be too long.

9.8 PREPARATION OF WELDS

Welds generally represent a small, but extremely important, part of a structure and yet they are often the most neglected when it comes to painting. In the majority of cases the first area of a painted structure to show paint breakdown is at the weld. During construction a weld may be accepted as structurally sound but may be considered unsatisfactory as a surface on which to apply paint, for which purpose it must be continuous and free from pinholes, sharp projections and excessive undercutting. Where possible all weld spatter should be removed, because not only will it protrude through most paint films, but it may also detach from the surface. Weld slag and deposits should also be removed since such deposits may be alkaline and will saponify the binder of oil-based paints and encourage the under-penetration of others.

Welds should be ground or blast-cleaned to remove contamination and sharp projections. Also any holes etc must be filled, either by re-welding or with an appropriate two-pack epoxy filler or similar material. When grinding a weld smooth, it is undesirable to overgrind or flatten since this may weaken the weld itself. As an additional precaution to assist with corrosion protection it is also always advisable in subsequent painting to add an extra stripe coat of the primer to the weld area.

9.9 REFERENCES

(1) CHANDLER, K A and REEVE, J, *J.O.C.C.A.*, **49**, No 6 (1966)
(2) MALLORY, A W, *Materials Performance*, **21**, No 11, 15 (1982)
(3) SCHWARB, L K and DRISKO, R W, *Materials Performance*, **20**, No 5, 32 (1981)

10 Paint coatings

Paints essentially consist of solid particles, called the pigment, dispersed in a liquid binding medium. The paint additionally contains solvents and thinners to control viscosity and provide suitable application properties. There are also other additives that will be considered later.

The medium or binder dries by one of a number of possible reactions, e.g. oxidation, to provide a dry protective film. The binder largely determines the protective quality of the coating. The pigment provides colour and opacity and to some extent influences the protective qualities of the film. Zinc as a pigment in zinc-rich paints is an exception as will be considered later. Paints are formulated to provide various properties such as suitable drying time, abrasion resistance, etc. From the protection standpoint, the ability to prevent rusting of the steel substrate over long periods, i.e. durability, is of prime importance. There are factors such as cost, ease of application and ease of maintenance, all of which are important and must relate to durability. A range of different paints is made, each having some advantages compared with another, but equally having disadvantages. All paints are, to some extent, a compromise as is their selection for different purposes.

Unlike alloys—which can generally be specified to provide materials with reasonably known properties—most paints do not fall into this category. There are paint specifications and standards and these will be considered later, but for most purposes, paints are denoted by fairly broad generic terms. It is, therefore, necessary to consider some of the main groups with an indication of their properties. As most paints are denoted by the binder, this is the best way to categorise them. A few paints are commonly described by their pigmentation, e.g. red lead. Strictly, this does not provide much indication of the type of paint. Although the term 'red lead paint' was generally taken to mean the oil-based paints, e.g. British Standard BS 2523 'Lead-based priming paints', this pigment is now used with many other binders, e.g. chlorinated rubber, and the properties of the paints are different.

With the exception of zinc noted above, pigments do not affect protective properties to the same extent as binders, but they still play an important role, particularly in priming paints.

Before considering paints in detail it is important to appreciate the various coatings that form the paint system.

10.1 PAINT SYSTEMS

Generally, steel protection is not achieved with a single coat of paint. Although many modern formulations are capable of providing thick films, it is generally advisable to apply at least two coats of paint to reduce the influence of any defects such as pinholes in the film. More often, protective coatings are produced by three or more coats of paint. Typical paint systems for marine conditions will be considered later but as the formulation of the various coats in the system often differ, the general concepts will be considered before discussing types of paint.

The basic elements of a paint system are (i) priming coat, (ii) undercoat, and (iii) finishing coat. Sometimes the terminology varies and some authorities use the term undercoat to indicate the priming coat as well. Each type of coat has a specific function, as follows.

10.1.1 Priming coats

This is the coat applied to the substrate to be protected, usually of steel but it may also be a metallic coating, e.g. zinc. In some situations, wash primers or etch primers may be applied before the actual primer, to improve adhesion. These are special materials, to be discussed later, but not in the class of primer considered here.

The primer acts as the foundation of the protective system so it must wet and adhere to the surface well, otherwise the whole system may become undermined and blister or flake. In many paint systems there is no essential difference between the undercoat and the primer but in oil-based and oleo-resinous systems, i.e. those containing a reasonable proportion of natural oils, the pigment in the primer may be what is commonly called 'inhibitive'. Such pigments, by various chemical actions, produce species that reduce the corrosion rate of steel. Some— such as red lead—work by virtue of a reaction with oil to produce soaps, so they would not be expected to act inhibitively in non-oil binders such as chlorinated rubber, although they are in practice used in such binders. Other inhibitive pigments such as metallic chromates depend upon their solubility in water to operate. Again, they would be expected to be less efficient in binders other than oil. When a great deal of steel was cleaned by wirebrushing rather than blast-cleaning so that primers were applied over rusty steel, these inhibitive pigments served to reduce the corrosion that inevitably occurred under the primer. Red lead in oil was a particularly good primer of this type but dried very slowly. Such primers are still used, particularly for maintenence work, but rarely in situations where new steelwork is painted in the factory, which should be the situations for coatings to be applied in aggressive marine environments. Apart from adhering well to the substrate, priming coats must be formulated so that the undercoats adhere well, i.e. good intercoat adhesion. This may mean, with certain types of binder, that limitations are placed on the period between applying the primer and the next coat of paint, to ensure that the primer is dry but does not harden, to the point where adhesion problems arise. Although, strictly speaking, the primer is the coat in contact with the substrate, an additional priming coat may be applied over the first true priming coat.

10.1.2 Undercoats

The undercoat is used basically to build up the thickness of the coating and it is usually formulated to provide properties somewhat similar to the finishing coat but not those of gloss, colour, etc. Intercoat adhesion is a necessary requirement. The pigments used are usually similar to those in the finishing coat.

10.1.3 Finishing coats

This coat provides the main protection from the environment and is formulated accordingly. The pigments are non-inhibitive and inert, such as rutile, titanium dioxide, with others to provide the required colour. Often lamellar pigments such as micaceous iron oxide (MIO) are used to provide added durability but these tend to provide a grey colour to the paintwork. Although other colours can be obtained, with MIO pigments, this may lead to a loss of durability.

An essential requirement for a paint system is that all the paints used are compatible. This should be checked with the paint suppliers.

10.2 PIGMENTS

Pigments are usually added to paints to provide colour, opacity and to improve durability. Most pigments are reasonably inert but some used in primary paints have some inhibitive properties. A brief description of the commonly used pigments is given below.

10.2.1 Inhibitive pigments

Red lead
This is probably the best known and most effective inhibitive pigment when used in drying oil binders. Although it is used in binders other than oil it is by no means as effective in such paints. Some authorities hold the view that red lead has no inhibiting effects unless in an oil binder. Because of its toxicity, its use is being severely limited and is prohibited by some organisations.

Metallic lead
This is also inhibitive and has been used widely in paints for inland structures. It is less reactive than red lead but is also toxic.

Calcium plumbate
It is claimed that paints containing this pigment are particularly useful for application to galvanised coatings to provide good adhesion. However, paints containing calcium plumbate pigments can give an alkaline reaction which may affect some paints applied over them.

Zinc chromate
Inhibitive but when used in conventional primers tends, because of its high solubility, to provide limited protection if not overcoated. A less soluble type is used for blast primers; it is considered to be toxic.

Zinc dust
This is used with zinc oxide or in zinc-rich paints to provide coatings that give some sacrificial protection (see 'Zinc-rich paints').

Zinc phosphate
A comparatively new pigment which is now widely used in priming coats as a replacement for toxic pigments, e.g. lead. Its inhibitive properties are considered to be poorer than many of the other pigments used for priming coats.

10.2.2 Non-inhibitive pigments

Rutile, titanium dioxide
Used in finishing coats for white or light colours and with the addition of other pigments it is used to obtain a range of colours.

Red oxide
This has no inhibitive properties but has been widely used in primers.
Sometimes it is mixed with zinc chromate because of its cheapness.

Aluminium
A useful pigment for heat-resisting paints. It is also used as a barrier
pigment for finishing coats but may react with atmospheric pollutants,
resulting in aluminium salts and local film breakdown. It is used in
bituminous coatings, to increase durability particularly in the
atmosphere.

Micaceous iron oxide (MIO)
MIO is widely used in some parts of the world in protective paints for
structural steelwork. It is claimed that the lamellar type of pigment
improves the durability of finishing paints. Many pigments of this type,
however, are not lamellar, and the advantages provided by such
pigments are probably less than those of the lemallar type.
    Other pigments are also used for various colours and special purposes,
e.g. graphite.
    Extenders such as barytes, talc, china clay and asbestine are used to
improve opacity and to provide some of the advantages of pigments at a
lower cost.

10.3 PAINT ADDITIVES, SOLVENTS AND DILUENTS

Although the binder and pigment are the essential ingredients of paint,
other materials are also used. These include driers, organic salts of metals
such as cobalt, lead and manganese, which are added to improve drying
properties.
    Anti-oxidants are used to prevent skimming in containers. Surface
active agents are used to help disperse pigments in paint and other
additives include thickeners and anti-settling agents.
    Solvents and diluents are used to enable satisfactory application of
paints or to obtain a higher solids content at a given viscosity. The type of
solvent will depend upon the binder and it is important to ensure that the
correct type is used.

10.4 BINDERS

There are various ways of classifying binders and one suitable way is as
follows:

(i) Drying oil type, including oil, modified oil, and alkyd.
(ii) One-pack chemical-resistant, including vinyl and chlorinated rubber.
(iii) Two-pack chemical-resistant, including epoxy, polyurethane and coal tar or pitch modifications.
(iv) Bituminous coatings—strictly they fall into classification (ii) but are used for specific purposes where cheapness is important.
(v) Silicates.

These binders provide the basis for the generic terminology of paints, with the exception of zinc-rich paints, which because of the high proportion of pigment in the paint are generally categorised by the pigment.

10.5 TYPES OF PAINT

10.5.1 Drying oil types

This group includes not only traditional drying oils such as linseed oil but also those modified with synthetic resins such as alkyds, phenolic, epoxy esters and urethane oils. They are usually applied to a thickness of about 35 μm in the dry film. Drying oils tend to be slow drying although this is improved by modification with synthetic resins, which also improve their durability. They dry by oxidation and paints based on them are generally applied in a series of comparatively thin coatings. Such paints are the most common in general use. They are used in marine situations but not for the most aggressive conditions.

The true oil-based paints are no longer widely used. They are prepared from treated natural oils such as linseed and tung and dry by oxidation. They are slow drying but have good wetting properties—a property of value in primers applied over rusty steel. Primers based on oil with inhibitive pigments such as red lead are probably still the best paints for manually cleaned steel. They are, however, toxic and many organisations will not allow their use. Furthermore, they have to be brush applied.

Generally, oils are modified with resins to provide quicker drying, more resistant paint coatings. The more commonly used paints of this type are:

Phenolic varnish
Alkyd
Modified alkyd
Epoxy ester

For priming paints the pigment is usually inhibitive. There is a growing trend in the use of zinc phosphate pigment in such paints. In undercoats and finishing coats, non-inhibitive or inert pigments such as rutile titanium dioxide and micaceous iron oxide (MIO) are used.

(i) *Phenolic modified paint*

These contain natural oils blended with phenolic resins; tung-phenolic is a common binder. The paints dry mainly by oxidation but they dry faster than oil paints. They have a better chemical resistance than oil paints but have poorer gloss retention than alkyds (see below).

(ii) *Alkyd paints*

Alkyd paints are the most commonly used of the drying oil types. Alkyd is itself a generic term covering a range of products. Basically for the protection of steelwork, various oil modifications are produced to provide suitable properties. The term 'oil-length' is used to indicate the relative proportions of oil to resin in the binder. The advantage of a long-oil alkyd containing a high proportion (65–75%) of natural oils, is in its ability to wet a surface. Binders containing less oil, e.g. medium-oil alkyds, dry more quickly than the long-oil variety.

Alkyd paints are easily applied and dry reasonably quickly, usually overnight, at 25 °C, and coats are tack-free in two to eight hours at this temperature. They have good resistance to general atmospheric conditions but are not resistant to alkaline situations or to stronger solvents. They retain their gloss well and are available in a range of colours, which makes them suitable as decorative coatings. They can be applied by all the standard methods although some, formulated for fast drying, are best applied by spray.

Although widely used in a range of atmospheres they would not usually be recommended for aggressive marine conditions or for situations of continuous dampness.

A number of modified alkyd paints are also used. One of these, silicone-alkyd, is a good deal more durable than the general alkyd paints, although considerably more expensive. They combine the heat and durability properties of silicone with the application advantages of the alkyd. These paints have good weathering properties and a high gloss retention. Despite their additional cost they may well prove to be an economic alternative to standard alkyds for some marine purposes.

Other modifications include vinyl-alkyds, chlorinated rubber-alkyds, phenolic-alkyds and acrylic-alkyds. Such modifications tend to have properties somewhwere between those of the component parts. Chlorinated rubber-alkyds have been used for a number of situations, particularly as a 'travel cost' for chlorinated rubber paint systems to avoid excessive damage of the coating during handling and transport. Such modifications may lead to a lower performance than anticipated, especially where the alkyd additions are not clearly indicated, so the user is under the impression that he is using a chlorinated rubber paint.

(iii) *Epoxy ester*
This, despite its name, is quite different from the two-pack epoxies (considered in Section 10.5.3). The epoxide resins are esterified with fatty acids, the presence of which reduces the resistance of the binder to about that of an alkyd. These paints have some advantages over alkyds, particularly for adverse drying conditions, and are fairly widely used as primers.

10.5.2 Single-pack chemical-resistant paints

Polymers such as chlorinated rubber, vinyl and vinyl copolymers are soluble in certain solvents and dry by evaporation, to provide a protective film. Generally plasticisers are added to improve the mechanical properties of the film. Paints made from these binders can be softened by the original solvents and this is useful when considering maintenance painting. Unlike paints from oil binders or two-pack binders (see below), there is no basic chemical reaction, so such paints are sometimes—as a class—referred to as non-convertible whereas the other two main types of binder are known as convertible.

A problem with this type of paint is that of solvent retention, which can result in brittle films over a period of time or, if too much plasticiser is used, softening under hot climatic conditions. Furthermore, holes—called 'vacuoles'—can form in the film as a result of the drying mechanism, although it has been claimed that vacuoles do not affect the protective properties of the film.

A new addition to this class of binder has been termed acrylated rubber and it is claimed to have superior properties particularly so far as solvent retention is concerned.

Chlorinated rubber paints are formulated with additions of resins, plasticisers and solvents. Additionally various pigments are used, including inhibitive types for primers. The finishing coats cover a range of colours. Properly formulated paints can be applied by all the conventional methods and have good resistance to many chemicals.

The paints dry by solvent evaporation and can consequently be attacked by solvents. This means that problems can arise if coatings containing strong solvents are applied to chlorinated rubber coatings. Care must be taken with paint systems to ensure that paints such as two-pack epoxies are not applied over chlorinated rubber paints, although the reverse procedure can be used.

The paints can be applied satisfactorily by spraying, but what is termed 'cob-webbing' may sometimes occur if the viscosity is not correct. Paints formulated for brush application can be applied to a dry film thickness of 25–75 μm but special formulations are available to provide thixotropic materials capable of being applied to dry film thicknesses up to 250 μm. Chlorinated rubber coatings are suitable for many marine applications, not only for steel but also for concrete. Compared with two-pack chemical resistant paints, maintenance is usually easier because of the good adhesion arising from the solvent nature of the paints. They should be applied to properly cleaned steel surfaces but are somewhat more tolerant to inadequate surface preparation than are the two-pack epoxies and urethanes. They are not considered suitable for continuous service above about 65 °C. Problems can arise with the stacking of steel coated with chlorinated rubber if the paint has not dried sufficiently. Reactions may occur between the coatings on the stacked sections causing damage to the coating.

Vinyl paints are a blend of vinyl chloride and vinyl acetate copolymers, sometimes modified with acrylic resin. Additionally they may have non-saponifiable plasticisers. They are similar to chlorinated rubber paints but are generally less tolerant of poor surface preparation and have a higher water-vapour permeability. They are considered to be more durable when exposed to intense sunlight and to have better gloss and colour retention than chlorinated rubber systems. Pigmentation is similar to that of the chlorinated rubber paints. Suitable for many marine applications, but for sea water immersion, the Steel Structures Painting Council recommends that after blast-cleaning a wash primer should be applied before application of the vinyl paint system. Generally they are available in a range of colours; their advantages are similar to those of chlorinated rubber paints.

10.5.3 Two-pack chemical resistant paints

These binders are prepared by mixing two components just before application. Unlike paints made from other binders, paints of this type harden or cure by chemical reaction within the material itself. This development has led to a range of coatings that can be applied to provide thick protective films that cure comparatively quickly. As the process of drying is basically a chemical reaction, it is temperature and time dependent. This means that once a coating material is mixed it has only a limited life so it must be applied within these limits. Furthermore, at low temperatures drying times may be prolonged, so they should not be applied below the manufacturer's recommended temperatures.

These binders provide the possibility of a wide range of very protective coatings, the properties of which can be varied by using different curing agents.

Paints made from such binders are widely used for marine conditions, particularly for the most aggressive situations. Epoxy resins are the most common group in this class, the other of significance being polyurethanes. Coal tars and pitches are added to both for improved water resistance and to produce cheaper coatings for certain conditions.

These materials generally contain solvents and can be applied by the usual method, e.g. spray. There are, however, also solvent-free coatings, mainly epoxies, that are used for some applications. These can be broadly divided into powder coatings and solvent-free coatings. The powder coatings are sprayed and heated in much the same way as thermoplastic coatings (see Section 10.72(c)). The solvent-free epoxy coatings are applied by a two-component hot airless spray. The two components are heated in small tanks to provide sufficient viscosity for spraying. They are then supplied separately to the spray gun and mixed and immediately sprayed.

At the application temperature, curing begins immediately, so some skill and control of application is essential to produce sound coatings.

There would clearly be an advantage if solvent-free coatings could be applied without heating and developments are taking place in this area. Some products for spraying at ambient temperatures are available. Generally, solvent-free coatings can be applied to high film thicknesses, e.g. 300–1000 µm. Greater thicknesses, up to 5 mm, can be obtained with special types of product.

The term *two-pack epoxy coatings* covers a number of different paints with somewhat different properties, these depending mainly on the reaction mechanisms brought about during the curing process. These coating materials or paints consist of two components, one the epoxy resin and the other a 'hardener' or 'curing agent' containing amino groupings. These are mixed before use and after application the paint cures to a hard durable coating. Mixing of the components in the correct proportions— which vary with different products—is essential.

Aromatic amines provide good water resistance to the coating and cure at low temperature. Aliphatic amines cure more rapidly and have good solvent resistance but may provide poorer adhesion. They can be added as adducts, i.e. products that have been partly reacted with the epoxy resin before use.

The various curing agents modify the application and drying characteristics of the coating so a range of coatings is possible.

Epoxies are durable, hard and have good adhesion properties when applied to clean steel. The important aspects of epoxies are (i) the 'pot-life', i.e. the maximum period between mixing and application and (ii) the curing temperature. The curing temperature and the speed of reaction at different temperatures is important. Some epoxies cannot be cured at low temperatures, say below about 5 °C, and others may react too quickly at tropical temperatures. Clearly, the paint manufacturer's advice should be sought to ensure that the correct material is used for a specific purpose. Epoxies have excellent resistance to marine environments, including immersion in sea water.

They are available in a range of colours and, although highly durable and initially glossy, they tend to chalk in outdoor exposure conditions.

Because of their hardness—particularly as this increases with ageing—adhesion problems may arise at maintenance painting and some abrading action on the coating may be required.

Modification of epoxies with coal tar, to produce *coal-tar epoxies,* provides a cheaper product than the epoxies but one with very good properties, particularly for marine situations. They are two-pack materials and are produced in two forms: (i) with the coal tar, epoxy resin and solvent in one pack and the curing agent as the second component or (ii) with the coal tar, solvent and curing agent in one pack and the epoxy resin in the other. The two forms produce different ratios for mixing of the two components and (ii) is considered by some authorities to provide a more accurate mix.

Different curing agents are used but the polyamide cured coal tar epoxy is considered to be slightly more flexible and generally to have improved moisture resistance. This will, however, depend upon other factors in the formulation.

Coal-tar epoxies can be applied by all the conventional methods although spray application is generally recommended. The amount of coal tar in the paint varies with different manufacturers but one specification for a polyamide cured product recommends about 30% with 25% filler and 18% epoxy resin, the other ingredients being curing agent, etc. Coal-tar epoxy coatings are black or sometimes a dark red so they cannot be used where decorative aspects are important. They are particularly suitable for protection of steel immersed in sea water and also have excellent resistance to atmospheric environments, although they tend to chalk fairly quickly.

Two-pack urethane coatings have similar properties to the two-pack epoxies. There are, however, other one-pack materials variously known as *moisture-curing urethanes* and *polyurethanes.* They should not be confused with the two-pack materials. Some paints called polyurethanes are, in fact, modified single-pack materials with properties quite different from those to be considered here. The single-pack materials cure or dry by reaction with either moisture or oxygen in the atmosphere, whereas the two-pack materials have two components which are mixed before use. There is a range of two-pack materials, some modified with vinyls. For readers who wish to study the range and properties there is a useful NACE Technical Committee Report[1] dealing with this.

Urethanes provide hard, tough, abrasion-resisting coatings which by changes in formulation can produce a range of finishes from high gloss to semi-matt, with a wide choice of colours. A high standard of steel surface preparation is required for urethanes and problems can arise when applying them to other coating materials; checks for adhesion should be carried out on a small area before complete coating is undertaken. It may be necessary to carry out light blast-cleaning where urethanes are to be applied to aged urethane coatings.

Urethane materials are excellent for many marine purposes but in view of the range and type of products available, specialist advice should be sought before using them.

Two-pack urethane pitches and *coal-tar urethanes* are used for much the same purposes as coal tar epoxies and they are broadly similar. Generally the type of curing agent influences physical rather than chemical-resistant properties. Although the materials can be applied by all the conventional methods, spraying is recommended using a pressure pot and a gun with an external mix tap. Coal-tar urethanes are reported to have improved flexibility and impact resistance compared with coal-tar epoxies but generally are not so hard.

As the coatings are black or brown they are not used for decorative purposes.

10.5.4 Bituminous coatings

Bituminous coatings, generally hot applied with or without fillers are widely used for pipelines buried in soil and for submarine pipelines (which are discussed in Chapter 17). These coatings are usually thick, over 300 μm, and are not considered to be paints. These are, however, bituminous coatings that are applied in the same way as paints and for some purposes can be considered as cheap one-pack chemical-resistant paints. They are sometimes called 'black paints' (tar based), e.g. in British Standard 1070.

The paints are either natural bituminous or coal-tar petroleum distillates in suitable solvents, which evaporate leaving the bituminous coating. At reasonably high thicknesses they are useful for steel immersed in water but tend to crack and sag in hot weather when exposed to the atmosphere. They can be pigmented aith aluminium flake to improve their performance in air. Bituminous coatings are basically barrier coatings with no inhibitive properties.

There are a number of standards and specifications for bituminous coatings including BS. 1070 and BS. 3416.

10.5.5 Zinc pigmented paints

Many paints are pigmented with zinc powder or zinc oxide but there are two special groups of zinc-containing paints. The two groups— irrespective of the terminology used to describe them which may vary— are fairly easily described. Each has a high proportion of zinc dust in the paint (80–90%) and group (i) has an organic binder while group (ii) has an inorganic binder, i.e. a silicate. They are important paints for marine purposes and because of the high proportion of zinc dust in the dry film provide a measure of cathodic protection under suitable circumstances. Although they are most often used as primers, they can be used as a multi- coat system, provided the grey colour is acceptable. Application of zinc- rich paints is sometimes called 'cold galvanising' and sometimes they are grouped with zinc metal coatings. Although their performance is often very good, they should be differentiated from metallic coatings because there is some binder in the protective coat and unlike hot-dip galvanising, there is no alloying with the steel substrate. Furthermore, they are far more influenced by application procedures and surface preparation than hot-dip galvanising which has a much greater control built into the coating process.

The two groups of zinc pigmented paints have somewhat different properties so they will be considered separately.

(i) *Organic zinc-rich paints*

A number of different binders are used to produce a range of paints. The most commonly used for marine purposes is the two-pack epoxy but other binders include chlorinated rubber, polystyrene, styrene butadiene and epoxy esters. A high level of surface preparation of the steel substrate is required to obtain the best performance from the paints and good mixing is essential. Although the paints can be used to provide a complete protective system, e.g. a three-coat system, say 275 μm thick, they are more widely used as primers for other types of paint systems. They are excellent for many marine environments but there is some difference of opinion among authorities regarding their suitability as primers under immersed sea water conditions. As a conventional primer, zinc-rich paints are usually applied to a thickness of about 75 μm, but there are specially formulated blast primers that are applied to a thickness of only about 25 μm.

(ii) *Zinc silicate coatings*

The binders for the zinc-rich paints considered above are organic, in common with all the other binders so far discussed in this chapter. However, there is a class of inorganic binder, based on silicate, which is generally mixed with zinc dust to provide the *zinc silicate coatings*. Silicate binders are either aqueous solutions of alkali silicates, usually sodium, potassium or lithium silicate or alcohol solutions of alkyl esters of silicate. The latter are commonly termed 'alkyl silicates' and the former 'alkali silicates'. Although the alkyl silicates are based on an organic medium the organic silicates hydrolyse with the volatilisation of alcohols leaving an inorganic dried film.

Zinc silicate coatings are hard and cement-like with a number of favourable properties.

There is a range of different coatings that falls within the generic term 'zinc silicate'; these may have varying quantities of zinc, but for use in aggressive marine atmospheres there must be sufficient zinc to justify the term 'zinc-rich paint'. The inorganic nature of the dry film confers excellent resistance to solvents and provides the alkyl silicates with good high-temperature resistance (up to about 400 °C). They are also abrasion resistant and the coefficient of friction of the dry film is about 0.5 compared with about 0.1 for many organic coatings. It is, in fact, about the same as a blast-cleaned steel surface, so is one of the few coatings that can be used on surfaces connected by high friction grip bolts.

Zinc silicate paints are extensively used as primers for marine situations but they are also used without topcoats, particularly for tanks, where their solvent resistance is of great advantage. They are not recommended without further coatings for conditions outside the pH range of about 6–10.

Surface preparation is particularly important with zinc silicate coatings and generally the highest standards, e.g. Sa 3, would be recommended. The profile height should be between 25 and 50 μm. Zinc silicate paints should not be applied to blast-cleaned steel which has remnants of old organic coatings adhering to it. Although so-called 'mud cracking' was at one time a problem with these paints with dry film thicknesses in excess of 100 μm, this is not a major occurrence with modern paints. Spray application is generally recommended to provide a dry film thickness of about 75 μm. 'Dry spray' may be a problem as the solvents evaporate rapidly, and attention must be paid to application, particularly in hot conditions. Sometimes curing is accelerated by spraying with water but this is not generally necessary. Because they cure by reaction with atmospheric moisture, high humidity during application may actually be an advantage.

If there is a delay in overcoating, zinc salts may form on the surface and this may impair adhesion, so it may be necessary to sweep blast the surface to remove them before the application of a further coating. Sometimes vigorous brushing with a bristle brush is a method employed in preference to blast-cleaning, which reduces the thickness of the coating to some extent. Most of the commonly used chemical-resistant paints can be applied to zinc silicate paints. Alkyds are not, however, generally recommended for direct application over them. A number of materials, some single pack, are called zinc silicate but these will not generally compare in properties and performance with the two-pack zinc-rich type of silicate coating.

10.5.6 Other materials used as primers

Apart from the primers already mentioned, there are three other materials, two mainly used for metallic coatings, that are also used for application to the substrate before applying the full protective system.

(i) Blast primers (which also include prefabrication and holding primers) are not true primers. Generally, when considering the protective qualities of a system any additional thickness arising from these materials is not usually taken into account. They are primarily used to protect blast-cleaned steelwork from the atmosphere during the period before the final protective system is applied. Prefabrication primers are used specifically to protect steelwork before and to some extent during fabrifaction. A number of different materials is available but the most commonly used are:

   (a) One-pack, phenolic-modified polyvinyl butyral (similar to etch primers (see below)).
   (b) Two-pack epoxy resins, generally with iron oxide possibly combined with inhibitive pigments.
   (c) Two-pack epoxy resins with high proportions of zinc, i.e. zinc-rich or zinc silicate based materials (not widely used where welding is involved because of problems with fumes).

(ii) Wash primers, also called etch and pre-treatment primers, are solutions of vinyl resins with phosphoric acid additions for etching. These materials are mainly used as pre-treatments for zinc or aluminium and produce very thin films, under 20 μm. Modifications with phenolic resins is considered to provide improved resistance to moisture, which may produce adhesion problems if overcoating with the paint system is delayed for some time. Although a good basis for many paints, intercoat adhesion problems have been reported with some wash primers, e.g. with chlorinated rubber paint.

(iii) A material called 'T' wash was developed by the British Rail laboratories specifically for galvanised steel. The following formula has been published in BS 5493:

Phosphoric acid SG 1.70	9.0% by weight
Ethyl cellusolve	16.5
Methylated spirit	16.5
Water	57.0
Copper carbonate	1.0

It is a blue solution that turns the zinc surface black. If this does not occur then the pre-treatment has not been effective.

All the materials noted above are available as proprietary products.

10.5.7 Other organic-based coatings and fillers

There are a number of materials based on waxes and greases which are not strictly paints but are used to protect steelwork. A particular group of oil and grease products is used for temporary protection against corrosion and they will be considered separately.

There is a variety of formulations for waxes and greases but essentially all these materials are applied in very thick, comparatively soft, films to provide barrier protection. They sometimes contain inhibitive pigments and are useful for inaccessible areas and where decorative aspects are not important. US Maritime Administration Specification 52-MA-602, 'Compounds Rust Preventive' and some materials in British Standard BS 1133, The Packaging Code Part 6 'Temporary Protectives', cover some of these types of material. Specification 52-MA-602 covers a number of types, including the following.

Type C is primarily for hot application to certain faying surfaces and inaccessible spaces. It has about 20–25% of inhibitive pigment including chromate and is useful for filling spaces where steel is in contact with damp materials.

Type D produces a hard wax-like coating.

The effectiveness of these wax and grease-type materials depends upon the actual situation in which they are used. They often provide a very thick barrier film and surface preparation of steel before use is not highly important. They do, however, result in difficulties where steel has to be inspected and generally are not suitable for areas used frequently by personnel. On the other hand they are cheap and easily applied.

Some authorities claim excellent results from them; others consider that their performance does not equal that of conventional hard film paints.

Mastics

These are very thick coatings usually over 5 mm in thickness used in a manner somewhat similar to wax and grease paints to fill inaccessible places, overlaps, crevices, etc, and also for water-proofing. They are commonly based on bitumens, synthetic resins and synthetic rubbers. They often incorporate a filler material and generally are formulated with a viscosity that precludes normal application methods used for paints.

Most mastics are supplied as proprietary products and include:

Two-pack polysulphide
One-pack polysulphide
Epoxy polysulphide
Silicones
Butyl

A useful table of mastics with their properties is provided in the Engineering Equipment Users Association Handbook No 31 (also reproduced in BS 5493:1977, Table 7).

10.5.8 High-temperature paints

Protection of high-temperature surface is not specific to marine environments but in some ways the problems of protection at higher temperatures are more difficult under marine conditions. Many paints can be formulated to satisfactorily protect against reasonably high temperatures, but they are not necessarily particularly good protective coatings against the ambient marine conditions during cool periods. Advice should be sought before using paints for high temperature purposes because the actual conditions of operation, such as maximum temperature and its duration and the actual temperature cycle, will have an important influence on the choice of paint. Furthermore, hot gases can affect performance both at operating temperatures and during periods of cooling.

Sprayed aluminium metal coatings are problably more staisfactory than paints but zinc silicate and silicone materials (aluminium pigmented) may be useful for some situations.

10.6 HIGH-DUTY COATINGS AND PLASTICS COATINGS

For some purposes, in marine situations, the ordinary paints are not considered to provide sufficient protection and more resistant materials are used. Typical examples are the protection of steel in splash zones and exterior coatings for submarine pipelines. These materials are also used for tank linings, an application that is not specific to marine situations. When using these materials a very high level of surface preparation of steelwork is called for—usually Sa 3 (Swedish Standard).

Materials of this nature are usually applied by specialist firms and many of the coatings are used to resist specific chemicals. Some of the materials used are:

Reinforced bitumen–asphalt
Epoxy (solventless)
Fluorocarbons
Furfuryl (alcohol)
Glass
Neoprene and Hypalon
Phenolics (baked)
Polyesters
Polyethylene
Polyvinyl chloride (PVC)
Rubber
(concrete is also used)

Some of the materials are produced with reinforcements, e.g. glassfibre, and some are wrapped in an extruded tape form, e.g. polyethylene. From the standpoint of marine corrosion, three materials are of particular interest and are considered further, i.e. neoprene, polyethylene and fuse-bonded epoxies.

For readers requiring more information on many of these materials, a useful reference is 'Coatings and Linings for Immersion Service', TPC Publication No 2 (National Association of Corrosion Engineers, Houston, Texas, USA).

10.6.1 Plastics coatings

There has been a continuing increase in the use of plastics coatings to protect steelwork. The widest use for marine purposes is with components and comparatively small fabrications. However, some large steel fabrications have also been coated with plastics including pipes and the legs of oil platforms.

A wide range of plastics coating materials is available, the most popular being PVC (polyvinyl chloride). Other commonly used coatings include polyethylene, nylon, fluorocarbons such as PTFE (polytetrafluoroethylene) and powder epoxies. Virtually all the commonly applied plastics materials are chemically resistant to marine environments but they will not necessarily resist all conditions as coatings. This will depend upon the thickness of the coating, its resistance to damage and abrasion, its lack of defects such as pinholes and resistance to ultra-violet light, etc. Plastics coatings act purely as barriers to the environment and while they can be very effective in this respect, small pinholes or areas of damage can result in severe local attack on the steel substrate.

10.6.2 Methods of application

The commonly used methods of application for plastics coatings are as follows:

(i) *Dipping into liquid plastisol*
Pre-heated, primed steel is dipped into a tank containing cold liquid PVC. The heat causes a reaction between the polymer and plasticiser and a layer of plastisol is deposited on to the primed steel. The steel is removed from the tank and cured, usually in an oven at a controlled temperature.

(ii) *Dipping into a fluidised bed*
This method is frequently used for polyethylene, nylon, 'Penton' and PVC. Suitable powders are fluidised by passing low-pressure air through the tank. Steel—after cleaning, heating and, where required, priming— is dipped into the bath where the powder adheres to it. It is then reheated to fuse the powder into a smooth coating.

(iii) *Spraying*
This method is generally more expensive than dipping but can be used for large areas, e.g. tanks, that are too large for dipping. Electrostatic spraying is also used. A high voltage is applied to the nozzle of a spray gun, which produces a charge on the powder. If the steel or object to be coated is earthed then the powder is attracted to it, so coating it. This method has been widely used for epoxy powders and, unlike other spraying methods, it envelops the whole of the surface, not just the area facing the gun.

Flame-spraying can be carried out on site. Powder is agitated and forced through a flame near the nozzle which produces a molten material on the steel surface. This is then fused by passing the flame gently over the coating. Skill is required to produce good coatings that are not carbonised or embrittled.

Comments on some of the commonly applied plastics materials are given below.

PVC (polyvinyl chloride)
Generally, the addition of a plasticiser may slightly impair its corrosion-resistance but it is still chemically resistant to most marine conditions. In practice the properties can be varied and a range of colours can be produced by the addition of suitable pigments and polymers and by altering the amounts of plasticiser and solvent.

Steel must be thoroughly cleaned by pickling or shot-blasting and it is then usually coated with suitable adhesive primers before coating application by dipping or spraying. Thick coatings over 12 mm can be applied in one dip by the fluidised-bed method, provided that the steel components have a suitable heat capacity. The powder coating method can be used for thinner, more decorative coatings. Generally the plastic is contained in the bath in the form of a liquid plastisol. PVC coatings can also be applied by spray.

PVC coatings have been used for tank linings but they are most widely used for equipment, components, fencing, etc. They are also the most commonly applied coil coatings for sheet steel (see later).

Chlorinated polyethers
Generally known by the proprietary name 'Penton'. This can be applied from a fluidised bed or by spraying directly on to blast-cleaned steel and then stoved. It is not generally used for marine applications but is employed for highly resistant coatings on pumps, pipelines, and for chemical plant.

Nylon
This is not as corrosion-resistant as many other plastic materials but it can be used at temperatures up to 120 °C and has good solvent resistance. It is used as an erosion-resistant coating for pumps and other similar plant. It is not used for general marine applications.

Polyethylene
Polyethylene of the high-density type is used for tank linings and for coating submarine pipelines. It can be extruded on to pipes as a form of tape but it can also be applied from powders by spraying or dipping.

Fluorplastics
Fluorplastics such as PTFE (polytetrafluorethylene) and PTFCE (polytrifluoromonochloroethylene) are widely used to produce anti-stick coatings (PTFE) and are also highly corrosion-resistant coatings. PTFE is very resistant to acids, alkalis and solvents in the solid form, but

coatings have to be defect-free to provide the same resistance. One coat of PTFE applied over cadmium plating has been used for carbon steel nuts and bolts and in laboratory tests has shown very good resistance to salt spray. It also has the advantage of preventing seizure of the components.

Thermosetting materials
The above coating materials are thermoplastics. A number of thermosetting materials, such as epoxies, polyesters and polyurethanes, can also be applied to protect steel. Epoxy coatings, e.g. by fusion bonding, have been used for a number of applications.

Coil coatings
A specialised form of plastic coating is that applied to steel sheet. Such products are marketed under a variety of proprietary names, often by steel-producing companies. Plastic coatings are generally applied to sheet steel that has been zinc coated either by electro-deposition or hot-dipping. 'Coil coating' is the term used for the continuous process in which coatings are applied to steel strip usually in liquid form, although some plastics materials are applied as laminates.

A wide range of coatings is available, particularly for use on domestic appliances, motor cars, etc, but for buildings—particularly those exposed in marine atmospheres—PVC plastisol is most commonly used (80–200 μm thickness), for sheet which can be supplied in a range of profiles. Other coatings include fluor-carbon and silicone modified polyesters.

Before specifying these materials for cladding, roofs or for other purposes in marine atmospheres, advice should be sought from the manufacturers because they may have a limited life in some situations where temperatures are over about 25 °C and salt-laden atmospheres may also affect their durability.

10.6.3 Performance of plastic coatings

Many plastic materials are chemically resistant to the conditions encountered in marine situations. However, the plastic coatings act purely as barriers and any pinholes or other defects in the coatings will lead to rusting of the steel substrate and consequent failure of the coating. One advantage of zinc on coated sheet is to reduce the rusting of the steel below such defects. Clearly, thickness and integrity of the coating are important and a number of tests can be carried out to check the quality of the coating. For coatings to be exposed to marine environments, spark testing for pinholes is advisable. Repair of coating at any such defects is then carried out.

The effect of ultraviolet light, temperature, etc, varies on different plastic materials and advice should be sought on the best choice for any specific application.

A particular problem concerns welded fabrications. If site welding is required, then protection of the welded area also has to be carried out on site. This is true of all protective coatings but it is a greater problem with plastic (and metallic) coatings than with paints. Special techniques have been developed, including localised heating. In many situations, however, liquid plastics materials similar to paints are used at welds and these do not usually have the same protective properties as the works-applied plastics coatings.

10.7 WRAPPING TAPES

Although wrapping tapes do not fall into the same group as the paint and plastics coatings considered in this chapter, for convenience, they are dealt with here.

Petrolatum wrapping tapes are widely used for buried pipes but they have also been used for jetty pipelines and for other parts of jetty structures[1], e.g. piles.

One system used for marine purposes consists of the following stages:

(i) Manual cleaning by chipping, scraping and wirebrushing to remove all loose rust, paint, etc.

(ii) Priming with a soft petrolatum paste containing water repellant surface active agents. This is liberally applied by hand, using a glove, or by roller or a short bristle brush.

(iii) Petrolatum tape is spirally wrapped over the primed steel, allowing a 55% overlap.

(iv) Generally, an outerwrap is then applied. This can be one of the following:

(a) Cold applied bitumen tape using a nylon carrier with a PVC backing. This is applied spirally, and a circumferential band of polypropylene strapping with fixing buckles is used when a new tape is required and at the final turn of the tape.

(b) PVC tape but, at sub-zero temperatures, polyethylene tape is used where a low damage situation is anticipated.

(c) Circumferential application of a heavy-duty ultraviolet stable high density polyethylene meshed netting.

(d) In severe situations, a preformed glass reinforced resin jacket may be bolted round the pile. The jacket is usually 2–3 mm thick.

Wrapping tapes based on materials other than petrolatum are also used. The main problem with tapes is that while they can easily be applied to regular shapes such as tubes, problems arise on more complex sections such as I-beams.

10.8 PAINT APPLICATION

Paint application has a significant effect on paint performance and should be controlled, so far as is practicable, to ensure that the dry paint film has the correct qualities to attain maximum protection. It is comparatively easy to control paint application under workshop conditions but far more difficult in field situations, particularly in marine environments. Good application of paint can be achieved only by using skilled operative with sound supervision. Matters that are in themselves straightforward, e.g. mixing of paints, can—if not properly carried out— lead to defective coatings. Furthermore, as there are a number of stages involved in applying a coating, each must be carried out to the highest possible standard. Paint application under the conditions likely to prevail on offshore and marine structures is often an unpleasant and difficult task but the standards achieved will have a marked effect on the lives of the coating.

10.8.1 Methods of application

The standard methods for application of paint to structural steel are (i) brush, (ii) roller, (iii) conventional air spray, (iv) airless spray. Other methods, e.g. stoving and dip-application, are also used for components and factory-produced units. Some very high viscosity materials may be trowelled on to the surface to produce a very thick coating. Special methods are used for the hot application of solvent-free coatings.

(i) *Brush application*
This method is less widely employed nowadays particularly for the types of paint used for the more aggressive marine situations. It is the slowest of the methods in general use and cannot be used for some types of paint. Nevertheless, it has certain advantages over the other methods.

(a) It requires no masking of adjacent areas that are not to be painted.
(b) The loss of paint during application is a good deal less than with spraying methods.
(c) It can be used in restricted spaces.
(d) It has advantages for the application of priming paints on surfaces

that have some contaminants and moisture present because brushing can, at least to some extent, displace these contaminants.

(e) No costly equipment is required.

As already noted its main disadvantage is the rate of application, which compared with other methods is slow. It is a method that is still widely used for maintenance painting, particularly where the parts to be painted are a small area of the total structure. It is also generally employed for stripe coats and for some inhibitive primers.

Clearly, good quality brushes are required and these must be properly cleaned after use. Animal or nylon bristles are used for brushes; often a brush comprised of 50% of each of the two types of bristle is used for industrial paints.

(ii) *Roller application*

Roller application is a useful method for large flat areas provided the paint has satisfactory flow-out properties but it is not suitable for all paints. Some authorities consider that certain types of pigments in paints, which can cut the fibres of the roller material, should not be rolled on because of the possibility of contamination of the paint film; micaceous iron oxide is in this category. Generally, for areas where paint cannot be properly applied by roller, e.g. around nuts and bolts, brush application is used. Careful attention must be paid to the thickness of paint that is applied by roller as it is possible to apply comparatively thin films with some paints.

Although mohair is generally considered to be the best material for rollers, other materials such as lambswool and various plastics materials are used, often as foams.

(iii) *Conventional air spary*

Conventional spray painting involves atomisation of the paint by a rise in air pressure. Fine droplets are then deposited on the steel surface and these merge into a continuous film. For some solventless types of coating, the paint may be pre-heated to about 80 °C to control its viscosity. This allows thick coatings to be applied without sagging.

Air spraying is efficient and reasonably fast but there is some wastage of paint. Furthermore, time must be allowed as part of the operation for setting up and cleaning the equipment. This may prove to be uneconomic for small areas of steelwork, particularly if different types of paint are being used.

(v) *Airless spray*

In this method higher pressures are used than in air spraying and the paint itself is forced through a small nozzle and atomised as an elliptical jet. It is faster than air spraying with a lower wastage of paint. Any blockage of the jet by dirt particles can lead to uneven coverage. The rate of discharge is controlled by the size and type of jet.

10.8.2 Rates of application

The overall time spent on coating a given area of steelwork by the different methods is only partly determined by the actual speed of application. Time to set up and clean equipment will be greater for spraying compared with brushing. Masking may be necessary with spraying and generally paint wastage is higher with this method compared with brushing. Furthermore, the shape and sizes of steelwork influence the rate of application and costs. Spraying and the roller method are most efficient for large areas with easy access, whereas for small inaccessible parts, brush application may be advantageous.

The table below based on experience, gives a broad indication of the areas that, on average, an operator would be expected to cover in a working day:

Method	Area covered/day (m^2)
Brush	100
Roller	200–400
Air spray	400–800
Airless spray	800–1200

10.8.3 Control of painting application

It is, of course, much easier to control paint application in a shop but some site painting will always be required either for touch-up of new work or for maintenance of existing structures. It is always more difficult than shop painting and it is, therefore, essential to provide the best possible protective film initially. This will serve as a sound basis for future maintenance painting. To achieve this high initial protection means that certain controls are required on both the application and on the workshop conditions. The exact requirements will vary with the types of paints being used and controls may need to be more stringent for the higher-duty paints such as epoxies.

These controls are briefly considered below.

(i) *Temperature*

Control of temperature is important for a number of reasons. It influences solvent evaporation, drying time and viscosity, but it is particularly important with two-pack materials where the temperature has a marked influence on the curing time. Temperature also has a secondary influence on relative humidity (see below). Although, in temperate climates, problems generally arise because of the requirements to raise temperatures by heating workshops, in some hot climates problems will develop because of temperatures being high. With some paints this will lead to problems of too rapid curing of some two-pack material, which have been formulated for different conditions, and solvent retention. Paints should be formulated to take account of the conditions where they are being used. In very hot, sunny environments, it may be advantageous to coat dark coatings with a white coating of some type to improve the curing conditions.

In workshops, heating should be by indirect means and methods that produce combustion products inside the shop should be avoided.

(ii) *Relative humidity*

The relative humidity of the air indicates the amount of moisture in the air at different temperatures. Moisture influences painting operations in a number of ways. It may, at high relative humidities, lead to moisture on the surface to be painted and it may affect the curing of paints. In the case of zinc silicate paints, a certain degree of humidity is required to ensure correct curing. The relative humidity requirements for particular types of paint coatings should be given in coating specifications and must be adhered to.

When painting outdoors, trends in relative humidity and temperature may be more important than the instantaneous values that are measured. The relation between relative humidity and the dew point must be taken into account to ensure that surfaces are not too moist for the application of paints. Some of the more conventional oleo-resinous paints can tolerate a certain amount of moisture on the surface but many of the two-pack chemical resistant paints are sensitive to the presence of surface moisture. Although various moisture detecting devices have been developed, they tend to be too sensitive for use in painting, i.e. they cannot easily be calibrated to indicate whether the surface moisture is at a level suitable for painting.

In BS 5493:1977 'Code of Practice for protective coating of iron and steel structures against corrosion', it is advised that coatings should not be applied to surfaces where the relative humidity is such that (a) condensation is present on the surface or (b) it will affect the application and/or drying of the coating. Furthermore, when a *rising* relative humidity reaches a value that would give rise to (a) and (b) then painting should not be started or if already started, should be suspended; work should not be resumed during the time the relative humidity remains at or exceeds that value. The limiting value of relative humidity has to be selected in relation to a number of factors including the thermal inertia of large sections and the probable changes in temperature. It is considered advisable to ensure that the steel temperature is maintained at not less than 3 °C above the dew point.

(iii) *General conditions in workshop*

Workshops must be properly ventilated to keep the concentration of fumes and vapours to a safe level, i.e. below TLV (Threshold Limit Value). Apart from safety and health requirements, a high concentration of solvents may affect the drying of the paint film.

Lighting should also be of a high standard as should general cleanliness, etc.

(iv) *Site application*

For site application of paint, it is possible to control the working conditions by the use of covers but generally the painting periods are chosen to suit the weather conditions. In some areas of the world, e.g. the North Sea, this means that really suitable conditions for painting are limited to certain months of the year. A problem with site application under marine conditions arises from the salt droplets that settle on surfaces. These lead to difficulties in surface preparation and painting. Salt deposits can cause adhesion problems between coats of paint. Furthermore, the presence of chloride salts leads to surface moisture at relative humidities as low as 40%. In practice, the only solution to such problems is to choose painting periods carefully and to maintain a constant check on the weather conditions. Often it may be necessary to paint under conditions that are far from ideal. In such cases there must be a strong probability that the life of the protective system will be lower than that of a similar one applied under good conditions.

10.9 TEMPORARY PROJECTION

Large steel sections are usually either (i) stored and transported in the as-rolled condition and cleaned and coated at their final destination or (ii) cleaned and primed before being transported to the erection site. They are not usually coated with temporary protectives within the definition in

The British Standard Packaging Code BS 1133. This defines 'temporary' in the sense that a coating can easily be removed by a simple solvent, such as white spirit, not as an indication of the degree of protection afforded to the steel. Consequently, when considering temporary protectives, priming coats of standard paints even where they are less than 25 μm in thickness, e.g. blast or weld-through primers, are not included.

The application of oils or greases is the most commonly used method of providing temporary protection for steel. A wide range of products is available. They are usually marketed as proprietary products, but most of them fall into a number of fairly well-defined categories, as specified in BS 1133, Section 6.

  (i) Solvent-deposited type, producing a hard film (designated TP1). These consist of film-forming ingredients dissolved in solvents to give low viscosity liquids at room temperature. On evaporation of the solvent, a thin, tough, abrasion-resistant film is obtained. Such films can be applied by dipping or by spraying to protect simple items either of high finish or with machined surfaces. They are not particularly suitable for batches of items that might stick together. Typical uses for this type of protective would be steel rod and bar or simple tools.

 (ii) Solvent-deposited type, producing a soft film (TP2). Similar to (i) but the films are soft and not initially resistant to abrasion, although in some instances improvements may occur on ageing. Generally, though, this type is not suitable where serious abrasion is likely to occur. It can be used for single items such as small tools and for interim protection of assemblies.

(iii) Hot-dip type (TP3). Unlike (i) and (ii), these are normally applied by dipping in a bath containing the melted material at about 70–90 °C and a thicker coating is achieved. The thicker film provides longer-term protection and is more resistant to damage than (i) and (ii).

(iv) Smearing type (TP4). This is a soft-film type that is smeared on and contains greases, usually containing an inhibitor, which combine the functions of lubricants and protectives. This type can be used for bare parts of painted items as well as for working components such as bearings.

 (v) Brushing or swabbing type (TP5). This is a semi-fluid, soft-film type, generally used where hand application is the only feasible method.

(vi) Non-drying oil type (TP6). Products in this group are lubricating oils containing inhibitors. They often serve a dual purpose as lubricants and protectives. They are generally used for internal protection but can be used for short-term protection of small parts.

(vii) Strippable type, hot application (TP7). These produce a thick coating 1.25–2.55 mm (0.05–0.10 in) thick and resist mechanical damage as well as providing corrosion protection. Unlike the other temporary protectives (i)–(vi), which require solvents to effect their removal, these can be stripped manually. They are more expensive than the others but ease of stripping makes them convenient for many products.

(viii) Strippable type (TP8). Similar to (vii), they can be applied by spraying or dipping at room temperature, providing a thinner coating than (vii).

Strippable types of coatings differ from the other temporary protectives in that they seal the component with a skin that does not adhere to the steel surface. Consequently they can be stripped manually without using solvents.

The choice of protective will depend on many factors, e.g. the complexity of the article, the number to be treated, the possibility of mechanical damage, the amount of handling, the environments and periods of exposure and the associated packaging that is required.

10.10 REFERENCES

(1) LOGAN, A G T, *Proc. UK National Corrosion Conf.* (1982), Inst. Corr. Sci. Tech., London

Other useful references

Painting Steelwork, CIRIA Report 93, Construction Industry Research and Information Association, London (1982)

British Standard 5943:1977, Code of Practice for Protective Coating of Iron and Steel Structures Against Corrosion

British Standard 2015:1965, Glossary of Paint Terms

Coatings and Linings Handbook, National Association of Corrosion Engineers, Houston, Texas, USA

Steel Structures Painting Manual, Vols 1 and 2, Steel Structures Painting Council, Pittsburgh (1982)

11 Metallic coatings

Under most conditions non-ferrous metals are more corrosion-resistant than carbon steels. Generally, however, they are a good deal more expensive and do not necessarily possess the required properties for constructional purposes. The advantages of both steel and non-ferrous metals can often be combined by applying a coating of the more corrosion-resistant material to the steel. Metal coatings may be applied for pure decoration, e.g. silver, or for a combination of decorative and corrosion-resistant purposes, e.g. chromium plating. For marine situations, however, metal coatings are applied to improve corrosion-resistance or for a specific engineering purpose such as hardness. Although many different types of coatings may be used, the most commonly applied to steel used for constructional purposes are zinc, aluminium and cadmium; zinc being the most widely used. Other metals and alloys, e.g. 'Monel', are increasingly being used for critical areas in marine environments. These are, however, not generally used for structural sections.

The method of applying the coating has an influence on a number of important properties such as cost, porosity, thickness and adhesion to the steel.

11.1 METHODS OF APPLYING METAL COATINGS

The common methods of applying metallic coatings to steel are (i) hot dipping, (ii) electroplating, (iii) hot spraying, and (iv) diffusion.

Other methods are employed to a lesser extent; these include cladding, vacuum evaporation, immersion plating, plasma spraying and welding an overlay. The most common application method for constructional steelwork is by hot-dipping and the most common metal applied in this way is zinc, i.e. hot-dip galvanising. Alloys other than steel are also coated with metals, e.g. copper with tin, but for marine situations, the metallic coating of steel is the most important.

233

11.1.1 Hot-dipping processes for coatings

The application of metallic coatings to a bare alloy by hot-dipping is the oldest and in some ways the simplest method. In practice the method is limited to coatings for zinc, aluminium, tin and lead. The method is most generally used for coating ferrous metals, particularly carbon steels.

The principles of the hot-dipping process are straightforward and cover the following stages.

(i) Cleaning of the alloy to be coated by degreasing, and removal of scale, most usually by acid pickling, although blast-cleaning may be preferred for cast and malleable iron. Blast-cleaning may also be used to achieve thicker coatings of zinc on steel.

(ii) Fluxing by immersion in a special aqueous pre-fluxing bath or by immersion through flux on the molten metal in the bath. Both methods may be used in some circumstances.

(iii) Dipping into the molten bath. Small components may be fixed to a jig or placed in a perforated basket. Large steel sections are dipped directly into the bath. Where the section to be coated is longer than the bath, a double dipping operation may be carried out.

(iv) Treatments of coatings are sometimes carried out immediately after the steel has been removed from the bath. This involves reducing the metal adhering to the steel by wiping, centrifuging or by air-blasting while the coating is still molten.

The coating is formed by direct reaction between the basic alloy and the molten coating metal, and this results in the formation of a series of alloy layers with nearly pure metal at the surface. The amount and nature of the alloy layers can be influenced by additions to the bath and the composition of the alloy to be coated, e.g. silicon in steel may affect the thickness and properties of zinc coatings. Sheet steel is hot-dipped by a continuous process and the treatment varies somewhat from that outlined above. This process is used to produce pre-coated steel or strip that is marketed as a standard product. Wire can also be coated by hot-dipping, in a continuous process.

Of the coatings produced by hot-dipping, only those of zinc and aluminium will be considered in further detail as tin, lead and terne coatings (lead-tin) are not generally used for engineering purposes in marine situations.

(i) *Hot-dip galvanised coatings*

The temperature of the zinc bath in hot-dip galvanising is about 460 °C. Higher temperatures, up to 560 °C, are sometimes used for smaller items, such as nuts and washers, to facilitate centrifuging. For these higher temperatures a ceramic-lined pot is required. The structure of the coating after reaction in the bath consists of a series of Fe-Zn alloy layers with the percentage of iron increasing from the surface to the basis steel. Nearly pure zinc occurs at the surface. *Figure 11.1* shows the layers in a typical cross-section of galvanised steel. The thickness of the alloy layers will vary with the type of steel and other factors. The thickness of the coating is influenced to some extent by the condition of the steel surface, the composition and temperature of the zinc bath, the immersion time and the size of the steel being galvanised. Generally, however, the thickness of the coating applied in a particular bath cannot be controlled or altered to any marked extent by variations in the operating conditions. Blast-cleaning the steel before dipping and variations in the composition of the steel are ways of altering the coating thickness obtained with a particular bath.

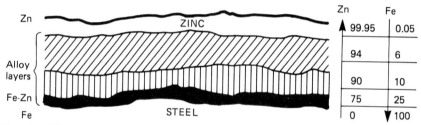

Figure 11.1 Diagrammatic cross-section of a hot-dip galvanised coating, showing approximate percentages of zinc and iron in the alloy layers

When this zinc solidifies, the surface tends to take on a metallic lustre and surface crystals may be formed giving a 'spangle' finish. This type of finish is not necessarily produced with silicon-killed steels but there is no difference in the corrosion-resistance of the coating, which often tends to have a darker matt finish. The presence or absence of 'spangles' provides no indication of the quality of the coating.

Generally, the overall thickness of the coating is greater with silicon containing steels. There is not, however, a simple linear relation between the silicon content of the steel and the thickness of the coating. Immersion time in the bath, the temperature of the bath and the silicon content of the steel—all influence the thickness.

Other alloying elements in the amounts present in structural carbon steels do not have a marked influence on the galvanised coatings, although carbon levels above about 0.3% may lead to increased thickness of zinc coatings. The sulphur level in ordinary steels has virtually no effect on the coating unless it is concentrated at the steel surface. Some free-cutting steels contain about 0.2% sulphur and these may prove to be difficult if not impossible to galvanise.

The composition of the zinc bath is not usually varied for ordinary hot-dipping. Usually electrolytic zinc (99.95% zinc) is used, with small additions of lead (1% max.) and possibly aluminium (under 0.01%) to improve the coating appearance.

Although this book is concerned primarily with the corrosion-resistance of zinc coatings, it is worth noting some points about the physical properties of the coatings. Hydrogen embrittlement, which can occur with plated steels (see Section 4.6) does not generally occur with hot-dipped galvanised steel because any hydrogen absorbed during pickling is removed by immersion in the bath at a temperature of about 460 °C. Hardened steels can become embrittled due to hydrogen diffusion into the steel and they should be tested before galvanising. Again hardened steels should be checked to ensure that intercrystalline cracking is not likely to occur. This can result from penetration of zinc between the boundaries of the grains in the steel during dipping. Ordinary carbon steels are not attacked in this way.

(ii) *Fabrication and design of steelwork for hot-dipping*

Some care has to be taken with steel constructions that are to be hot-dip galvanised to ensure that all parts of the steelwork are coated properly and that distortion is avoided. There are also certain safety precautions to be observed.

Venting is important for enclosed spaces. During pickling, acids may penetrate into the spaces, e.g. through small gaps or pores in welds. When the steel is immersed the temperature of the bath is such that the liquid may vaporise and the pressure may be sufficient to burst the steel construction. Discussions should be held with the galvaniser to ensure adequate venting where this is required.

Distortion can arise if steel of widely differing thickness is combined in the same construction. This arises because of uneven heating in the bath and variation in the expansion characteristics of the materials. Thin material, below about 3 mm, may need stiffening to avoid distortion. Acid traps must be avoided because, if during pickling acid is retained, say at crevices or joints, subsequent galvanising may lead to the sealing of the crevice with zinc. The acid may then corrode the steel at the interface. Generally butt welds are preferable to lap welds. Where lap welds with a large area of contact are used, it may be necessary to provide a vent hole to avoid the risk of explosion.

Pockets where zinc can collect should be avoided and if necessary holes should be provided to ensure that the zinc can flow to all parts of the construction.

(iii) *Hot-dip aluminising*

Although steel strip is coated by hot-dip aluminising, the amount of other types of steel treated by this method is very small compared with hot-dip galvanising. This arises, in part, from the nature of the hot-dip aluminising process.

Aluminium has a higher melting point (600 °C) than zinc and this usually necessitates operating the bath at temperatures over 700 °C. Very rapid reaction occurs between the steel and molten aluminium, resulting in increased dross formation. Furthermore, because of this rapid attack on iron, it is necessary to use a more expensive ceramic-lined tank for the process. Fluxing is also more difficult with aluminium than with zinc and the oxide that forms on the aluminium surface (Al_2O_3) is difficult to remove and particles of it tend to become entrapped in the coating.

The alloy layer (Fe_2Al_5) tends to be thick and very hard, which leads to problems of bending and forming. An addition of about 3% silicon to the bath reduces the thickness and hardness of the alloy layer. Beryllium also has this effect on the alloy layer and in much smaller quantities than for silicon, but toxicity and cost have hindered its general use.

Hot-dip aluminising is mainly used for small components and for continuous strip.

Provided that the aluminium adheres well to the steel, its corrosion-resistance is generally better than that of zinc and its heat-resistance is much better. Usually, however, aluminium coatings for structural purposes are applied by spraying.

11.1.2 Diffusion processes for coatings

Diffusion coatings are produced by reacting a coating metal with the base alloy, usually steel, so that there is diffusion of the two metals, so providing an alloy protective layer on the base metal. Although many coating metals can be used for ferrous metals, chromium, zinc and aluminium are the only ones commonly used for corrosion-resistant purposes. The term 'cementation' is sometimes used for the method where powders of low-melting-point metals such as aluminium or zinc are used.

Diffused zinc coatings on steel are termed 'sherardised', and this method is used to coat comparatively small components such as nuts and bolts. The components after cleaning and pickling are packed in zinc powder or dust, usually with silicon or sand as a diluent and rotated slowly in a steel drum at a temperature below the melting point of zinc, e.g. 400 °C.

The coating process continues for some hours depending on the size of the components and the thickness of coating required. The coating is an

Fe-Zn alloy and has a matt grey appearance. An advantage of this method arises from the very small dimensional changes that occur. This is particularly useful for threaded components such as nuts and bolts.

Aluminium is also coated by a diffusion process similar to sherardising. This is also a cementation process, called 'calorising'. It is used mainly to provide coatings for high-temperature resistance.

Chromium coatings can also be produced by a diffusion process called 'chromising'. Such coatings can be produced from powders or from chromium halide vapours, sometimes a combination of the two methods. Chromised components are not widely used for purely corrosion-resistant purpose in marine environments. They are more frequently used for oxidation-resistance or for special purposes, e.g. prevention of inter-granular corrosion of ferrous and nickel-based alloys in certain environments.

Other metals can be used for diffusion coatings but will not be considered here as they are not related to marine corrosion. A good summary of diffusion processes and coatings is provided in *Corrosion*, Vol 2 (ed L L Shreir).

11.1.3 Coatings applied by spraying

The basic principles of spraying metals on to substrates to produce coatings is straightforward. A metal in the form of a powder or wire is fed through a nozzle with a stream of air or gas and is melted by a suitable gas-oxygen mixture. The melted particles are then sprayed on to the surface to be coated by compressed air or other gas. Wire is now more commonly employed than powder for corrosion-resistant coatings. There is no alloying between the sprayed particles and the surface being coated when application is by gas pistol.

Another method, less widely used for corrosion-resistant coatings, is arc spraying. In this process the basic principle is that two wires insulated from each other are fed to a point where an arc is formed, immediately before the jets of compressed air. The arc-melted metal is broken down to a spray of fine droplets which are projected on to the surface. This method is not used to any extent for zinc coatings and only to a limited extent for aluminium coatings. Because of the high temperatures produced by arcing, some incipient welding may occur between the sprayed particles and the surface being coated. This improves adhesion and is advan-tageous for sprayed aluminium coatings where good adhesion is less easily achieved than with zinc.

Surface preparation
Because the adhesion of sprayed coatings is basically mechanical, it is

necessary to provide a clean roughened surface for the process. This is achieved by blast-cleaning with angular grit. A high standard of cleanliness is required and while cleaning to Swedish Standard Sa $2\frac{1}{2}$ may be sufficient for zinc, a higher standard, Sa 3, is probably required for aluminium.

The surface preparation before spraying is an essential part of the process because this influences adhesion to a marked degree.

11.1.4 Electrodeposited coatings

Electroplating is the most common method of applying metallic coatings for general purposes. For structural marine purposes, the most commonly applied electrodeposited coating is zinc. Cadmium is used to some extent and other metals, such as nickel and chromium, are also used for special situations. The technology of electroplating is well developed and a range of solutions and processes is used, but these will not be discussed in any detail here. Basically, electroplating is similar to the corrosion process in the sense that two electrodes, an anode and a cathode, are immersed in a bath containing an electrolyte. The steel or metal to be plated is the cathode of the cell and the coating metal is the anode. The electrolytes used in plating baths are often complex, containing various salts and compounds in addition to salts of the metal to be coated. Often proprietary solutions are used for plating. Although wire and strip are electroplated in a continuous process, small components such as nuts and bolts are generally plated in batches, usually in a revolving barrel. The cleaning and pre-treatment of steel before plating is essential for the production of good coatings. There is usually no alloying with the substrate.

11.1.5 Other coating application methods

Coatings are applied by methods other than these described above but they have only a limited application to marine situations. Other methods include mechanical or electroless plating, vacuum evaporation and plasma spraying.

Cladding can also be used where a sheet or plate of one metal is rolled on to another. This method is used with stainless steel and other corrosion-resistant alloys rolled on to carbon steel to provide a cheaper product than solid stainless steel for some chemical applications. Corrosion problems may, of course, arise at edges or at damaged areas. Cladding of reinforced bars with stainless steel has been used to reduce corrosion in concrete structures exposed to particularly aggressive conditions. Aluminium has been used to clad aluminium alloys for some marine situations.

Welding sheets of more resistant alloys to steel have also been used widely in marine situations. A typical example of this concerns protection at the splash zone of offshore platforms where Monel 400 is wrapped round the tubes and welded in place to provide a high level of protection. Corrosion-resistant alloys can also be applied by a variety of methods to carbon steel to provide improved corrosion-resistance at the surface. Methods that have been used include, explosive bonding and weld overlays. Weld overlays have been used extensively for marine applications and a recent paper provides information on this method used for Inconel Alloy 625[1]. Basically, the coating metal is produced on the steel surface in the same form as welded metal, i.e. a cast structure, and will, therefore, have some differences in properties compared with the wrought metal. This method has been used for the seal areas of ships' propeller shafts where corrosion-erosion is a problem. It has also been employed for other marine applications such as pump parts, trunnions on lock gates and sub-sea transformer boxes.

11.2 CORROSION PROTECTION BY METALLIC COATINGS

The corrosion protection afforded by metallic coatings depends mainly on the coating metal and the thickness. The method of application has less influence that these other factors although it may have an effect on the type of corrosion that occurs. The coatings that are mainly used to protect steel in marine environments—zinc, aluminium and cadmium—are all anodic to carbon steel. Other coatings such as nickel are cathodic. There is a fundamental difference in the protective action of the two types of coating. At damaged areas or where there is a pore in the coating, the anodic ones will protect the steel, whereas those that are cathodic will intensity the local attack on the steel substrate. In atmospheric situations, zinc performs as an anodic coating to a much greater extent than either aluminium or cadmium, although aluminium tends to provide a greater degree of sacrificial protection when immersed in sea water than it does in the atmosphere. The anodic property of these metals, particularly zinc, is important. It reduces corrosion at scratches and other damaged areas and protects the cut edges of sheet material. Basically, though, the metal coating has to withstand the marine environment to provide an economic solution to the protection of steel.

The method of application has both a direct and indirect effect on corrosion performance because the thickness of coatings is related to the application method.

The most straightforward method is electrodeposition where, in effect, a layer of the metal protects the steel. Usually it is fairly thin, about 25 µm, and reacts with the atmosphere or sea water in a manner similar to that of the wrought metal until the coating is corroded away, leaving bare steel. The actual form and rate of corrosion for the different metals will be considered later. Initially hot-dip coatings behave in much the same way as electroplated coatings but the coating structure is different and as the virtually pure metal corrodes away, an alloy layer is uncovered. This continues to provide about the same level of corrosion-resistance as the pure metal but, with zinc, the metal most widely used in hot-dipping, the Fe-Zn alloy layer produces a rust-stained appearance. Although it is not really similar to the rust formed on carbon steel, it can be mistaken for corrosion of the steel substrate, which is in fact being protected. The period before such rust staining occurs will depend upon a number of factors not least the thickness of the alloy layer. With zinc diffusion coatings, this darkening or rust staining occurs soon after exposure because the whole coating is essentially an Fe-Zn alloy layer.

It should be emphasised that Fe-Zn alloy layers are protective despite their appearance and for some purposes, e.g. on steel sheets, are deliberately formed because they provide a sound basis for paint.

Other metals and alloys such as chromium and nickel are used for reclamation of engineering components and for coatings requiring specific mechanical properties; the only two coatings used for marine structures are aluminium and zinc.

Spraying is the only method of obtaining aluminium coatings on structural sections and zinc also is sometimes applied by spraying. The deposited coating is porous and consists of small particles of about 100 µm in diameter that have a flattened appearance resulting from the action of striking the surface. The pores tend to fill with corrosion products on exposure and in the case of zinc this can lead to problems if the sprayed coatings are painted, particularly if the zinc coating is exposed to a marine atmosphere for a time before painting. Where the coatings are exposed bare, the corrosion products do not cause such problems. Often they are insoluble, in which case the coatings can protect for a long time. In other situations, as with all zinc coatings, the corrosion products are soluble and this results in a reduction in the period of protection. The aluminium particles are surrounded by oxide coatings and the aluminium coatings tend to be somewhat more porous than zinc sprayed coatings. The oxide coating is not, however, continuous around the particles and the aluminium coating has a reasonable conductivity. In most atmospheric environments, zinc provides more sacrificial protection to steel than does aluminium but under immersed conditions, in sea water, aluminium provides effective sacrificial protection at scratches, etc.

In the early stages of exposure of sprayed aluminium coatings, the moisture penetrating the pores may lead to slight corrosion of the basis steel because the oxide covered particles act cathodically to the steel resulting in some rust staining. This is, however, only a short-term initial reaction and the pores fairly rapidly become filled with insoluble aluminium corrosion products. These corrosion products may be stained with the initially produced rust and though this can be unsightly, it does not affect the protective value of the aluminium coating to any significant degree.

Sprayed aluminium coatings can be used for oxidation resistance up to about 900 °C but the coating has to be sealed and heat-treated.

The protection afforded by both aluminium and zinc sprayed coatings depends upon the thickness and the relationship between thickness and time is reasonably linear. Generally the coatings are applied to a thickness of about 100 μm but thicker coatings can be applied to increase the period of protection.

Sprayed coatings are most advantageously applied to fairly large plain areas. They are less suitable for small components and sections with a large surface area/volume. They have been used for many structures including bridges, sheet piles and buildings.

11.2.1 Additional protection to sprayed coatings

The rough nature of sprayed surfaces leads to the collection of dirt and corrosion products which becomes unsightly. To improve the appearance and to add to their overall protective value, it was common practice a few years ago to paint them, e.g. the Forth Road Bridge in Scotland was zinc-sprayed and painted. This method was considered to be sound, provided that the paint coating was applied to the zinc coating before exposure to the atmosphere. Zinc is a reactive metal and problems were experienced in some cases where the zinc coating had corroded to some extent before being painted. The paint coatings failed prematurely and this resulted in a move away from the painting of metal coatings. The general view nowadays is that sprayed coatings should be sealed rather than painted.

A range of sealers is available and these materials should be applied immediately after spraying. In BS. 5493, it is recommended that pre-treatments based on two-pack polyvinyl butyral, pigmented with zinc tetroxy chromate, should be applied to sprayed coatings before sealing. Sealers include the following types of material (either pigmented or clear):

Two-pack epoxy
Two-pack polyurethane
Blend of vinyl chloride/acetate
Copolymers

The sealers are applied until absorption is complete; they are *not* an additional coat of paint.

11.2.2 Painting of galvanised surfaces

In some situations hot-dip galvanised coated steel is exposed with no additional coatings. At many inland atmospheres, protection of the steel for over 20 years is possible with galvanised coatings of sufficient thickness. However, in more aggressive environments, additional paint coats are applied. There is evidence to show that a paint system applied to galvanised steel will provide improved performance compared with a similar system applied to blast-cleaned steel.

There can, however, be problems with adhesion of the priming coat, particularly to newly galvanised steel. General recommendations have been made in a publication jointly issued by the Paint-makers Association of Great Britain and the Galvanizers Association[2]. These are summarised in *Table 11.1*. The final finishing coats will be chosen to suit the environmental conditions, but the chosen primer must be compatible with the finishing system and special coatings for the purpose are becoming available[2].

11.3 CORROSION DATA FOR METALLIC COATINGS IN MARINE ENVIRONMENTS

There is a large volume of data on the performance of zinc metal and zinc coatings in a range of atmospheres. This is summarised in a study commissioned by the International Lead Zinc Research Organisation and called 'Zinc: Its Corrosion Resistance'[3]. Some of the data is reproduced below. Less data is available on the performance of aluminium coatings but it clearly indicates the excellent overall performance of these coatings. In particular, the large scale tests on flame-sprayed metals carried out by the American Welding Society provides soundly based data over a period of more than 19 years for both aluminium and zinc sprayed coatings. Aluminium-zinc coatings have been tested both as a duplex layer where zinc is applied to the steel substrate and aluminium is applied to the zinc (a system still used in some shipyards), and as an alloy of the two metals. There is now a commercial

Table 11.1 General recommendations for painting galvanised steelwork (based on Reference (2) on p 252)

Galvanised surface	New	Weathered	Variable condition[3]
Surface condition	Clean[1]	Zinc corrosion products present dull, grey appearance—no rusting of steel substrate	Surface variable but no rusting of steel
Surface preparation	Dry and clean lightly abrade if necessary	Remove loose corrosion products Wash with clean water to remove soluble zinc salts Dry clean surface	As with weathered surface—abrade
Pre-treatment	Etch primer or 'T' wash	Usually nil	Etch primer or 'T' wash
Typical primers[2]		Zinc phosphate Zinc chromate Metallic lead Calcium plumbate	

Note (1) If chromate treatment has been applied, consult paint manufacturer.
     (2) Generally, with calcium plumbate primers, no pre-treatment is required.
     (3) If the steel substrate has rusted, consult paint manufacturer.

process for applying an alloy coating consisting of 55Al 43.4Zn 1.68Si to steel sheet. This was developed by the Bethlehem Steel Corporation in the United States and is marketed under proprietary names. Thirteen year data from tests on these materials will be considered and compared with zinc and aluminium coatings[4] in Section 11.3.3.

11.3.1 Zinc coatings

A general indication of the corrosion of zinc coatings is given in *Figure 11.2* based on BS 5493:1977. Other data for marine atmospheres has been summarised[5] and a corrosion rate of 5 μm/y is considered to be a general figure although some investigators quote figures of 2.5 μm/y as a general figure. This would provide the following lives for typical coating thicknesses produced by the various application methods.

Application method	Thickness (μm)	Life of coating (y)
Spraying or grit blasting before galvanising	200	30–50
Spraying or galvanising	100	15–25
Galvanised sheet or sherardising	25	4–5
Electroplating	5	0.25–1

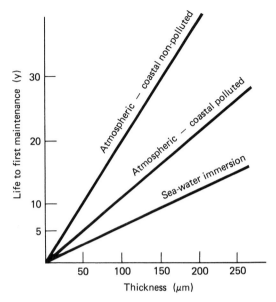

Figure 11.2 Typical lines of zinc coatings in marine environments (based on BS 5493:1977)

This data indicates the influence of thickness on the life and in particular on the temporary nature of protection afforded by electroplating and the limited life for sherardising. It follows that both should be painted to obtain reasonable lives in marine atmospheres.

The term 'atmospheric marine environment' provides no more than a broad indication of the nature of the conditions. In some marine environments, lower corrosion rates than those quoted above have been obtained. On the other hand, higher rates may be experienced, particularly close to the sea itself. At Kure Beach, North Carolina, 25 m from the sea, corrosion rates of 6.0 µm/y have been measured and at Galeta Point Beach, Panama, rates of 15.9 µm/y have been reported. This is unusually high but at tropical surf beaches, where salt droplets remain on the wet metal surface for prolonged periods, and where inhibiting magnesium salts are not present in the sea water, severe corrosion can occur.

Where the environment can more correctly be described as marine-industrial, e.g. where a plant is operating in a coastal situation, the corrosion may be determined more by the industrial pollution than by marine contamination. In such situations, corrosion rates may be double those anticipated for non-polluted marine situations.

In tests carried out on zinc-coated sheet[4], the corrosion rates 25 m from the ocean, compared with 250 m from the ocean, were about

4 µm/y and 1.5 µm/y, respectively. Under fully immersed conditions in sea water corrosion rates between 10 and 25 µm/y have been measured. Probably a rate of about 20 µm/y is a reasonable general figure. In tests carried out by the American Welding Society[6], corrosion rates were not determined, but the general performance of sprayed zinc coatings and sprayed aluminium coatings is discussed below. Readers are advised to study the actual report for details of the performance of the coatings and a description of test sites and techniques.

Although the performance of the coatings, particularly the aluminium, was very good in these tests, improved sealants have been developed since the test programme was started and some authorities consider that an etch primer prior to sealing is not always necessary.

At the end of 19 years' exposure in three general atmospheric marine environments, described in the report as 'salt-air', and at one environment described as 'severe marine', the overall results of the tests on sprayed zinc coatings can be summarised as follows:

Marine environments (including Kure Beach 250 m from the sea)
Coatings of 80 µm thickness exhibited corrosion as pinpoints of white rust on the zinc but the steel substrate was not attacked.

Where a sealer coat had been used this showed signs of attack or even virtual failure, but the zinc coating was intact. The surface condition was similar for thicker zinc coatings up to 0.38 mm thickness.

Severe marine environment (Kure Beach 25 m from the sea)
The coatings of 80 µm thickness had not protected the steel for this period. The steel was 75% exposed on the front of the specimens and rusting. It was 10% exposed on the back and rusting. With coatings of similar thickness (80 µm) but sealed with a wash primer followed by an aluminium vinyl coat, the steel substrate had corroded only at the edges of the specimen although the sealer coat had broken down and zinc corrosion products were evident. However, where a two-coat chlorinated rubber sealer coat had been used, this had broken down and 90% of the steel substrate was rusting on the fronts of the specimens and 20% rust was reported on the backs.

On specimens with thicker coatings of zinc, the effect of the chlorinated rubber sealer coat appeared to produce results inferior to both unsealed zinc spray and the coatings treated with the wash coat and aluminium vinyl. On the 150 µm coatings, no rusting occurred on either bare zinc or those sealed with the aluminium vinyl but the base metal was 5% rusted on the chlorinated rubber sealed specimens.

No attack of the steel occurred on any of the specimens where the thickness of the sprayed zinc coating was greater than 150 μm, although there was evidence of zinc corrosion products on some specimens.

Sea water immersion
As would be anticipated, over 19 years' test, the sprayed zinc coatings were no longer protecting the steel and in some cases the steel itself had been severely corroded. The chlorinated rubber sealer had not been particularly effective. At zinc thicknesses of 0.30 mm and over, although the zinc had corroded, there was no attack on the steel. Where the chlorinated rubber sealer coat had been applied, 25–50% of the steel was rusting.

Alternate exposure to atmosphere and sea water
Coating thicknesses up to 0.23 mm did not protect the steel for the 19 year period of the test but coating thicknesses of 0.30 mm and over did so, although the zinc exhibited thick white corrosion products and pitting.

11.3.2 Aluminium coatings

Test results for aluminium coated sheets indicate a low rate of corrosion in marine environments. In tests carried out at Kure Beach (25 m from the ocean), the corrosion rate over 13 years was under 0.5 μm/y (compared with about 4 μm/y for zinc coatings). At 250 m from the ocean, the corrosion rate was about 0.20 μm/y.

Other tests, mainly on sprayed aluminium coatings, have demonstrated their durability. The American Welding Society tests carried out at the same time as those on the zinc coatings, quoted above, demonstrate their superiority over zinc. The results can be summarised as follows:

Marine environments
Even with coatings of 80 μm thickness, the aluminium without a sealing coat protected the steel substrate for the 19 years of test, although there was black discolouration and some modification of sealers where they were used.

Severe marine environment
With no sealer coat, the 80 μm coating of aluminium protected the steel although there was some slight rust staining on the backs of the specimens and some evidence of corrosion of the aluminium. Where either the wash coat—aluminium vinyl or the chlorinated rubber sealer coats—were

used, there was no attack on the steel. There was some discolouration of the sealer coats.

Sea water immersion

In the tests, the aluminium coatings of 80 μm thickness were not tested without sealing coats but with this additional protection there was some rust, 3–5% on the specimens. With a coating thickness of 150 μm, the sealed aluminium showed some evidence of corrosion of the coating but no corrosion of the steel substrate. Aluminium coatings (150 μm) without the sealers did not completely protect the steel and red rust staining was evident over 50% of the surface of some specimens. Sealed coatings of aluminium protected the steel. At coating thicknesses up to 0.46 mm, there was rust staining of some specimens on unsealed aluminium.

Alternate exposure to atmosphere and sea water

The aluminium coatings at all thicknesses from 80 μm with sealer coats, protected the steel but without the additional sealing, the sprayed aluminium coating tended to blister with some attack on the steel under the blisters. This occurred with all coating thicknesses up to 0.46 mm although there appeared to be little loss of steel.

11.3.3 55% Al-Zn coatings

As would be expected, coatings of this composition performed at a level between aluminium and zinc. The results from tests on sheet metal coated with this alloy have been published[4]. Comparative tests on zinc and aluminium coatings were also carried out and these are shown in *Figure 11.3*, based on Reference 4. However, the results on the three types of coatings are not directly comparable; full details on the test should be studied before using the results as design data.

11.3.4 Cadmium

Cadmium coatings are used mainly for small components. The corrosion rate in marine environments has been established in various test programmes[7],[8] and has been stated to be about 1.3 μm/y. It is generally less than zinc in marine atmospheres but higher in industrial ones. It follows that the nature of the marine conditions would have a bearing on the corrosion rate.

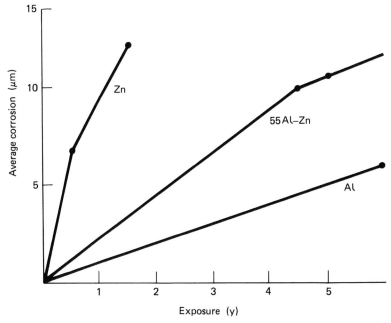

Figure 11.3 General performance of zinc and aluminium coatings in marine atmospheres (based on Reference 4)

11.4 SELECTION OF METAL COATINGS

A summary of metal coatings used for steel is given in *Figure 11.4*. The selection of metal coatings falls into broad groups:

(i) As the main protective coating for structural members.
(ii) As a coating for smaller components such as fasteners.
(iii) For non-ferrous metals and special purposes, e.g. valves.
(iv) For critical areas in marine environments.

(i) The choice for structural members is made between aluminium and zinc. Although aluminium coatings provide better protection than do zinc coatings in marine atmospheres, other considerations have to be taken into account. Aluminium can be applied only by spraying, whereas there is a choice of hot-dipping with zinc. Furthermore, the standard of surface preparation is probably more important with aluminium than with zinc. In atmospheric conditions, zinc tends to protect at scratches and damaged areas to a greater extent than does aluminium. Hot-dip processes also have some advantages over spray methods because the probability of obtaining a high standard of coating is higher. Therefore, for many purposes zinc is preferred to

Application method		Metallurgical bond	Metals generally applied			General coating thickness (μm)
			Structural sections	Strip	Components	
Metal / Alloy layer / Steel	Hot dipping	Yes	Zn	Zn Al Al-Zn	Zn Al	100 – 200
Metal	Spray	No	Zn Al	0	Zn Al	100
Metal	Electro deposition	No	0	Zn	Zn Cd	5
Alloy layer	Diffusion	Yes	0	Zn	Zn Al	25
Metal	Cladding (rolling)	No	Stainless steel on steel Aluminium or Al alloys	0	0	Varies
Metal / Alloy layer	Weld overlay	Yes	0	0	High alloys on steel	Varies
Metal	Envelope sheet cladding	No	High alloys e.g. Monel	0	0	Varies

Figure 11.4 Summary of metallic coatings applied to steel

aluminium although if the coating process is properly controlled, aluminium is likely to provide a greater measure of protection than zinc. A disadvantage of hot-dip galvanising is the size limitation on its use. Large dipping baths are available and, furthermore, double-dipping can be carried out which allows the coating of sections of much greater length than the baths. However, the method is clearly less flexible than metal spraying.

(ii) For smaller items, such as fasteners, zinc is generally the preferred choice not least because coating facilities whether by hot-dipping, electroplating or diffusion are readily available. Aluminium is not widely used as a coating for fasteners and other small components except for high temperature purposes. An alternative coating to zinc for components used in marine environments is cadmium, which corrodes at a lower rate but is usually only applied at coating levels of 25 μm or less. It is, however, a good deal more expensive than zinc. Cadmium is used for instrument components and in aircraft, in part because it is easier to solder, but not for general marine engineering purposes. However, it is commonly used for high-strength steel fasteners used with aluminium alloys. Nickel and chromium coatings are used, although not primarily for their corrosion-resistance, e.g. for wear-resistance. Both coatings are cathodic to steel so they must be pore-free for long-term service in marine atmospheres.

(iii) Metal coatings are also used to coat non-ferrous metals for specialised purposes, e.g. cadmium, and sometimes zinc, may be used on copper alloys to reduce bimetallic corrosion effects where such alloys may be in contact with aluminium or magnesium alloys.

(iv) Corrosion-resistant alloys, e.g. Monel, are being used for critical marine areas. The best known example is at splash zones on offshore structures where sheets of Monel are wrapped round tubular members and welded into place. Other alloys such as cupro-nickels can also be used for similar purposes.

The other development is in the welding of comparatively expensive alloy overlays on small components for particularly aggressive situations. Clearly there is a wide range of materials that can be used in this way. The use of high-alloy material in these ways is likely to find increasing applications as an alternative to paint and plastics coatings for comparatively small areas where the additional expense can be justified.

11.5 WELDED AREAS

Problems may arise with metallic coatings at field welds. These have to be cleaned and coated on site and clearly it is not possible to carry out any of the usual methods of application except possibly spraying. Although this may be done in a few cases, generally the weld areas are painted, usually with a zinc-rich paint but, good though such paints may be, this does tend to reduce the overall advantage of metallic coatings in situations where long-term protection without maintenance is required.

Weld areas of hot-dip galvanised structures can be protected by the application of a fusible stick (somewhat similar to solder) over the cleaned area; this method is not widely used.

11.6 GENERAL COMMENTS

The potential life of metallic coatings in marine atmospheres has been demonstrated in tests but the same standard of coating application is not always achieved in practice. There is a clear correlation between thickness of coating and its life, so any areas that are not properly coated will fail prematurely; this is a problem on the edges of steel sections where the coatings are often comparatively thin. Damaged areas can also lead to breakdown of a coating and in the atmosphere, aluminium protects steel to a lesser degree than does zinc at such areas. For sprayed coatings, a high level of blast-cleaning is essential, particularly for aluminium if long-term protection is to be obtained.

Hot-dip galvanising is a process which—to some extent—has controls built into it and for that reason is less likely to cause problems than sprayed coatings, which have something in common with paint in that a high level of quality control and application supervision is essential to achieve long-term performance.

Two useful booklets on inspection of coatings are available. They cover the requirement for the production of coatings of a high standard, which is essential for long-term protection in marine environments. They are *Inspection of Sprayed Aluminium Coatings* (Association of Metal Sprayers, Birmingham), and *Inspection of Zinc Sprayed Coatings* (Zinc Development Association, London).

11.7 REFERENCES

(1) POWELL, G A and DAVIS, R V, *Proc. UK National Corrosion Conf.* (1982), p 71
(2) *Recommendations for Painting Galvanised Steel*, Paintmakers Association of Great Britain, London, or The Galvanizers Association, London
(3) SLUNDER, C J and BOYD, W K, *Zinc: Its Corrosion Resistance*, Zinc Development Association, London (1971)
(4) TOWNSEND, H E and ZOCCOLA, J O, *Materials Performance*, **18**, 10 (1979)
(5) SHREIR, L L (ed), *Corrosion*, 2nd ed, Butterworths, London (1976)
(6) *Corrosion Tests of Flame-sprayed Coated Steel, 19 Year Report*, AWS C2, 14–74, Amer. Welding Soc., Miami, Florida, USA
(7) BIESTEK, T, *Proc. First International Congress on Metallic Corrosion* (1962), Butterworths, London
(8) CLARKE, S G and LONGHURST, E E, *Proc. First International Congress on Metallic Corrosion* (1962), Butterworths, London

Attaining the potential performance of coating systems in marine environments

12

It is difficult to predict accurately the performance of alloys in all circumstances, particularly in marine environments. Localised corrosion may occur because of changes in the environment that could not be foreseen in the early stages of design. Velocities and temperature of sea water cannot always be clearly defined when alloys are being selected and the materials in one part of a construction can have a marked effect on those in other parts. Furthermore, different design teams may be working on different parts of a plant or construction; consequently, the metallurgical, engineering and corrosion requirements for any situation are rarely so carefully designed and implemented so as to avoid all problems in practice. This is equally true of coatings. In fact, there are probably more problems with protective systems than with other materials because they involve so many different parties and processes in the overall coating programme. Although coating failures rarely lead immediately to serious structural problems, they do result in considerable financial burdens from loss of production and additional maintenance costs.

The process of coating steelwork, particularly by paints, involves a knowledge of the various types of materials suitable for particular environments and the correct methods of storing, preparing and applying such materials. However, to attain the potential performance from such materials, particularly in aggressive marine situations, it is essential that all stages of the protective process are carried out properly. A weakness in the early stages of the coating process will undoubtedly lead to inbuilt weaknesses in the whole system, out of proportion to the original defects.

There are three important ways of reducing the likelihood of problems and ensuring that the probability of success is at a high level. These are:

(i) Proper consideration of the conditions and the choice of an adequate coating system to withstand such conditions. This also involves careful attention to the design of the construction to ensure that

253

steelwork can be properly coated initially and maintenance repainted subsequently. Additionally, any design features likely to reduce the performance of coatings, e.g. water traps, should where practicable be altered to avoid such problems.

(ii) The preparation of clear and concise coating specifications.

(iii) The application of adequate quality control measures.

Chapters 5, 9, 10 and 11 of this book deal with many of the requirements of (i). This chapter deals with specifications and quality control for coating systems. The main factors that determine the protection afforded by a coating system is summarised in *Figure 12.1* of these factors, all but the environment can be controlled by selection, specifications and inspection.

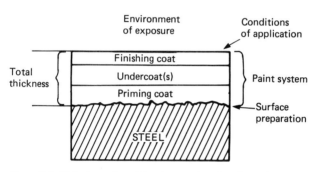

Figure 12.1 Main factors determining the protective value of a coating system

12.1 SPECIFICATIONS

The term 'specification' has a number of meanings in relation to coating systems. It is often used synonymously with the term 'standard' to indicate a document that has been prepared by a national body or a group representing a specific industry or governmental authority to cover certain requirements, e.g. British Defence Specifications or Steel Structure Painting Council specifications. These documents are generally prepared by a group of specialists to provide minimum standards for materials or processes. Usually such documents include methods of checking such requirements, e.g. analytical or test methods. Such standard, specifications or codes of practice—as they are sometimes called—are important as representing accepted and agreed levels of requirements. The term 'specification' also has a clear meaning in relation to contract documents where it has legal implications, but it is also used to denote a set of documents prepared to communicate requirements regarding the quality of materials and workmanship for a specific project. It is in this context that the term is being used here.

The function of a coating specification can be summed up as follows:

(i) To provide the detailed requirements for materials, workmanship and processes for the coating application.
(ii) To provide a complete technical document for the material suppliers, contractors and all other parties involved in the work.
(iii) To provide an accurate basis for costing and tendering.
(iv) To reduce to a minimum disputes and misunderstandings regarding the coating requirements.
(v) To provide a document for communication between all those concerned in the work.

As the technical coating specification will contain information required by many parties, it should be given wide circulation. It certainly should not be treated as a confidential document to be seen by as few people as possible. It must be prepared to provide a concise and clear statement of all the requirements. It should be devoid of ambiguity. Any tendencies in that direction should be strenuously avoided as ambiguities may well lead to delays, increased costs and eventually a system of a lower standard than anticipated. There is also the possibility that the matter will end up with litigation to determine the actual meaning of the specification.

The main areas to be covered by a coatings specification are noted below. Although the general pattern is indicated, it is not possible to cover all the clauses in detail. Furthermore, some coatings—particularly those that are factory applied or of a specialised nature—will require a somewhat different approach.

12.1.1 Surface preparation of steelwork

This is probably the most important single item of the specification and it must indicate quite clearly the requirements. Clauses such as 'Remove all rust and scale by thorough wire brushing' are pointless and although 'Removal of loose ruse and scale by wirebrushing' is more realistic there is still a vagueness in the sense that 'loose' is not very precise. A clause on the lines of 'Remove all rust and scale by suitable methods' is in one sense acceptable but the methods are left to the contractor and if he decides to keep his tender price low by using manual methods such as wirebrushing, problems will amost certainly arise in marine situations.

The following clauses should be suitably covered in the specification:

(i) Degreasing with appropriate methods specified: British Standard CP3012 or Steel Structures Painting Council SP1.
(ii) Removal of rust and scale.

There are a number of standards covering the requirements for cleaning steel and these should be specified, e.g. Swedish Standard Sa $2\frac{1}{2}$, blast-cleaning. Clauses such as 'Blast clean to Swedish Standard', without any further details is of little or no value. If the initial state of the steel, e.g. whether it is badly pitted, is important, then any limitations should be stated. The Swedish Standard covers four grades of steel surface condition prior to cleaning, and this may be a useful reference point.

In some situations, particularly under immersed sea water conditions or under conditions of condensation in marine situations, salts such as ferrous chloride remaining in pits after dry blast-cleaning may adversely affect coatings and their removal may be considered as a requirement in the specification, e.g. by washing or by the use of wet blast-cleaning.

The profile produced by blast-cleaning may be important and if so should be specified. A general 'rule of thumb' requirement is that the profile should not be more than about one-third of the total thickness of the paint system. So for a total film thickness of 220 μm, the maximum profile after blast-cleaning would be about 75 μm. The general method of blast cleaning may be specified and where required any limitations on abrasive type and size.

Sometimes a standard panel is blast-cleaned to the required standard and used as a reference; this may be specified.

Where steelwork is to be metal sprayed after blast-cleaning, suitable standards can usually be referred to, say, British Standard BS 2569 'Sprayed Metal Coatings; Part 1 Protection of Iron and Steel by Aluminium and Zinc Against Atmospheric Corrosion'. Where acid pickling is employed for the surface preparation of the steel, it is usually as part of a process for electroplating or hot-dip galvanising and may be included as part of any standard used for the overall process. The Steel Structures Painting Council Specification SP-8 covers pickling and may usefully be quoted in appropriate situations.

Flame-cleaning may be specified for maintenance painting, but not usually for new work. The Steel Structures Painting Council Specification SSPC-SP4 'Flame-cleaning of New Steel' has been discontinued in the latest copies of their Painting Manual, but is still available and may be considered as a reference document.

Because of its variability, there are difficulties in producing really satisfactory standards for manual cleaning, e.g. scraping and wire-brushing. The Swedish 'St' Standard and The Steel Structures Painting Council Specifications—SSPC-SP2 'Hand Tool Cleaning' and SSPC-SP3 'Power Tool Cleaning'—refer to manual cleaning methods although SP2 and SP3 are based on the Swedish Standard requirements. These standards refer to photographs of rusted steel that has been manually cleaned. They are virtually the only standards available but in practice the photographs may prove to bear only a limited resemblance to many types of rusted steelwork. This should not, however, prove to be a serious problem with new steelwork for marine conditions because manual methods of cleaning are not likely to be used for important structural parts. For maintenance work, however, where the standard of cleaning of rusted steel may require to be specified, there appear to be no really satisfactory methods other than 'removal of all loose rust, scale and other contaminants'. Other points that may be considered as specification requirements include the following:

(i) Exclusion of steel tools for situations where sparking may lead to a risk of explosion. Bronze tools may be specified.
(ii) Avoidance of burrs, gouging, etc, arising from lack of care in use of power tools.

12.1.2 Coating systems

The specification of the coating system must cover more than a broad indication of the type of coating, e.g. 'Chlorinated rubber system'. It should cover the following general requirements:

(i) Blast primer (where required).
(ii) Priming coat.
(iii) Undercoat(s).
(iv) Finishing coat.

together with the dry film thickness requirements for each coating and the total dry film thickness (dft) for the system.

Details of remedial action for damage during handling, transport, etc, should be included. There are varying opinions on the detail of the coating systems that should be included in the specification. Some specifying authorities list the coatings in detail, including proprietary products, whereas others prefer to indicate the type of paint, leaving the contractor to make the final choice. There is sometimes a compromise with a list of, say, three manufacturers, any of which may supply the paint. Some authorities hold the view that too much restriction on the contractor's choice may remove some of the liability inherent in legal contracts. It is advisable to specify areas where additional 'stripe' coats are required, e.g. edges and welds.

Where metallic coatings are used, the thickness of sprayed coatings can be specified but generally hot-dip galvanised and electroplated coatings are covered by standards which can be included in the specification. The thickness of diffusion coatings, e.g. sherardised, cannot usually be specified with any precision.

12.1.3 Paints

Specifications should cover storage, preparation for use and, where required, testing procedures. Where specific paints have been listed, it is advantageous to indicate acceptable alternatives to ensure a continuing supply of paint in case problems arise with supplies of the specified material.

12.1.4 Application of coatings

This should cover the following general requirements for paints:

(i) Method of application, e.g. airless spray.
(ii) Working conditions, e.g. temperature, relative humidity, lighting in paint shop.
(iii) Handling and storage.

The data sheets supplied by paint manufacturers should be referred to and the paints should be applied in strict accordance with the manufacturer.s requirements. For other types of coating such as metal coatings and specialised plastic coatings the application requirements are often not specified by the user. Such coatings are either marketed as part of the product, e.g. coated strip and tubes, or the process itself is so carefully controlled that the only real checks of its effectiveness is to carry out suitable tests on the coatings themselves. Specifications for wrapping tapes are usually in accordance with the manufacturer's instructions.

12.1.5 Handling, transport and storage

A good deal of damage can occur on painted steelwork during its handling and transport from the paintshop or fabricator's works to the final erection site. It is, therefore, important to specify the procedures that will reduce damage to a minimum and—where it does occur—to ensure that the paintwork is suitably repaired to, as far as is practicable, its original state. The specification should cover the following aspects:

(i) Suitable slings (*Figure 12.2*), protection for chains, and where appropriate special harnesses.
(ii) Suitable drying of coatings, particularly where solvent-evaporation type coatings such as chlorinated rubber are used. If such coatings are carelessly handled or coated steelwork is stacked before a suitable time has elapsed to allow adequate drying, considerable damage may occur.
(iii) Correct stacking of painted steelwork.

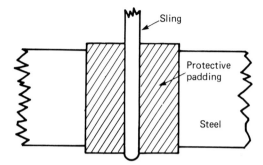

Figure 12.2 Protection of steelwork against damage by slings

(iv) Where appropriate welding of special lifting lugs on to large steel constructions to minimise damage (*Figure 12.3*).
(v) Suitable wrapping of tubes, sections and components.
(vi) Proper protection and supports where steelwork is to be carried as deck cargo and proper lashings and stackings for hold cargo.

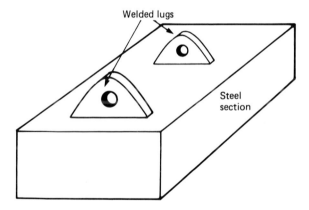

Figure 12.3 Lugs welded to heavy section to avoid damage during handling

Sound specifications are essential if protective coatings are to achieve their potential. However, no matter how carefully specifications are prepared, they will serve little purpose if there is not strict adherence to the requirements that are included. It is, therefore, necessary to have proper quality control procedures.

12.2 PERFORMANCE SPECIFICATIONS AND GUARANTEES

The above discussion has concerned what are often termed 'Method Specifications', i.e. the methods for the different process requirements are specified. An alternative form of specification is based on the required performance from the coating system and is called a 'Performance Specification'. It is quite common for some form of guarantee to be involved with performance specifications. For example, the specification may call for a six-year life for a protective system, with a guarantee to cover remedial work should the system fail to meet the life requirements.

Performance specifications clearly have a number of attractions:

(i) The total protective costs are known for the period covered by the specification.
(ii) There is no requirement to produce a detailed method specification.
(iii) There is a certainty of performance in the sense that any premature failures are dealt with at no cost to the client.

In practice, however, the system has not always worked in such a straightforward manner. Many such specifications do not, in fact, provide the complete 'guarantee' of performance but may cover only a percentage of the costs determined by the point at which failure occurs. For example, if the period covered is six years and failure occurs after four years, then often only two-sixths of the costs of repair of the system are covered, the client meeting the other 66% of the total. Clearly, this will be determined by the nature of the agreement but rarely does the 'guarantee' cover costs involved other than the actual repainting, e.g. loss of production or inconvenience caused by the requirement for early maintenance.

Other problems also arise with performance specifications:

(i) The 'life' offered is often below the anticipated potential life of the system.
(ii) Problems arise in determining the criterion of failure at which the agreement for repainting will come into force.
(iii) The responsibility for failures arising from mechanical damage is often a problem in relation to the agreement.
(iv) Additional costs arise because of the insurance element in this type of specification.
(v) There may well be a delay in completing the painting of the structure because the painting contractor may be prepared to work only under virtually ideal conditions. Often this has resulted in the 'guarantee' clause being waived on certain, often critical parts, of a structure.

(vi) Because of the number of parties involved in protecting steelwork, delays often occur in settling claims made after the premature failure of a coating.

Although performance requirements may operate satisfactorily in some situations, clearly the agreements must be couched in unambiguous terms to avoid the problems noted above.

12.3 QUALITY CONTROL OF THE APPLICATION OF COATINGS

Quality control of coatings application is necessary to obtain the highest standard of performance on any coated steelwork. Although it is not a requirement that is specific to steelwork exposed in marine situations, because of the aggressive conditions that have to be withstood, it has become a matter of considerable economic importance in this particular field. The developments in offshore technology have resulted in the formation of many specialised companies (particularly in the United Kingdom) that carry out quality control work on coatings, or 'inspection' as it is generally called.

Quality control of materials and processes is commonly applied and can be considered as an essential requirement for optimum performance. Clearly, the correct production, heat treatment and forming of alloys is a necessary prerequisite for satisfactory materials used in plant and structures. Many standard tests exist to check composition, tensile strength, impact resistance, etc, and welding inspection is considered to be essential for important elements in construction.

It is, however, more difficult to specify control requirements for coatings. By its nature, the coating process has to be controlled at every stage and it is often difficult, if not impossible, to deduce from examination of the completed work what had occurred at an earlier stage. Furthermore, unlike most industrial processes, coating is susceptible to operator error and adverse environmental influences at every stage of the operation; often these are 'hidden' by the later stages of the work. Clearly it is easier to control coating operations that take place in the shop and in situations where a minimum reliance is placed on the operator, e.g. the painter. Hot-dip galvanising is in many ways 'operator-proof' and, while faults can occur, they are less likely than with paints or sprayed metal coatings.

To determine the approach to quality control, it is necessary to consider some of the factors that lead to coating failure. In this context failure is taken to mean that the coating fails to attain its potential performance. Although coatings do sometimes fail fairly rapidly, more often poor application procedures tend to reduce the life of the coating, sometimes by a significant amount.

12.3.1 Causes of coating failures

Coatings may fail prematurely for many reasons and it is sometimes difficult to determine the cause. Often, there are a number of related problems, none of which alone might be sufficient to cause early breakdown but which together lead to a serious reduction in coating life and sometimes a premature failure such as flaking or complete loss of adhesion.

Some common causes of failure are listed below according to the stage of the coating operation where the problems may have arisen.

(i) *Surface preparation*

(a) Use of manually-cleaned steel in situations where thorough cleaning is essential to achieve good performance.
(b) Use of steel that is so deeply pitted that it cannot be properly cleaned by the chosen method or has serious laminations and shelling that have not been treated.
(c) Inadequate blast-cleaning so that the level of salts, rust and scale is too high for the service conditions on the coating system.
(d) Not attaining the standards of cleaning specified.
(e) Dirt and grease on surface.
(f) Inadequate storage and handling of steelwork after blast-cleaning so that it re-rusts.
(g) Inadequate treatment of welds (one of the most common causes of paint failure).

(ii) *Paint*

(a) Incorrect storage and use after expiry of shelf or pot life.
(b) Inadequate mixing, particularly of two-pack materials and paints with heavy pigments.
(c) Use of incorrect or excessive amounts of thinners.
(d) Poor manufacture of paint, e.g. incorrect formulation.
(e) Use of paint unsuitable for particular conditions, e.g. paints formulated for temperate climates may not always be satisfactory in the tropics.

(iii) *Application*

(a) Use of defective or unsuitable equipment.
(b) Incorrect use of equipment.
(c) Inadequate working conditions, e.g. from the standpoint of temperature, humidity, lighting.
(d) Incorrect drying periods between applications of successive coatings.
(e) Poor application procedures.
(f) Low film thickness.
(g) Poor supervision leading to omission of coats, etc.
(h) Lack of experience of special requirements for some materials.

(iv) *Handling, storage and transport*

(a) Excessive damage to coating.
(b) Inadequate drying of coatings so that during storage they are affected by reaction between the coatings on different parts of steelwork.
(c) Poor stacking of steelwork.

There are other causes of failure, including incorrect formulation or faults in production of paint materials. Generally, however, the problems arise from the choice of protective systems for specific situations and inadequate cleaning of steelwork and poor application of the systems.

The basic purpose of quality control procedures is to ensure that specifications are correctly followed. The specification document should in fact cover quality control and inspection procedures so that the level and type of control and responsibilities involved in these procedures is clearly understood.

12.3.2 Inspection

Inspection is a part of quality control and covers the requirements of many parts of the specification [1] such as surface preparation of steelwork, application of coating system and handling and storage during and after coating application.

Strictly, the choice of protective systems, design audits and the preparation of specifications are not part of inspection, although many inspection companies carry out this type of work as part of an overall consultancy-inspection requirement. Some companies even become involved in a semi-supervisory role in the sense that their inspectors are employed by clients to act with a wide-based authority over the cleaning and coating operations. Such arrangements go beyond the strict interpretation of quality control and inspection although, in practice, they may well achieve a high level of coating performance.

In this particular section, though, quality control and inspection will be considered as being divorced from production requirements. Essentially, their role is to provide the client, in the case of third-party organisations, or the management where in-house inspection is used, with the necessary information on a technical level to enable sound decisions to be made. Clearly, experienced and competent inspection personnel will attempt to improve matters by drawing to the attention of production staff possible difficulties that may arise rather than allowing such problems to occur. It is, however, important to appreciate that too much blurring of the dividing line between quality control and production requirements may prove to be anything but beneficial.

Although inspection is commonly carried out by specialised organisations, it is always the responsibility of the contractor to ensure that specifications are followed.

12.3.3 Instruments used for quality control of coatings

To assist in the control of surface preparation and coating application, a number of instruments is available to check various requirements.

(i) *Surface preparation*

There are two main requirements to be checked, surface cleanliness and profile. Although instruments for measuring surface cleanliness have been developed, they are not widely used and generally assessment is by pictorial standards based on Swedish Standard SIS 0559 00 are preferred.

Various methods are employed to measure or assess the surface profile. The most commonly used are comparators and direct reading instruments based on dial gauges.

Various comparators are available. These basically consist of specially prepared gauges with a number of segments each with a different profile pattern. The profile to be examined is compared, usually, visually with the segments on the comparator gauge.

The other widely used method of determining profiles is by direct measurement with a dial gauge. The profile may also be measured by taking a replica of the surface with a special plastic tape and then measuring this with a specially adapted dial gauge. This method has the advantage that a record is retained on the replica tape.

Another method of profile measuring instruments is based on a stylus that is moved over the surface to produce a trace of the surface roughness. The profile can also be measured with an optical microscope by focusing first on a peak then an adjacent trough of the steel surface and noting the difference on a vernier scale. This method is not used for general quality control work.

(ii) *Coating thickness*

This is an important element in quality control and there is a range of instruments to measure coating thickness. Wet film thickness of paint coatings can be measured by a simple 'comb' device which is placed in the wet film and the position of the wet paint on the steps of the comb indicates the paint thickness.

There is also a small wheel with an eccentric inner cam and markings on the circumference, which can be rotated in the wet paint to provide a reading of thickness.

There are many types of instruments for measuring the dry film thickness of both paint coatings and metal coatings. Comparatively simple magnetic gauges include those where the coating acts as an air gap to measure the change in magnetic attraction. Such instruments were widely used at one time but more accurate types are now available. These include an instrument in which a permanent calibrated magnet is mounted at one end of a balanced arm, with a tension spring connected to the arm and a scale ring. The instrument is placed on the painted surface and the scale ring is turned to increase the tension of the spring until the magnet lifts off the surface. The thickness is read from the scale.

In electromagnetic instruments, an alternating current is applied to a coil and an oscillating magnetic field is established around the coil. The changes in magnetic field when a piece of coated steel is introduced can be used to determine the coating thickness. Many of these instruments supply digital readings of thickness.

All the above instruments can be used only for coatings on a ferrous base and careful calibration is essential. Eddy current instruments are also produced for the measurement of coatings on non-ferrous substrates. There are also instruments that can be used to measure dry film thickness destructively. A small V-scratch, with one type of instrument, and a small hole drilled through the coating with another, are examined microscopically and the coating thickness measured directly. These instruments can also be used, in many cases, to check individual coating thicknesses. Furthermore, they can be used on a range of surfaces, e.g. steel, non-ferrous and concrete.

(iii) *Porosity*

The porosity of paint coatings up to about 500 μm in thickness can be measured by a simple wet sponge device. A wetting agent is applied over the coating by means of a small probe with a sponge at one end. The wetting agent seeps through any small pinholes in the coating causing electrical contact and the completion of a circuit in which is incorporated an audio or visual signal.

For thicker coatings and for detecting not only holidays, i.e. pinholes, but also flaws within the coating, e.g. small bubbles, a high voltage instrument is used. The principle of operation is based on the variation in the dielectric strength of a coating in which there are flaws. The instrument is used after calibration by passing a probe over the surface of the coating. At flaws a spark will penetrate the thinner area and will thus be detected.

(iv) *Climatic conditions*

The climatic conditions at the time of paint application can be measured by a variety of instruments. These include devices for electronically measuring the steel surface temperature and instruments for measuring relative humidity and the dew point.

12.4 REFERENCE

(1) BAYLISS, D A, *Ocean Management*, **7**, 299 (1981)

13 The selection of coating systems for marine environments

13.1 INTRODUCTION

There are no simple rules for selecting systems for the protection of steelwork in marine situations or, for that matter, in other environments. There is a multitude of systems and materials to choose from and clearly there is no general agreement on the optimum system for a particular purpose. There are in fact widely diverging views on the best coating systems for the protection of some types of marine installations. This arises because of experience with certain types of coating, either good or bad, and the influence of manufacturers of coating materials, who market their products with considerable expertise and convince designers and specifiers of the merit of their particular product. Consequently some specifiers tend to use one type of protective system while others prefer alternative types recommended by other companies.

The basic systems from which the choice will be made are:

(i) Paint coatings.
(ii) Metallic coatings.
(iii) High-duty coatings, e.g. neoprene.
(iv) Plastics coatings, e.g. PVC.

Under immersed conditions these may be used with or without cathodic protection; of course, cathodic protection alone may be used. Generally, paint coatings will be considered as the first choice and—only if they are not suitable—will the other types of coatings be considered. Paint coatings are usually the cheapest when initial cost is the sole criterion but are not necessarily so over a long period when maintenance costs are taken into account. Metallic coatings may, in many situations, be as cheap as paint coatings initially and sometimes last longer.

If paint is first considered, what are the factors that will influence the choice? There are many, not all of equal importance and not all applicable to every situation but the following list, not complete, indicates the range of factors:

  (i) Cost and probable life to first maintenance.
 (ii) Suitability for the particular service conditions.
(iii) Application properties, taking into account skill of the operatives and the number of coatings required.
 (iv) Moisture tolerance during application, particularly important for some situations.
  (v) Adhesion properties.
 (vi) Elasticity to allow for steel expansion.
(vii) Abrasion and impact resistance.
(viii) Resistance to other chemicals that may be in the vicinity of the structure.
 (ix) Resistance to sea water and sea salts.
  (x) Toxicity of paints.
 (xi) Shelf life of paints.
(xii) Mixing, not only for two-pack materials but heavily pigmented one-pack paints.
(xiii) Pot life, for two-pack materials.
(xiv) Time to cure or dry at relevant temperatures.
 (xv) Resistance to bacterial effects.
(xvi) Surface preparation requirements.
(xvii) Ease of maintenance.
(xviii) Colour and general appearance requirements.

Clearly there are many other factors that should be considered but just one of the above can eliminate a coating that has most of the required properties. Other factors will only be then of importance in the final choice between two or three possible paints.

A basic factor in the list is the 'life' of the coating system. The term 'system' is used rather than paint because sometimes more than one type of paint will be used to build up the protective system. In practice, the probable life is the one requirement that cannot easily be determined. Physical properties such as abrasion-resistance and properties such as curing temperature and shelf life can usually be obtained without great difficulty from the paint manufacturers. However, the important aspect, i.e. how long it will last, is much more difficult to determine. Attempts have been made to provide tables showing the lives of protective systems and they serve a useful purpose but at most are only a guide to what can be achieved. Because of the complex nature of the coating process, with a possibility of errors at virtually every stage, the intrinsic protective properties of a paint film (assuming there is such a thing) will be achieved only by following a very clearly defined set of rules, written into concise specifications nad a high level of quality control. It is, therefore, unusual—and understandably so—for paint manufacturers to provide data on the life of their products. They may, however, have useful performance data, which can be used to assist in selection. All data must, however, be carefully studied and checked to ensure that the system in question and the situation in which it has been used are genuinely similar to the one under consideration. Data from tests can be useful but this requires very critical appraisal of the methods by which it was obtained and the presentation of such information (see Chapter 21). In particular, performance data, as opposed to physical properties, obtained by laboratory tests, should be treated with caution. Although such tests are useful for a number of purposes, they rarely provide reliable data on 'lives' of systems.

First hand experience is, of course, particularly useful but not always available. Where it is, again it must relate to the situation under consideration. A coating system that performed well on a pier or jetty in the United Kingdom would not necessarily do so on a similar structure in the Arabian Gulf.

Eventually, based on data, experience and possibly in-house tests, a broad indication of the life will be obtained and provided that cost and other factors are satisfactory a short list of possible systems can be drawn up.

It is then necessary to consider how these coatings relate to other requirements, although in practice this would be carried out at the same time as appraisal was being given to the factors inherent in the coating system.

Another list can be drawn up of such factors and some of the more important ones are:

(i) How difficult will maintenance be? Will access be difficult and expensive? Will there be a very limited period for maintenance? Will the weather conditions be necessarily adverse?

The answers to such questions may indicate that even at a high cost, a very protective system should be chosen. On the other hand, it may be that maintenance will be fairly straightforward and can be arranged to coincide with good weather conditions. Consequently the original life of the system would not be a determining factor although, of course, a long-term system might still be the most economic.

(ii) How important is the structure?

All structures are important but, relatively, some more so than others. Will corrosion problems with a structure lead to considerable loss of production, with the ensuing financial losses? Will there be safety problems?

(iii) Will the choice of certain coating systems delay the completion and use of the structure and will this be unacceptable?

Delays of any sort are likely to be costly but on some contracts the possible increase in costs may lead to a choice of coatings that can be applied within the required time scale, even though they may not have the preferred durability. In such cases, of course, high maintenance costs and problems are likely to arise. Nevertheless, if this is the management decision, then the coatings must be selected accordingly. The attention of those making decisions should be drawn to such potential consequences.

(iv) Are the required coating facilities available?

Clearly some facilities are not readily available in some parts of the world, e.g. galvanising baths are not always available. Moreover, even blast-cleaning facilities may not be conveniently situated. It may, therefore, be necessary to make changes in the system to allow for part of the work to be carried out in a situation away from the site, with the attendant problems of transport and storage. In some cases this may lead to a reconsideration of the whole coating system.

(v) Is the design of the structure and its fabrication likely to influence the choice of coating?

Bolted structures pose less problems than those that are of welded construction from the standpoint of corrosion protection because there is no requirement for site cleaning and protection of welds. Where high friction grip bolts are used, the protective system on the faying surfaces must meet the slip requirements but this is not usually a serious difficulty.

(vi) Is the preferred system compatible with cathodic protection and is it likely to involve any safety hazards during maintenance?

Generally there are no problems in choosing systems that will be compatible with cathodic protection but the possibilities of cathodic disbondment must be taken into account. Aluminium castings may be unacceptable for some oil-producing situations.

Clearly, this list is not exhaustive. Comparatively straightforward matters such as colour, general appearance and requirements to use locally produced materials may be the limiting factors of choice. The point to be made is that the selection of protective systems is often far more complex than just choosing the one that is likely to be the most durable. Furthermore, the protective system specifications must cover the many minor variations that are likely to occur over a large structure.

With these points in mind, the next section provides an indication of the types of coating systems that are used for various marine structures and installations. In some situations, paint coatings are not considered able to provide the required degree of durability and plastics-type coatings, concrete and alloy sheet material are all used.

In Chapters 17, 18 and 19 an indication of the types of coating used for offshore structures, ships, submarine pipelines and steel piling are given. It is interesting to note that tests carried out on coatings for sheet piling have shown a disappointing performance from materials that might have been considered to be eminently suitable based on previous test data. This illustrates the undoubted problems that designers and specifiers face when choosing protective systems for aggressive marine environments.

13.2 SYSTEMS FOR GENERAL MARINE STRUCTURES AND INSTALLATIONS

Many marine structures are exposed to conditions somewhat similar to those offshore. Generally, however, conditions are less severe and maintenance can be carried out more easily. In the atmospheric zone, coating systems similar to those considered for offshore purposes may be used but for many structures, however, where the conditions are by no means so aggressive, less durable coatings may be chosen. Clearly, the systems will be determined by the nature of the environment and often oleo-resinous coatings perform quite adequately a comparatively short distance from the coast, or even on coastal installations where conditions are not too severe. Some indication of the choice of coatings is given below; the thicknesses of the coatings will have an important influence on the lives before maintenance. It is difficult to provide hard and fast rules but from BS 5493:1977 the conclusion can be drawn that in air generally for an increase in thickness of 25 μm in the total thickness of the paint system, an additional 12–18 months durability can be anticipated. This is probably correct where abrasion and damage does not affect the coating. The effect of increased thickness would be expected to be less than the figure noted above in more aggressive situations or with very thick coating systems.

(i) Near the coast but in reasonably non-aggressive atmospheres, the following type of system would be suitable:

Blast-clean to Sa 2½ (Swedish Standard)
Priming coat of oleo-resinous type
Two or three coats of alkyd paint of a suitable colour

Alternatively, a one-pack, chemical-resistant paint system such as chlorinated rubber or vinyl might be preferred. On the other hand more resistant types of coatings, suitable for more aggressive situations might be preferred to achieve long periods between maintenance.

Plastic coatings are commonly used for street furniture, rails, etc, as are stainless steels. These should preferably be of the molybdenum-containing type, e.g. 316, to reduce rust staining.

(ii) For more aggressive atmospheric conditions, chemical-resistant types of paint either one or two-pack will generally be considered. These include chlorinated rubber, vinyl, epoxy and urethane. In all cases the steel would be blast-cleaned to at least grade Sa $2\frac{1}{2}$ (Chapter 17).

Hot-dip galvanised and sprayed aluminium (sealed) coatings could also be considered. Galvanised coatings would usually be painted.

(iii) The protection at the tidal zone of jetties, piers, etc, will in some situations follow that for offshore structures (Chapter 17). However, the conditions near the coast are often less aggressive and systems such as the following might be considered:

    (a) Wrapping tapes on suitable shaped sections, e.g. tubes.
    (b) Sealed sprayed aluminium (150 μm).
    (c) Coal tar epoxy system (400 μm).
    (d) One coat two-pack epoxy primer.
        One coat two-pack epoxy.
        One coat chlorinated rubber (350 μm).
        The chlorinated rubber coating is considered to be helpful for maintenance by some authorities.
    (e) Bituminous coating preferably pigmented with aluminium (250 μm). For short-term protection or in reasonably non-aggressive, accessible situations.

(iv) For immersed conditions, the systems chosen will be determined by the nature of the structure and the difficulties of maintenance, which in many situations is extremely difficult:

    (a) Cathodic protection with or without coal tar epoxy system (see Chapter 16 for cathodic protection).
    (b) Coal tar epoxy system (400 μm).
or   (c) Aluminium flake filled bituminous system.
        A comparatively cheap system, where removal of steel unit for maintenance is practicable, e.g. rafts.

When maintenance in a dry dock is practicable, the systems given for ships' bottoms can be used.

Paint and painting specifications

More detailed advice is provided in SSPC Manual, Vol 2, in which specifications covering all the above paints are given. BS 5493:1977 also provides useful advice but, apart from reference to a few British Standards, does not include paint specifications. Paint manufacturers provide paints, usually under proprietary names, for all the above requirements.

14 Maintenance painting

The approach to maintenance painting in marine environments will be influenced by the type of structure and the nature of the conditions to which it is exposed. The maintenance painting of offshore structures and ships is discussed in Chapter 17 and Chapter 18 respectively. The general approach to maintenance of coatings is the same for onshore situations but the problems are not usually of the same magnitude. Furthermore, the protective systems employed for coastal installations may differ from those used offshore. The initial surface protection, selection and application of the protective system is the foundation for the performance over the life of the structure. It is therefore right that a large measure of attention should be paid to this aspect. However, the attitude to maintenance is likely to determine the overall life-cycle costs for protecting a structure, whether off-shore or on land.

The economics of protection are complicated by financial factors such as the tax systems operating in different parts of the world and by requirements of a non-technical nature. It is, therefore, difficult to discuss the economics of maintenance in broad terms. Nevertheless, the total cost, irrespective of the method of calculating it, is basically the initial cost plus the total of all the individual maintenance costs discounted to present-day values. The overall pattern will be determined by the individual costs for each painting requirement and the total number of times maintenance is required during the service life of the structure. This will be considered later.

14.1 THE NEED FOR MAINTENANCE

All structures deteriorate with time and eventually require maintenance. The only exception is where the corrosion of the steelwork is within the limits required to maintain structural integrity. This may be the situation for temporary or comparatively short-life requirements, or where a form of corrosion allowance is integrated into the degisn of the structure. Both these approaches must be considered with some caution because corrosion is not necessarily, or even usually, uniform over a structure and often it is higher in certain critical areas.

Apart from structural integrity the other reason for maintenance is appearance. This cannot be ignored but sometimes considerable costs may be incurred in improving this in situations where the coating is providing adequate protection. This is clearly a matter for the organisation responsible for the structure but it is not always appreciated that features concerned with appearance, such as gloss, colour retention etc, often deteriorate quite rapidly compared with the protective factors. For example, two-pack epoxies are very resistant to marine atmospheres but may chalk fairly quickly. This does not necessarily seriously affect the protective properties so the continual application of coatings to retain the appearance may be somewhat uneconomic overall and, perhaps, coatings such as urethanes should be considered as alternatives in some situations.

14.2 FACTORS AFFECTING MAINTENANCE

A number of factors influence the breakdown of coatings in marine situations. These include:

(i) The total environmental conditions, taking into account whether the steelwork is exposed to salt spray or immersed in sea water; the pattern of condensation, which may be more acute on undersides or in sheltered areas; the temperature, both the maximum experienced and the fluctuations that occur. Abrasive conditions may also have a significant effect on coating performance.

(ii) The quality of the initial protection on new structures, or the efficacy of earlier maintenance work on older structures. This obviously must take into account the surface preparation, type of coating and thickness applied.

(iii) The type of structure. The function of a structure often changes over a period of time. Additions are made, often to increase the load-carrying capability. Various services such as pipes may be added without necessarily taking into account the maintenance problems that may occur through the lack of access to the main steelwork arising from a proliferation of additional components.

(iv) The design and purpose of the structure may have an influence on maintenance requirements, particularly where there are problems of access for scaffolding and the work itself is likely to cause difficulties with surrounding structures, buildings and plant.

All these factors should be taken into account when planning repainting. Where deterioration has reached an advanced stage, through lack of proper maintenance over a period of time, then it may be necessary to carry out structural repairs. Depending on the overall condition, attention should be given to the advantages likely to accrue by blast cleaning and repainting the whole structure.

14.3 THE APPROACH TO MAINTENANCE

In the short term, except possibly in the most aggressive situations, neither the operation of a plant nor the structural stability of steelwork will be seriously affected by lack of coating maintenance. It is, therefore, often tempting in times of financial stringency to delay maintenance. This may be reasonable as part of an overall strategy but can be a very unsound policy if carried out in an 'ad hoc' way. Delays in painting beyond a limited period usually result in considerably increased expenditure because, unlike metal coatings which tend to degrade in a fairly linear way, organic coatings tend to reach a point where breakdown increases rapidly with time. *Figure 14.1* illustrates the pattern of breakdown of conventional, i.e. oleo-resinous, coatings in a marine atmosphere. This is based on a number of investigations that have been carried out and while the actual times to breakdown will vary with the

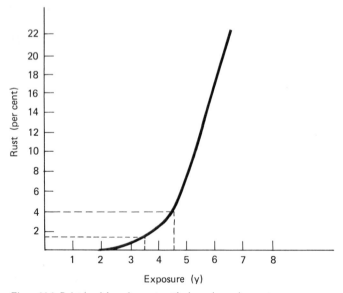

Figure 14.1 Paint breakdown in an atmospheric marine environment

specific conditions of exposure and the type of protective system, the general pattern tends to be similar.

The criterion of breakdown before maintenance painting is important and there are differences of opinion among various authorities regarding the situation at which repainting will be most effective. Ideally, it should be carried out before any significant breakdown of the paint occurs but in practice this may not be economic. The continual application of coatings with increasing thickness beyond that required for protection is not always a sound procedure. Therefore, it is common practice to consider a situation where a small amount of breakdown has occurred as being an appropriate time for repainting, e.g. 0.5% of rusting. To assist in determining the stage at which painting should be carried out, a series of photographs has been produced[1]. These are numbered with the prefix 'RE', e.g. RE6, and may be used to specify the stage at which maintenance work should be carried out.

Often, however, the photographs do not provide a good match with the condition of the paint. It may, therefore, be preferable to use an area criterion such as percentage of rust or blisters. In some situations where appearance is the main criterion, loss of gloss, chalking, etc, will be preferred.

Reference to *Figure 14.1* will show that beyond a certain point, in this case about $3\frac{1}{2}$ years, a delay of a year would lead to an increase from about $1\frac{1}{2}$ to 4% in the amount of rust. As the area to be treated and painted is inevitably greater than the area of rust, this might be a difference of, say, 10% and 20% of the area to be treated. So, such a delay would increase maintenance costs very considerably. Longer delays would clearly be even more costly. Furthermore, at a certain stage of breakdown, which will vary with different systems and the type of structure, patch painting of localised areas of breakdown becomes uneconomic; a figure of greater than 10% has been quoted[2].

There must be a balance of the costs of regular short-period maintenance against more expensive longer-term repainting. This will be influenced by costs of access and the potential problems that may arise if corrosion of the steelwork is allowed to continue for prolonged period. It should be appreciated that while the corrosion of steelwork in marine atmospheres—other than the splash zone—is generally between 75–125 μm/y, in critical areas, e.g. at crevices, overlaps, poorly ventilated spaces and where salt water is trapped, the rates may be a good deal higher.

The single most important element in repainting is the surface preparation. This is so for any situation but it is especially important in marine situations. The rust that forms because of the lack of steel protection arising from the breakdown of the coating will contain considerable quantities of iron salts, particularly chlorides. If these are not removed then premature failure of the coating applied at maintenance will occur.

Furthermore, many of the chemically resistant paints such as epoxies will not adhere well to rust and even with paints containing oils, their performance on rust will be greatly reduced compared with application to blast-cleaned surfaces.

Blast-cleaning may be difficult and expensive in some marine situations but the problems will be even greater if steel is allowed to rust to the extent that it is exhibiting large areas of pitting. The time and cost of cleaning such areas to obtain an effective surface for repainting can be two or more times that for unpitted steel. Developments in surface preparation techniques, e.g. water-blasting and coated abrasives, might be expected to lead to improvements in the overall problems of cleaning of steel prior to repainting.

14.4 PLANNING OF MAINTENANCE

Planning is required if the most economic return on maintenance expenditure is to be achieved. Some form of periodic inspection of the structure, plant or building is required. This is usually done but often in a somewhat cursory way. It is not sufficient to carry out a largely visual examination of paintwork from a point some distance from the structure. It is really necessary to make a proper assessment on the condition as part of an ongoing plan to ensure that such valuable assets are properly maintained.

A thorough survey of the particular construction before planning maintenance is probably essential. In this way, the overall situation can be assessed and a proper plan of operation can be formulated. There are always restrictions on budgets and this is the method to achieve the most effective use of the finance available.

14.5 SURVEYS FOR MAINTENANCE

The survey should provide detailed data on the state of the coating system, the degree of corrosion and the influence of specific features on the performance of the coating. For example, if pipework is placed in a position that results in lack of an adequate coating on adjacent steelwork because of lack of access, this should be reported. The survey should be carried out in detail with close examination of the coating system. The extent of the survey and the time spent will depend on the type of structure and the state of the protective coating, but might well cover the following points:

(i) Thickness of coating system.
(ii) Adhesion of coating.
(iii) Removal of loose paint flakes for examination.
(iv) Deterioration of state of surface under loose paint, e.g. salts in the rust.
(v) Checking number of coats applied.
(vi) Commenting on areas where failure is particularly pronounced.
(vii) Photographic records of critical areas.
(viii) Measurement of loss of steel by corrosion to determine the thickness of section remaining.

The survey should provide a complete record of the state of the structure with regard to protective coatings and corrosion. The report will most usefully serve as a 'log book' for future surveys and records of maintenance carried out. It is then used as the basis for determining the repainting and other requirements, including the rectification of areas of unusually high breakdown of coatings or severe corrosion of the steelwork.

In some cases, it may be advantageous to organise trials to determine the best methods and treatments for maintenance. This may be particularly useful where there is no record of the coating materials that have been previously used, and will eliminate the choice of unsuitable paints. The final stage in the planning is to prepare clear specifications for the work including suitable quality control procedure to ensure that the work will be undertaken and completed in an efficient and competent manner.

Although this section has been concerned with paints, a similar approach should be adopted for metallic coatings. However, dependant upon the level of breakdown, it may be a requirement to remove all the remaining corroded metal coatings by blast-cleaning prior to the application of either a sprayed metal coating or—more usually—a paint coating system

14.6 TREATMENT OF PAINTED STEELWORK

The treatment of painted steelwork will be determined by the condition of the coating. With some paint coatings, light abrasion of the surface may be required to obtain good adhesion when applying further coats during repainting, e.g. on two-pack epoxies.

An indication of the approach that may be adopted for different states of breakdown is given below.

(i) Loss of decorative aspects, e.g. chalking, loss of gloss, etc, but no rusting.
Treatment: Wash down with clean water. Apply one or more coatings to provide a suitably acceptable appearance.

(ii) No rusting, but blistering or slight flaking of coating.
Treatment: Scrape and clean areas of breakdown. Wash down. Apply undercoat to such areas and—where appropriate—a final finishing coat.

(iii) Rusting, but some sound paintwork.
Treatment: Wash down. Remove rust preferably by blast-cleaning. Scrape and clean areas of breakdown where there is no rusting. Prime and undercoat areas of bare steel. Where necessary undercoat areas of paint that have blistered etc. Apply final coat.

(iv) Rusting or flaking over more than 20% of area.
Treatment: Consider blast-cleaning whole area and repainting with complete system.

In all the above situations the coating thickness must be sufficient to provide protection for the required period.

The treatment will, in practice, be determined by the overall purpose and life of the structure. Generally, the same paint system should be applied as initially used. Where there is no record of the original system, tests for compatibility should be carried out and advice should be sought. Serious problems can arise if paints of different types from those originally used are employed for maintenance. For example, two-pack epoxies should not be applied directly to chlorinated rubber paints.

Water-blasting is a convenient method of surface preparation for many marine structures. However, as supplies of fresh water may be limited, sea water is sometimes used in the process. This may be unavoidable but a final treatment with fresh water is probably essential to ensure an adequate performance from coatings applied to the blasted surface.

14.7 TREATMENT OF METAL COATINGS

Generally, special problems do not arise when painting galvanised or sherardised surfaces, but advice should be sought before painting sprayed metal coatings that exhibit signs of breakdown. Recommendations for painting galvanised steel are given in Section 11.22.

14.8 REFERENCES

(1)   Echelle Européenne de degrés d'enrouillement pour peintures antiroulle.
(2)   *British Standard 5443*:1977, Section 5, Maintenance.

15 Control and treatment of the environment

Corrosion occurs as the result of a reaction between a metal and its environment. In marine situations, corrosion is generally controlled by a form of insulation of the alloy from the environment. This may be by a coating or by using alloys that form a 'natural' protective surface film. However, the environment itself can be controlled and some of the methods used will be considered in this chapter.

The methods adopted will depend upon the nature of the environment, in particular whether it is atmospheric or aqueous. Generally, environmental control has a limited, albeit important, application to reasonably enclosed spaces or systems where such methods can be used economically.

15.1 ATMOSPHERIC ENVIRONMENTS

The two main factors that influence atmospheric corrosion at ambient temperatures are moisture and contaminants such as dust, sulphur dioxide and chlorides.

Generally, at low relative humidities, corrosion is not significant with constructional alloys. Hydrogen sulphide is an exception and may lead to the tranishing of some alloys. The relative humidity below which corrosion is inappreciable varies with the alloy and the contaminants present in the atmosphere but, generally, relative humidity control below 50% is satisfactory, although a lower level may be required in the presence of chloride salts. Often relative humidity can be reduced by heating store rooms; otherwise air conditioning equipment can be installed. In structures, e.g. box girders, the relative humidity is often controlled by employing desiccants, e.g. silica gel or activated alumina. These are readily reactivated after use by heating to 130–300 °C for silica gel and 150–700 °C for activated alumina.

Silica gel used at the rate of 250 g/m³ of void is effective for two to three years inside a reasonably well-sealed structure.

For some situations, where corrosion has to be prevented, a space can be completely purged of air and replaced by an inert gas such as nitrogen.

Methods employing desiccants are also used for packaging but there are also available 'volatile corrosion inhibitors', which, because of their volatile nature, inhibit corrosion at the alloy surface within a confined space. These materials are based on compounds containing inhibiting ions, e.g. nitrites and carbonates which sublime slowly, producing vapours that react with moisture to produce inhibition at the alloy surface. Volatile corrosion inhibitors (VCI) are amine salts which can be added to packages or enclosed spaces in equipment in the form of powder on trays or in cloth bags, or impregnated into paper. There is a range of inhibitors and they vary in their volatility. For larger spaces, it is probably advantageous to choose one with a relatively high volatility, so that it is quickly transferred into the water vapour, thus ensuring that condensed moisture is inhibited. Although VCIs are used for packages, they have not been widely employed for structures, e.g. box girders. It has been suggested that CHC (cyclohexylamine carbonate) could be used satisfactorily in large spaces if the openings are fitted with caps. Some VCIs will attack one group of alloys, while protecting others. They are most generally used for steel and advice should be sought from the manufacturers as to the type most suitable for a particular application.

When either volatile inhibitors or desiccants are used in packages, the nature of the packaging material and the effectiveness of reducing the ingress of air and moisture are important factors to be taken into account. A useful guide to packaging is British Standard Packaging Code BS 1133.

15.2 AQUEOUS ENVIRONMENTS

For corrosion to occur in near neutral aqueous environments, generally oxygen must be available and one method of controlling the environment is to remove oxygen. Again, corrosion can be controlled by making solutions alkaline. Both methods are used, particularly for water used in boilers. This is not, however, specifically related to marine corrosion, and a good summary is available in *Corrosion* (ed L L Shreir, Section 18.4). Both oxygen removal and alkaline treatments may be used in circumstances other than boiler feed water. For example, on some buildings the fire protection requirements are obtained by a system in which water is pumped through the tubular structural members. To control corrosion, such water is treated with suitable chemicals, e.g. sodium carbonate, to maintain alkalinity. Alkaline treatments are suitable for many ferrous alloys but not for all alloys. Such treatments will increase the corrosion of aluminium and its alloys.

Inhibitors are widely used as a corrosion control method in many situations. A very large number of papers and books have been produced on this topic and it is important for many processes related to the petrochemical industry. It is, however, a specialised subject and in this book only a few broad principles will be considered.

Apart from aqueous solutions such as supply waters and cooling waters, inhibitors are also used in acids for pickling (i.e. cleaning) of alloys and for operations concerned with oil-refining. In acid solutions, a wide range of substance is available to inhibit the corrosion of alloys, including many organic compounds. The inhibitors act in acid solutions by adsorption on to the metal surface, which is generally free of surface oxide film. The adsorbed inhibitor then acts to retard the corrosion process. The exact nature of the adsorption process may be complex and will not be considered here.

In aqueous, near-neutral solutions, the mechanism of inhibition is somewhat different from that in acids because there is generally some form of surface film on the metal. Furthermore, whereas in acids the cathodic reaction is hydrogen evolution, in neutral solutions it is oxygen reduction. Consequently, many of the inhibitors that can be used successfully in acids do not operate satisfactorily in aqueous solutions because they cannot be adsorbed on the metal surface. Generally, the inhibitors used in neutral solutions act by forming or stabilising protective films on the metal surface.

Inhibitors that form surface films of an insoluble salt include zinc, manganese and nickel salts, soluble calcium salts and polyphosphates in the presence of zinc or calcium. These salt films restrict diffusion of dissolved oxygen to the surface and so prevent oxygen reduction. For this reason they are usually called cathodic inhibitors.

Another group of inhibitors act by stabilising the oxide film on the metal surface; these include chromates, nitrites and benzoates. Because they act to protect the metal surface they inhibit the anodic reaction and are often termed anodic inhibitors.

In practice, the effectiveness of inhibition depends upon many factors, including the following:

(i) Concentration of inhibitors.
(ii) pH of solution.
(iii) Dissolved oxygen in solution.
(iv) State of metal surface.
(v) Temperature.
(vi) Nature of the solution.

A large number of inhibitors is available, many of them of proprietary compositions. Often they are specific to a particular alloy and solution.

Substances that act as an inhibitor for one alloy do not necessarily act in the same way for others; they may, in fact, accelerate corrosion. In some situations, certain inhibitors are only partially successful in retarding corrosion. There is, however, a considerable amount of information available on inhibitors from the manufacturers as well as in published papers. Before using inhibitors, specialist advice should be sought, but the following general points should be borne in mind:

(i) Inhibitors must be present above a certain minimum concentration to be effective. If the concentration is below this critical concentration, then it may lead to localised attack on the unprotected areas, causing pitting. Such inhibitors are often called 'dangerous' as opposed to 'safe' inhibitors which—when used below the critical concentration—allow only the uniform attack that would occur in the absence of an inhibitor. 'Dangerous' inhibitors tend to be those in the anodic inhibitor class. On the other hand, this class—in the correct concentration—is the most efficient for many purposes. Chromates and nitrites tend to fall into the 'dangerous' category.

(ii) Dissimilar metals in a system may require a more complex treatment generally by using inhibitors appropriate to each of the metals. It is, of course, necessary to ensure that the inhibitors chosen for one alloy do not have an adverse effect on another. For example, the action of a zinc galvanised coating in protecting steel at damaged areas may be ineffective in the presence of certain inhibitors, e.g. sodium nitrite.

(iii) The presence of certain ions in the solution may influence the effectiveness of inhibitors. Chlorides and sulphates are particular examples and when present increased concentrations of inhibitor may be required. For example, 50–100 ppm of sodium nitrite will inhibit the corrosion of steel in soft fresh water, but fifty times that concentration is required for sea water.

(iv) Generally, an adequate supply of oxygen is necessary for inhibitors to function properly. Furthermore, continued access by the inhibitor to all parts of the surface is required. Problems may, therefore, arise in crevices where the inhibitor may not be effective.

(v) Micro-organisms and bacteria may affect the efficiency of certain inhibitors and the addition of suitable bactericides to the inhibitors may be necessary.

(vi) Sodium nitrite in high concentrations can be effective in inhibiting the corrosion of steel in sea water. It has been reported that corrosion in 50% sea water can be inhibited by 10% sodium nitrite[1]. Mixtures of chromates and phosphates have been considered for desalination plants, apparently with some success[2].

Although inhibitors are widely used in the oil industry, particular problems can arise when the oil is being produced from deposits in the sea-bed. This is, however, a specialised topic outside the scope of this book, but useful information is available in the NACE publication 'Corrosion Inhibitors'.

A general guide to the effectiveness of some inhibitors in near neutral solutions is given in *Table 15.1*[3].

Table 15.1 General guide to the effectiveness of various inhibitors in the near-neutral pH range

Metal	Chromates	Nitrites	Benzoates	Phosphates
Steel	1	1	1	1
Cast iron	1	1	0	1
Zinc	1	0	0	—
Copper	1	2	2	1
Copper alloys	1	2	2	1
Aluminium	1	2	2	2
Aluminium alloys	1	2	2	2

Key: 1—effective, 2—partially effective, 0—ineffective. (After Table 18.1, *Corrosion,* Vol 2, ed L. L. Shreir, Butterworths, London (1976).)

15.3 SUMMARY

Inhibitors are widely used and can be very effective if used correctly. Many factors, however, influence their performance and in some situations either a low concentration or an incorrect choice of inhibitors where there are dissimilar alloys in a system may lead to intense local attack on one or more of the alloys. In particular, the effectiveness of many inhibitors is related to individual alloys and solutions and may not be applicable in other situations. Many tables of inhibitors and their performance in relation to specific alloys and solutions are published. Where these are based on laboratory tests, the data should be used with caution.

15.4 REFERENCES

(1) OAKES, B D *et al*, Proc. 26th NACE Conference 549–552 (1970), NACE, Houston, USA.
(2) HOAR, T P J. Soc. Chem. Industr. 69, 356–362 (1950)
(3) 'Corrosion Inhibition, Principles and Practice', Chapter 18.2, *Corrosion* Vol 2, (ed L L Shreir, Butterworths (1976)

16 Cathodic protection

16.1 INTRODUCTION

In view of the electrochemical nature of corrosion, it would seem logical for an electrochemical method to be devised to control or prevent it; cathodic protection is such a method. It is based on the reactions that occur in a simple electrolytic cell and provides a situation where an alloy, usually steel, is made the cathode or non-corroding electrode of the cell. The principles and concept of cathodic protection are straightforward. The method is not new; it was used over 150 years ago when Sir Humphry Davy proposed it as a method of preventing the corrosion of copper sheathing used to protect the wooden hulls of ships. Slabs of zinc were attached to the copper sheathing and this protected it from corrosion. The method was not widely used at the time, although it was employed to a limited extent when steel hulls replaced the earlier wooden ones.

The widespread use of cathodic protection has occurred in the last thirty or so years as technology has developed and practical problems have been solved. Cathodic protection is restricted to environments of suitable conductivity. It is particularly suitable for steels immersed in sea water but is also widely used in other waters and soils of suitable conductivity. It is not, however, a method suitable for the control of corrosion in air. Attempts have been made to develop special types of coating with suitable conductivity for use on steel, but these have not proved to be of practical value. Steel reinforcements in concrete can be protected in some atmospheric situations by cathodic protection.

The principles of cathodic protection are straightforward but the application to practical situations may not prove to be a simple matter. An understanding of the basic principles may assist engineers in determining when and where to consider the application of the method. In practice, however, it is usual to seek specialist advice.

Sections 16.4–16.8 of this chapter are based on notes prepared by Mr. Bryan Wyatt.

16.2 GENERAL PRINCIPLES OF CATHODIC PROTECTION

The principles of corrosion have been considered in Chapter 2 and the basic points discussed below have been dealt with in greater detail there. It has been shown that the chemical reactions occurring in the corrosion process can be sub-divided into two electrochemical reactions, one of which involves oxidation, i.e. the release of electrons, and the other, reduction, i.e. the consumption of electrons.

Typical anodic (oxidation) and cathodic (reduction) reactions are shown below for the corrosion of iron (or steel). In practice, of course, these reactions occur on the same surface of the iron or steel.

Anodic reaction: $2Fe \rightarrow 2Fe^{2+}(aq) + 4e^-$

Cathodic reaction: $O_2 + 2H_2O + 4e^- \rightarrow 4OH^-$

An alternative cathodic reaction occurs under some circumstances leading to the evolution of hydrogen rather than the formation of alkali (OH) ions. This does not affect the electrochemical balance or the basic principles involved. As can be seen, all the electrons released in the anodic reaction are consumed in the cathodic reaction. If electrons are supplied to the iron from some external source, i.e. a current is imposed, then the anodic reactions will be suppressed and the potential of the iron will be lowered. If the potential of the iron is lowered sufficiently, then eventually no current will flow between the anodes and cathodes on the iron surface and corrosion will cease. During cathodic protection, this situation is brought about and—provided sufficient current is supplied— full protection will be achieved. If insufficient current is supplied, then some corrosion will occur, but less than the amount experienced without cathodic protection.

Reference to the potential-pH (Pourbaix) diagram in Chapter 2 indicates that, in near neutral solutions, steel can be prevented from corroding by lowering the potential to below about -0.62 V although in practice the figure may be somewhat lower in many situations. Another way of explaining cathodic protection is in terms of the potential current (Evans) diagrams also considered in Chapter 2. *Figure 16.1* shows such a diagram.

In cathodic protection the whole metal surface is at the same potential, so local currents no longer flow; unlike the situations without cathodic protection where local currents can flow from one area of the metal surface to another.

In *Figure 16.1*, E_{corr} is the potential at which corrosion freely occurs, i.e. the anodic and cathodic reaction rates are equal, and is termed the *corrosion potential* and the corresponding current I_{corr} represents in electrical

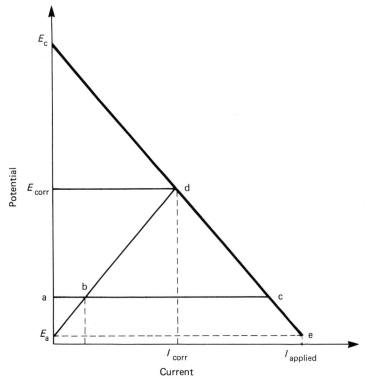

Figure 16.1 Polarisation diagram illustrating the principle of cathodic protection

terms the corrosion rate. To achieve protection, E_{corr} is depressed until it coincides with E_a when all the metal surface will be at the same potential and no corrosion currents will flow. $I_{applied}$ is the current required for complete cathodic protection, in effect to supply the additional electrons required to lower the potential.

If the current supplied from the external source is insufficient and the potential moves only from d to c (*Figure 16.1*), instead of from d to e, then b–c represents the external current and a–b the residual corrosion current. In such a situation, full cathodic protection will not be achieved. Although some corrosion will occur, the situation is not intrinsically harmful. On the other hand, if excess current is applied so that the potential is lowered below E_a, then not only will there be an unnecessary expenditure of current but other detrimental effects may occur. The applied current causes a cathodic reaction with either the evolution of hydrogen or the reduction of oxygen. This tends to make the solution alkaline in the vicinity of the protected metal and may cause problems with paint films that are subject to saponification. Consequently coatings

such as chlorinated rubber or coal tar epoxies—which are resistant to
alkalies— should be used in conjunction with cathodic protection. Over-
protection of aluminium with the formation of alkali can lead to corrosion
rather than protection of the metal.

In situations where hydrogen is evolved, other problems may arise. On
a coated structure, the hydrogen may blow off part of the coating. Where
high-strength steels are used, excessive hydrogen may enter the steel
leading to problems of hydrogen embrittlement (see Section 4.6).

16.3 THE PRACTICAL APPLICATION OF CATHODIC PROTECTION

The essential requirement for cathodic protection is to achieve a situation
on the surface of the metal to be protected where no local corrosion
currents occur. In electrochemical terms, this involves the supply of
electrons to the steel, which can be achieved in two ways:

(i) By applying a current through an external source. This is called the
    impressed current method.
(ii) By forming a galvanic cell using a block of a suitable metal, e.g. zinc,
     electrically connected to the steel. Provided the attached metal has a
     more negative potential than the steel, electrons will pass spon-
     taneously from it to the steel. This is called the *sacrificial anode method*
     because the attached metal is 'sacrificed' to protect the steel.

16.3.1 Impressed current method

The basic principle of the method is illustrated in *Figure 16.2*. The original
anodes and cathodes on the steel surface are indicated diagrammatically
by A and C. The current from a d.c. source is delivered through an
auxiliary anode, X. The areas of steel that were originally anodic reach
the same potential as the original cathodic areas, so the whole surface
becomes a cathode of the new cell, thus stifling corrosion.

A range of materials is used for auxiliary anodes and a useful guide has
been provided by Brand[1]. Some of the more common materials used for
marine purposes are considered below.

(i) *Scrap steel and cast iron*
These materials are cheap but bulky and have a limited durability as
anodes. They have largely been replaced for marine situations by other
materials, although they may still be used to a limited extent for sheet
piling and jetties.

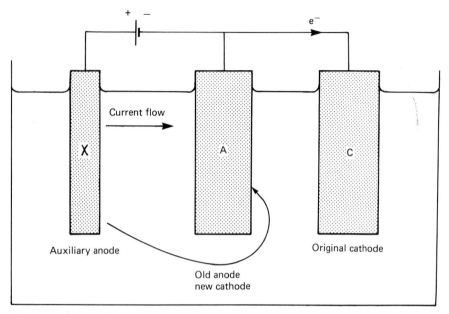

Figure 16.2 Corrosion cell with impressed current cathodic protection

(ii) *Graphite*
Graphite, particularly when impregnated with resin to reduce porosity, has been widely used in saline situations, but it is brittle and may fracture in sea water if subjected to mechanical shock. It is not particularly resistant to the velocity effects of moving water, e.g. erosion and impingement.

(iii) *High-silicon irons*
Usually denoted as 'HSI', these contain about 14.5% Si and have been used for anodes in sea water. However, they tend to pit and small additions of molybdenum (3%) or chromium (4–5%) considerably improve their performance. The chromium-containing silicon iron (HSCI) is the preferred alloy for sea water use.

(iv) *Lead alloys*
A number of different lead alloys have been used for marine conditions, but the Pb-6Sb-1Ag alloy is now generally used. This depends on the formation of a PbO_2 film to provide the long-term performance achieved with anodes produced from such alloys. For sea water service, platinum-activated lead alloy anodes have proved to be successful. Small platinum

wires are inserted into the surface and provide microelectrodes of Pt which stimulate the formation of a stable PbO_2.

(v) *Platinised-type anodes*

Platinum has many of the requirements for an ideal anode material. It is one of the most noble metals and also tends to passivate by forming a thin electrically conductive film. These properties provide it with a long, virtually permanent life. It is, however, a very expensive material, so is usually used in the form of an electro-deposited coating on titanium. Other methods of application of the platinum such as spraying and cladding are also employed. Tantalum and niobium are also used in preference to titanium for some systems where high voltages are employed.

16.3.2 Sacrificial anode method

No external current source is required with this method. Anodes of suitable metals of more negative potential than steel are attached to the structure to provide electrical contact. This produces a cell in which the steel is the cathode; the anode material corrodes, so must be replaced periodically. The essential requirement for the anode material is its ability to polarise the steel to a potential where it either does not corrode or corrodes at an acceptable rate.

Other requirements are reasonable cost, ability to be manufactured to suitable shapes and sizes, and even and reasonably uniform corrosion. Zinc, aluminium and magnesium are used in marine situations. All are sufficiently negative in potential to act as anodes to protect steel in sea water. However, in practice, the exact composition of these metals when used for anode manufacture are important. Certain alloying elements markedly influence the performance of the metals as anodes for cathodic protection. Sound quality control procedure including a full analysis of the materials are necessary requirements when using such anodes.

The composition must be such as to produce the following properties:

(i) A sufficiently negative potential to ensure cathodic protection in a particular environment.
(ii) Ability to continue to corrode during use and not to develop a passive or protective film on the surface.
(iii) A high anode efficiency.

(i) *Zinc*

High-duty zinc alloys are widely used for anodes in marine situations. Since the corrosion of zinc is comparatively low in sea water, the high efficiency (85–95%) of these alloys is maintained throughout the current density range.

Iron is the major harmful impurity and should be maintained below about 0.0014–0.005%, depending on the alloy used. Small additions of aluminium and silicon can be used to neutralise the effect of iron.

(ii) *Aluminium*

Aluminium depends for its corrosion-resistance on a protective oxide film. This film is detrimental to the use of aluminium as an anode material so the alloys used for sacrificial anodes include mercury or indium, which prevent the occurrence of the passive film on the anode material. Other elements are also added, e.g. zinc and tin to make the anode potential more negative.

(iii) *Magnesium*

Magnesium has the most negative electrode potential of the metals used for sacrificial anodes and is particularly useful in environments of high resistivity where its high current output per unit weight is useful. Magnesium is not widely used for sea water applications.

This chapter is primarily concerned with the cathodic protection of steel, but other metals can also be protected, e.g. copper with an iron anode.

16.3.3 Comparison of the methods of cathodic protection

Each method has advantages and disadvantages compared with the other. A comparative summary is given below. In some situations, both methods may be employed. This has been termed the hybrid method and is considered in Section 16.7.2.

Sacrificial anode method
The advantages of the sacrificial anode method are:

  (i) Can be used where there is no power.
 (ii) Initial costs cheaper.
(iii) Less supervision required.
 (iv) Installation is comparatively simple and additional anodes can usually be added if the initial protection proves to be inadequate.
  (v) The anodes cannot be fitted incorrectly, e.g. so that current is supplied in the wrong direction, causing additional corrosion instead of protection.
 (vi) Generally, over-protection does not occur.

Disadvantages are:

(i) The current available depends on the anode area, which on large structures or ships may require many anodes.
(ii) There is a limit to the driving voltage available and this is a good deal lower than for the impressed current system.
(iii) The requirement for a reasonably highly conductive environment is not usually a problem in marine situations but may be in some soils.

Impressed current method
Advantages of the impressed current method are:

(i) The high driving voltage is conducive to efficient protection of large structures.
(ii) Fewer anodes are required.
(iii) The method is capable of better control to provide optimum performance.

Disadvantages are:

(i) Continuous d.c. power must be available.
(ii) Mistakes can occur with the direction of connections; this could lead to corrosion rather than protection.
(iii) Supervision of a higher level is required.
(iv) Poor control can result in over-protection and the possibility of problems arising with coatings and high-strength steels.
(v) In aggressive environments such as the North Sea, physical damage is likely to cause more problems than with sacrificial anodes.

16.4 APPLICATIONS OF CATHODIC PROTECTION IN MARINE SITUATIONS

Cathodic protection is widely used to protect steelwork immersed in sea water and in estuarine waters. Offshore structures, harbours, jetties and ships are typical of the range of applications. The detailed design of cathodic protection systems will not be considered in this book. However, in later sections an indication is given of the general approach to be adopted.

16.4.1 General factors to be considered

Cathodic protection is generally used in conjunction with suitable coatings. The cathodic protection system, then, has only to protect areas where the coating has failed to protect the steel, e.g. at pinholes, pores and damaged areas. There is, however, one important exception. The underwater steelwork of the majority of offshore platforms used in the North Sea—and in other areas of the world—is not protected by coatings. This is an economic decision arising from a number of factors such as the overall costs involved in the protection of such a massive amount of steelwork and the delays that ensue because of the time required for coating. This has to be balanced against the greater cathodic protection requirement on uncoated steelwork and the much greater weight involved in using more anodes.

Incidentally, one large oil company does coat the underwater areas of its platforms in the North Sea.

The general factors to be considered include the following:

 (i) Superficial area to be protected.
 (ii) Estimate of current requirements.
(iii) Assessment of electrical continuity requirements.
(iv) Determination of resistivity of environment.
 (v) Type of cathodic protection, i.e. impressed current or sacrificial anode method.
(vi) Cost.

Further factors to be taken into account depend on the method of cathodic protection chosen, but include:

 (vii) Anode metal.
(viii) Number of anodes.
 (ix) Anode spacing.

16.4.2 Potential requirements

The current required to achieve cathodic protection has been considered in relation to the Evans' diagram (*Figure 16.1*). However, this current cannot be directly measured because the original anodic and cathodic areas on the steel are present on the same surface so no cable or ammeter can be used between them. The most satisfactory method of monitoring for successful protection is to measure the potential of the structure being protected in the electrolyte solution in which it is operating. As discussed in Chapter 2, the potential is obtained by the use of reference electrodes of known values. A variety of reference electrodes are used in practical laboratory and field cathodic protection and are presented in *Table 16.1*. Each has its own fundamental half cell potential which, by convention, is expressed with respect to a standard hydrogen electrode. Therefore every metal/electrolyte potential must be expressed with respect to the

Table 16.1 Common reference electrodes

Electrode	Potential (V, SHE) (V at 25 °C)	Temperature coefficient (mV °C)	Stable potential in sea water (V, SHE)	Protection potential for steel in sea water (V)	Comments
Calomel in					Normally restricted to laboratory use. Accurate, used for calibration of other references
0.1N KCl	+0.336	−0.06	Not applicable	−0.868	
1.0N KCl	+0.283	−0.24	(always used	−0.815	
Sat KCl (SCE)	+0.244	−0.65	with salt bridge)		
Silver/silver chloride	Varies with Cl⁻ concentration	−0.6	+0.240	−0.772	Normal, robust electrode for sea water use. Needs no salt bridge. Steel protection potential usually taken as −0.800 V wrt Ag/AgCl
Copper/copper sulphate (saturated solution)	+0.318	—	Not applicable (always used with salt bridge)	−0.850	Most widely used electrode, can be used in sea water if salt bridge solution often changed
Zinc	−0.763	—	−0.810	+0.278	Relatively inaccurate but long term reliable reference. Steel protection potential usually taken as +0.250 V wrt Zn

particular reference electrode used. With this in mind, the reference electrode can be considered simply and correctly as merely a convenient connector to the electrolyte.

Theory and practice both indicate that in clean aerated sea water at approximately 25 °C, full cathodic protection is achieved at steel/sea water interface potentials more negative than −0.80 V with respect to (wrt) a silver/silver chloride Ag/AgCl reference electrode. This is the generally accepted figure although values from −0.78 to −0.91 V (wrt Ag/AgCl) are found in the literature, the most negative being in polluted water. These figures should be compared with the steel/sea water interface potential under freely corroding conditions of approximately −0.5 to −0.6 V (wrt Ag/AgCl). In anaerobic conditions such as closed compartments, muds and clays, sulphate-reducing bacteria can be active. This is considered in Section 4.13. Such conditions can lead to rapid corrosion, characterised by deep pits and black, slimy corrosion products. Early work on buried pipelines indicated a protective potential of -0.9 V (wrt Ag/AgCl) or (-0.95 V wrt Cu/CuSO$_4$) [2],[3], and this is generally accepted. However, recent work [4] indicates that in marine sediments containing extremely active sulphate reducing bacteria, a steel/electrolyte potential of -0.95 V (wrt Ag/AgCl) may be required. As noted previously, when current is supplied in excess of that required for optimum protection, an increase in alkalinity occurs. As the alkalinity increases, the deposition of bicarbonate is encouraged and carbonates, normally as calcium carbonate, are precipitated from the sea water. At higher current densities magnesium hydroxide can be precipitated. These precipitates are known collectively as calcareous deposits and their importance in the reduction of current density requirements to achieve particular potential levels is significant. These films are electrically isolating, limiting oxygen diffusion to the steel surface and reducing current demand significantly [5-8] and are, therefore, beneficial.

Increasing current density and decreasing steel/sea potential beyond the optimum, results in increasing hydrogen generation. There is some valid concern, for high strength steels, that hydrogen produced in this way may limit the fatigue life of steel structures [9],[10]. Although a cautious design for protecting steel in sea water would attempt to limit steel/sea water interface potentials less negative than -1.10 V (wrt Ag/AgCl) some systems have resulted in apparent potentials as negative as -1.3 V (wrt Ag/AgCl) with no known damage to conventional mild or low alloy steels. Much of the laboratory data on which the theoretical concern for overprotection is based has been derived at current densities some orders of magnitude greater than those reached at equilibrium in actual structures in sea water.

In order to protect fully a marine structure, submarine pipeline, offshore platform, ship or harbour installation from corrosion below the water line, it is necessary to achieve the potential requirements over all immersed/buried surfaces of the structure. A significant part of the design will be to ensure current distribution from the anodes into the electrolyte and onto the structure is sufficiently even to achieve and maintain all steel/electrolyte potentials in that band between underprotection and overprotection. Often areas most difficult to protect due to features such as geometric complexity or unsatisfactory coatings are the most structurally critical areas, such as the nodal joints and conductor guide frames of tubular offshore platforms.

16.4.3 Current requirements

The required current density to achieve a particular steel/sea water or steel/mud potential is generally considered to be determined by the availability of oxygen at the steel/electrolyte interface. In sea water, prior to film formation and marine growth, the predominant factors are probably the sea water velocity and the oxygen concentration. Current density requirements can be deduced theoretically, at least prior to polarisation and film formation. The actual practical application of the numerical values may, however, be questionable. Film formation, the so-called calcareous deposits noted earlier, will limit oxygen diffusion to the surface and reduce current density demands substantially [5-8].

There is evidence that these films are more difficult to form in cold waters and that there are ranges of optimum current density under which the most effective film forms [7].

In saline muds, the current density required to achieve cathodic protection is probably determined by a more complex variety of relationships between the bulk electrical resistivity (well accepted as the predominant indicator of corrosivity for soils), oxygen content, level of sulphate reducing bacteria activity and the extent of water and water-borne ionic species transported through the mud.

For certain marine applications, specific combinations of materials and environments will result in particular current density requirements. These will be referred to later but it is worth noting that high oxygen availability due to high water flow rates or high turbulence will invariably result in high current demands for cathodic protection of non-coated or 'filmed' surfaces.

16.5 CATHODIC PROTECTION OF FIXED OFFSHORE STRUCTURES

All fixed offshore structures associated with the drilling and production of offshore oil and gas are provided with cathodic protection. Without effective cathodic protection, these structures would suffer general corrosion loss resulting in structural weakening and possibly perforation of members, but the most dangerous result of underprotection is, and has been, preferential corrosion at or near welds between structural tubulars with resultant increased concentration of stress and reduction in fatigue life. An increased likelihood of corrosion fatigue problems also arises in this situation (see Section 4.7). Effective cathodic protection will not correct poor structural design where the fatigue life is insufficient for the intended life of the structure. However, insufficient cathodic protection and the resultant corrosion can result in rapid and substantial damage to the structure that may affect a significant proportion of its designed fatigue life.

In many areas of the world, government and quasi-government bodies are now involved in setting minimum standards for cathodic protection of oil and gas associated offshore platforms. It is possible to question some of the specific recommendations endorsed by these bodies but they generally have improved the quality of offshore cathodic protection designs. However, a proportion of 'approved' systems have failed and there are some areas of the world where design and monitoring standards are far less strict, possibly due to less aggressive environmental conditions than those that exist offshore northern Europe.

In many instances harbour and jetty installations have much in common with fixed offshore structures with similar corrosion rates. The fact that some harbour and jetty installations are without cathodic protection is probably due more to a lack of familiarity with these techniques of corrosion control by designers and operators in this field than to any inherent lack of economic applicability. This section will concentrate on oil- and gas-related fixed structures. The principles and parameters are widely applicable to harbour and jetty installations which are referred to specifically in the next section.

16.5.1 The structure

Most platforms are welded steel tubular 'space frames' piled to the sea bed but several steel reinforced concrete 'gravity platforms' have been constructed where sea bed loadings have avoided the necessity of piling. For deeper waters and shorter-life fields, there are recent developments towards tethered buoyant platforms and for sea bed located control and production modules.

The most common form of fixed offshore structure in the petroleum industry will be considered. This is the welded steel tubular 'jacket' supported on driven foundation piles and itself supporting a substructure and decks upon which the equipment is placed. Normally the piles have no designed corrosion allowance and will require cathodic protection.

The jacket is the support for the deck-mounted topside equipment. This may include fire-fighting equipment, heat exchangers and process equipment—all of which may require sea water intake caissons and/or dump caissons. Other caissons and/or J-tubes or pulling frames may be provided to facilitate the pulling of cables, small diameter control and facilities pipework, large diameter gathering pipelines for remote wells or even large diameter import and export risers from the sea bed up to the platform topside facilities.

These items are often under different design control from the jacket and often do not appear on jacket drawings offered to the cathodic protection designer. If the caissons are not effectively electrically isolated from the jacket (and there is normally little merit in doing so), their external cathodic protection demands must be considered in conjunction with that of the jacket. If the caissons are for some reason electrically isolated from the jacket, their external surfaces should receive separate cathodic protection provision. In any case, separate cathodic protection should be made for the caisson internal surfaces, even if coated, unless the caissons are either designed for a limited life or made from corrosion-resistant materials. Similarly, items within the caissons may require corrosion protection.

(i) *Risers and conductors*

Risers, i.e. pipelines carrying gas or oil onto or off drilling, production or pumping platforms, are one of the most important operational components of a platform. They often have some of the most severe corrosion problems (see Section 17.7). Conductors, tubular members through which oil or gas wells are drilled and then through which production casing tubing is inserted and often grouted into place, are present in drilling and production platforms. For some projects the wall thickness of the conductors and/or the number of internal casings grouted within the conductors may be considered by the operators to negate the requirement for corrosion protection of the conductors. However, they are invariably in electrical contact with the jacket, above and below water conductor support framings, and therefore are present as a current demand upon the cathodic protection systems. Furthermore, the conventional conductor arrangement on platforms is for a number of closely spaced rows at one location. This has presented many current distribution problems, some due to geometric difficulties of current distribution to the conductors and their framings, and others due to inadequate provision of current to very high local surface areas.

(ii) *Ancillary steelwork*

The jacket will carry ancillary steelwork such as mud mats, grout lines, monitoring conduits, temporary or permanent buoyancy chambers, temporary or permanent pile guides, boat bumpers and construction and installation steelwork, e.g. walkways, pad eyes, valves and spindles etc. All of these should be considered as short- or long-term current demands on the cathodic protection system.

(iii) *Splash zone*

Almost certainly for the riser(s), possibly for the jacket itself and even for some caissons and conductors, there will be additional corrosion protection for steelwork in the splash zone. The selection of the protective system for the splash zone is interdependant upon the cathodic protection scheme; each will affect the other and require combined design input at an early stage (see Sections 13.2 and 17.2.2).

(iv) *Sealed members*
Some of the tubular members of the jacket may be flooded with sea water. Corrosion prevention for these areas should certainly be considered although cathodic protection is unlikely to be the chosen solution. The corrosion in sealed flooded members may be comparatively low because oxygen once depleted will not be replaced. Oxygen replenishment through non-sealed orifices or anaerobic bacterial corrosion in sealed compartments are both possibilities which can be overcome at relatively low costs by design, inhibition and biocide treatment. Consideration of these aspects may also be applicable to foundation pile internals and sealed caissons.

All these aspects require attention from the cathodic protection system designer; omissions will result in areas of underprotection, corrosion, increased inspection costs and, possibly, high-cost retrofits.

16.5.2 Design parameters

In some parts of the world, government bodies, or those acting as government agents, set standards and requirements for the cathodic protection of offshore structures. These bodies are often the Classification Societies, such as Lloyds Register of Shipping and Det Norske Veritas in the UK and Norway respectively, which have developed from primarily shipping interests into many areas of marine technology. The recommendations of one such authority in respect of required steel/electrolyte potentials and current densities are presented in *Table 16.2* and *Table 16.3*.

The data in *Table 16.2* is widely accepted in respect of the positive limit (least protected) requirement, although Reference 4 would imply some caution in the universal application of -0.9 V (wrt Ag/AgCl) to all anaerobic conditions.

The negative limit (most protected) values are far more arbitrary, based on contradictory data in respect of the onset of damage due to overprotection. It should be noted, however, that sacrificial anodes of aluminium or zinc are unlikely to cause negative potentials in excess of the recommended limits on bare steel of conventional strength in sea water. However, for high-strength steels the implied design 'band' of 0.15 V (or 0.05 V under anaerobic conditions) would be most difficult to achieve.

Table 16.2 Recommended potentials for protection of fixed offshore steel platforms (Det Norske Veritas TNA 703, 1981, Rev 1)

Metal	Reference electrode		
	$Cu/CuSO_4$ (V)	$Ag/AgCl$ (V)	$Zinc$ (V)
Steel aerobic environment			
Positive limit	− 0.85	− 0.80	+ 0.25
Negative limit	− 1.10	− 1.05	+ 0.00
Steel anaerobic environment			
Positive limit	− 0.95	− 0.90	+ 0.15
Negative limit	− 1.10	− 1.05	+ 0.00
Very high strength steel UTS > 700 N/mm^2			
Positive limit	− 0.85	− 0.80	+ 0.25
Negative limit	− 1.00	− 0.95	+ 0.10

For temperatures between 25 °C and 100 °C potentials more negative by 1 mV/°C.

Table 16.3 Recommended minimum design current densities for protection of fixed offshore steel platforms (Det Norske Veritas TNA 703, 1981, Rev 1)

Location	Current density (ma/m^2)		
	Initial	Mean	Final
North Sea (northern)	160	120	100
North Sea (southern)	130	100	90
Arabian Gulf	120	90	80
India	120	90	80
Australia	120	90	80
Brazil	120	90	80
Gulf of Mexico	100	80	70
West Africa	120	90	80
Indonesia	100	80	70
Buried pipelines	50	40	30
Risers (in shafts with flowing sea water)	180	140	120
Risers (in shafts with stagnant sea water)	120	90	80
Saline mud (ambient temperature)	25	20	15

For temperatures between 25 °C and 100 °C current density increased by 1 mA/m^2/°C.

The data in *Table 16.3* represents the most widely used conservative design approach. Some authorities consider that it overestimates the real requirements by a factor of 0.2 to 0.3 in the *mean* and *final* figures for all items except saline mud. Also if *initial* is taken as instantaneous non-equilibrium current density prior to film formation, it may well underestimate the real requirements by a factor of 1 to 20 in the *initial* figures[4],[11-19].

Possibly the most important comment on these recommended figures is that they take no apparent account of the increase in current density requirements due to higher water velocities during winter storm conditions. Typically current demand in the upper part of fixed offshore platforms (certainly to -30 metres and possibly below) will be some 25% higher than mean figures during winter storms[21],[22]. However, despite the possibly conservative nature of these figures, designs based upon them have been satisfactory in achieving and maintaining polarisation of structures without excessive anode consumption and with the apparent prevention of external corrosion.

The requirement of a cathodic protection system applied to fixed offshore structures is to fully protect all immersed steelwork from within days of the placement of the structure in the sea and throughout the design life. The steel/electrolyte potentials required will vary over the structure, as will the current densities required to maintain them. The current density required at any single location to achieve a particular steel/electrolyte potential will also vary with time as polarisation takes place and environmental changes modify the extent of this polarisation.

16.5.3 Cathodic protection with or without coatings

As noted previously, the underwater steelwork of most fixed offshore platforms is not coated. There are both advantages and disadvantages to coating steelwork that is to be cathodically protected. The performance of coated structures has—in practice—been good and they require less cathodic protection current to maintain the required levels of protection potentials. Furthermore, there is a greater degree of tolerance in the design of the cathodic protection system. For example, because lower current densities are required with coated steel, the positioning of the anodes is less critical than with uncoated steel. On the other hand, it must be appreciated that the coating will be damaged and generally deteriorate during service. There is at present no 'in-service' experience covering the anticipated life span (20–40 years) for such coatings. The recommendations of one of the Classification Societies concerning allowances for coating breakdown is shown in *Table 16.4*.

Full protection can be achieved with a properly designed cathodic protection system on a non-coated structure. The approximate costs (1983 figures) and weights arising from the coated and uncoated options are shown in *Table 16.5*, based on a design project for a notional deep-water structure[22]. Clearly, the data in the tables will not be appropriate for every situation, but it does indicate the following points of interest: (i) There is little difference in the overall cost between cathodically protecting bare and coated steel. (ii) The additional weight is considerably greater with bare steel.

Table 16.4 Recommended coating breakdown allowances for fixed offshore structures over 20/30 years (Det Norske Veritas TNA 703, 1981, Rev 1)

	Percentage breakdown		
Coating classification	*Initial*	*Mean*	*Final*
Thick film pipeline	1	10	20
Vinyl systems	2	20	50
Epoxy coal tar	2	20	50
Epoxy (high build)	2	20	50

Table 16.5 Cost and weight variations due to coating submerged steel (notional deep water structure[22]). Aluminium sacrificial cathodic protection system

Parameter	*Bare steel option value*	*Coated steel option value*
Anode net weight	230 kg	115 kg
Anode gross weight	292 kg	157 kg
Anode quantity (total)	5530	5132
Budget price per anode	£300	£160
Budget installation price per anode	£300	£250
Budget coating cost per m^2 (63 000 m^2)	—	£20
Total budget cost	£3 318 000	£3 364 120
Total estimated weight	1615 tonnes	806 tonnes + paint weight

Note (1) Coating breakdown assumed to be 44% average, 70% at the end of life, 35 years after installation.

(2) Smaller anodes are used in coated option to give adequate end of life current output and distribution.

16.5.4 Choice between impressed current and sacrificial anode method

The selection of the type of cathodic protection system for offshore structures appears in some cases to have been somewhat arbitrary, depending on the experience and preference of the designers or consultants.

A method termed the 'hybrid system' has been discussed by Wyatt[18]. With this system, sacrificial anode protection is used in the early months prior to the commissioning of the impressed current system. It has been claimed that the cost is only about 60% of that anticipated from a full sacrificial system. This approach was used—with reported success—for a major North Sea platform.

There have been problems with both methods of cathodically protecting structures. On some sacrificial anode designs, although the overall anode provision has been sufficient and reliable, poor distribution has resulted in areas of underprotection. On others, outdated and unreliable anode alloys have been used, again leading to poor performance.

The main problems with impressed current systems have arisen from inadequate mechanical designs leading to poor reliability and poor distribution from high current anodes, resulting in both over—and under—protection of substantial areas.

Generally, however, both systems have provided satisfactory service and the following conclusions have been drawn by one authority[22].

(i) For harsh environmental conditions such as are found in the North Sea and other high northern latitudes, the mechanical vulnerability of impressed current systems results in a preference for sacrificial anode systems for long-term cathodic protection.

(ii) By virtue of the relatively low individual anode current output available from sacrificial anodes, such a design will tend to have a degree of redundancy and advantages of distribution not available from conventional impressed current systems.

(iii) The major disadvantage of sacrificial anode systems is weight and not cost. *Table 16.6* compares weights and costs of alternative sacrificial or impressed current systems for a notional deepwater

Table 16.6 Comparison of cost and total weight between impressed current and sacrificial anode systems

	Sacrificial system	*Impressed current*
Total installed cost	£3 318 000	£4 600 000
Total weight	1615 tonnes	470 tonnes

Note (1) Impressed current system assumes: 50 A output anodes (structure mounted); cabling and conduits; transformer rectifiers; 20% sacrificial anodes for polarisation and protection before power available.
(2) Both systems are for bare steel.
(3) Mean current 8850 A.

platform which is uncoated below the water level. In severe environmental conditions, where consideration should be given only to impressed current systems with a high percentage of redundancy below the water level and where cable runs to structure-mounted anodes would be within substantial caissons or inside tubular members, impressed current systems of high reliability are of equivalent or greater cost than sacrificial anode systems.

(iv) For severe environmental conditions or where other requirements dictate high reliability, it is considered that weight restrictions are best met by coating the platform and thereby reducing the weight of anodes as indicated in *Table 16.5*.

(v) The costs of repair and maintenance of an impressed current system in a severe environment are high, reaching perhaps 20 times that of the initial cost for a full offshore replacement or 'retrofit'.

(vi) If the weight of a sacrificial anode system on a coated structure cannot be accommodated and impressed current is to be installed in a severe environment, a prudent owner/operator should budget for at least one major retrofit during a 20–30-year life.

The pessimistic view of impressed current reliability noted in the above conclusions is less likely to be applicable to offshore environments such as the Gulf of Mexico, the Arabian Gulf and the Far East, which are generally not so aggressive as the North Sea. However, problems have arisen in the Gulf of Mexico [23],[24] and the limited availability of trained personnel in the other two areas may lead to difficulties with the maintenance of the systems.

Wyatt has expressed the following views [22]:

'Any impressed current system provided for a fixed offshore platform should be a hybrid system with sufficient number, weight and distribution of sacrificial anodes to fully protect at least the critical areas of the jacket during the 1–3 years between placement of the jacket and full operation of the impressed current system and for subsequent periods when the impressed current system is switched off or failed. Even impressed current systems of high presumed reliability may be switched off for periods to avoid danger to divers working near to anodes, although this problem may well have been overstated [24]. A realistic, if slightly pessimistic view, would be that after the first three years the sacrificial system will be required for some 20% of the remaining design life.'

It is worth noting that the concept of locally protecting critical areas of the jacket such as nodes with no (or only partial) protection of the major areas, is not practical. Attempts have been made, and have failed, using large sacrificial anodes placed near nodes with no anodes elsewhere; the result is generally a marginally protected jacket. Generally, the sacrificial anode method is preferred for the protection of offshore structures in the North Sea. Although there is a saving in weight with an impressed current system, it does appear to be more vulnerable to damage and probably requires more design expertise than does the sacrificial method.

16.5.5 Design of cathodic protection systems

Although a detailed discussion on the design of cathodic protection systems for offshore structures in not appropriate in this book, some of the underlying requirements are worth considering so that engineers may gain a better appreciation of the design process.

The current and anode requirements for each part of the structure must be considered individually. Local current requirements will vary over the structure and *Table 16.7* indicates the variations anticipated on an actual deep-water platform.

Table 16.7 Local current density design figures (actual deep water platform)

Location	Extent (m)	Mean current density (mA/m^2)
Upper submerged	−11 to −35	175
Lower submerged	−35 to −110	150
Sea bed silt	−110 to −115	75
Sea bed	−115 to −180	25

Note (1) Initial current density provision for first three months of life approximately 30% greater than mean value.

(2) End of life current density equals mean value.

The current densities selected reflect the values anticipated from knowledge of sea water temperature, salinity, oxygen content, lateral flow rates, suspended solids and frequency and magnitude of storms, and the figures selected are considered to be generous. In a new environment designers generally consider that, with the agreement of the operator, generous current density allowances should be made to reduce the possibility of unknown factors resulting in an inadequate design with a resultant retrofit requirement. The result should be a cathodic protection system with a true life in excess of its design life; this excess life may even be utilised in extended platform operation.

(i) *Anode current output*

Various formulae are available to calculate anode to electrolyte resistance. Some are based on theoretical, others on empirical, derivations. It is generally assumed that cathode/electrolyte resistance is negligible, although this assumption may be invalid, particularly when current distribution is being considered. The detailed formulae will not be considered here but are discussed in a number of papers[25–29]. Basically, the anode resistance, R, is calculated taking into account the dimensions of the anode and the electrolyte resistivity. The anode current output is then calculated from Ohm's Law $I = E/R$, where I is the anode current output in amperes and E is the anode/cathode potential difference for sacrificial anodes or the anode/electrolyte driving voltage for impressed current anodes in volts.

For sacrificial anodes, it is conventional to use in Ohm's Law, for anode current output calculations, the design difference between operating anode potential and polarised cathode potential, which are typically -1.05 V and -0.80 V for zinc or Al-Zn-Hg anodes protecting steel at ambient temperatures in sea water. This approach is open to question as field potential measurements and electrical field theory both indicate that for close anode/cathode spacings, as normally encountered on offshore sacrificial anode installations, cathode potentials near to anodes will be closer to -1 V than -0.80 V. However, the combination of assumptions and formulae used are currently producing workable designs. It is necessary to design the anodes and the overall cathodic protection system to have sufficient current output to be able to provide the required degree of cathodic protection throughout the life. In particular, it is necessary for the system to have the capability of rapidly polarising the structure immediately after placement and for protection to be maintained for the entire design life of the system. This latter requirement is believed to have been omitted from some early designs, in that anode current outputs were calculated on the basis of the anodes being consumed by an arbitrary 40%. Schemes designed in this way may well have working anodes remaining at the end of the system life. However, it is probable that, due to the reduced size of the anodes at, say, 85% or 90% consumed, the current output available from them will be significantly below the design figure and insufficient to maintain adequate protection. It is worth noting that the initial current output of sacrificial anodes is determined by their operating potential, the polarised (or partially polarised) potential of the steel and the initial 'as-cast' dimensions of the anode. The 'end-of-life' final current output is determined by the same potential parameters and the 'end-of-life' anode dimensions which are themselves primarily determined by the anode insert. The insert, around which the sacrificial alloy is cast, provides mechanical strength, a suitable member for welding to the structure and the 'end-of-life' dimensions. Therefore, anodes of the same net weight of sacrificial material can be given greater initial and final current output characteristics by increasing steel insert dimensions; tubular inserts are the most effective and most commonly used for properly designed stand-off anodes.

(ii) *Anode parameters*

All cathodic protection anode materials and designs, whether impressed current or sacrificial, have specific parameters relating to their rates of consumption, utilisation and operating voltage. These determine the useful life that can be obtained from the anode for a given number of

Table 16.8 Sacrificial anode operating parameters

Material	Capacity (Ah kg^{-1})	Consumption Rate (g/Ah)	Anode/electrolyte voltage† (V)
In sea water			
US mil spec zinc	780	1.282	0.25
Aluminium-zinc-mercury	2830	0.353	0.25
Aluminium-zinc-indium	2550	0.392	0.30
Magnesium (standard grade)	1230–1500	0.733	0.70
Magnesium (high voltage grade)	1230–1500	0.733	0.90
In mud			
US mil spec zinc (ambient temp.)	730	1.370	0.25
US mil spec zinc (60 °C)	Unsuitable	Unsuitable	—
Aluminium-zinc-mercury (ambient temp.)	2500	0.400	0.25
Aluminium-zinc-mercury (60 °C)	1175	0.851	0.20
Aluminium-zinc-indium (ambient temp.)	2200	0.455	0.30
Aluminium-zinc-indium (60 °C)	1250	0.800	0.25
Magnesium (standard grade)	1230	0.813	0.70
Magnesium (high voltage grade)	1230	0.813	0.90

† To steel polarised to -0.8 V wrt Ag/AgCl.

ampere-hours output and the available anode/cathode potential for sacrificial anodes or the maximum anode/electrolyte voltage for impressed current anodes.

Typical figures for sacrificial anodes in sea water and saline mud are given in *Table 16.8.*

The utilisation factor is determined by the amount of anode material consumed when the anode can no longer deliver the current required. This factor should take account of both reduced size of the anode and/or disbondment of sacrificial material from the core at the end of life.

Typical utilisation factors used, and those recommended by one Classification Society, are 0.9 for long slender anodes and 0.85 for bracelet anodes. It should be noted that utilisation factors of these magnitudes are only achievable if the anode cores are designed specifically to achieve them. Many anode cores, particularly for bracelet anodes and flush-mounted anodes, appear to be unsuitable for utilisation factors in excess of 0.7. For impressed current anodes, typical rates of consumption, utilisation factors and operating voltage range are given in *Table 16.9.*

Of the anodes listed in *Table 16.9*, only the platinum-coated niobium and titanium anodes are widely used in large fixed offshore structure cathodic protection designs. Lead alloys have, however, been widely used for impressed current systems for semi-submersible drilling rigs. Silicon–iron anodes are not normally used for permanent systems in deep

Table 16.9 Operating parameters of impressed current anodes

Anode material	Anode/electrolyte voltage for use in current output calculations (V)	Typical anode current densities (A/m²)	Consumption rate (mg/Ah)	Utilisation factor for rod shaped anodes
Platinised or platinum clad niobium	100	250–1500	1–2	0.85 (0.5 to 0.85 for wire wound)
Platinised titanium	8.75 max (can be exceeded remote from cathode)	250–750	1–2	0.85
Lead-silver-antimony	100	250–1000	5–15	0.5–0.85
Chromium bearing high silicon cast iron	100	10–100	30–60	0.5–0.85

water due to their high consumption rates and low brittle strength. These parameters are used to determine the weight and/or size of anodes required to provide the current for the full life of the system. The anode arrangement selected must meet this life requirement, the current output requirements discussed earlier and distribution requirements.

Typical impressed current anodes for fixed offshore structures can be generally categorised into three types. These are relatively small cantilever anodes, flush-mounted anodes and remote seabed or buoyant anodes.

The cantilever anodes may be rigidly fixed to the structure with a dielectric coating on the anode mount and possibly on the structural tubular itself. Alternatively, the cantilever anode may be supported by an anode conduit or riser, through which the anode may be retractable. The anode end of the anode conduit may be relatively remote from structural tubulars and may itself have a dielectric coating.

Generally, due to the relatively small size of cantilever anodes, it is necessary to operate this type of anode at relatively high voltages often in the range 5–30 V and therefore the electrochemical and mechanical properties of platinum-coated niobium are often utilised.

Flush-mounted anodes are normally carried in a bracelet form dielectric carrier utilising a technique originally developed for tubular piled jetties. It is possible in these designs to have a greater surface area of anode exposed to the electrolyte and to utilise the lower cost and lower operating voltage of platinum-coated titanium, often in formed sheet, inlaid into the dielectric material. Lead alloys have also been used in this type of anode, particularly in the Middle East.

The third category of impressed current anode for offshore structures, the remote anode, can—by virtue of its remoteness— be designed to give a high individual anode output. A number of different designs have been employed, including platinum-coated niobium or, for larger areas, platinum-coated titanium anodes on buoyant or foundation supports.

(iii) Anode distribution

Incorrect anode distribution of both sacrificial and impressed current systems is considered by many authorities to be one of the most frequent causes of poor cathodic protection performance on fixed offshore platforms. Due to their low current output, generally in the range 3–6 A, sacrificial anodes can be adequately distributed with simple, if laborious numerical techniques. If, for instance, the local current demand is estimated to be 100 mA/m², an anode of 3 A mean current output will protect 30 m² of steel. If the structural member is relatively large, say 3 m diameter, a single anode placed in the centre of this 30 m² area is not much more than 3 m from the extremities of the cathode it is protecting. For small-diameter tubulars, attenuation calculations or simple mathematical modelling may be necessary to ensure that the selected anode size can protect the cathode half way to the next anode, taking account of the additive effects of that anode.

The distribution of impressed current anodes becomes more difficult the greater the individual current output.

For a 50 A structure-mounted anode, for example, the calculations necessary to ensure proper distribution can probably be limited to fairly simple mathematical modelling to ensure local steel/electrolyte potentials near the anode are not excessively negative and that the attenuation between adjacent anodes is acceptable. These relatively small impressed current anodes can then be allocated on a 'member-by-member' basis in a similar manner to that described for sacrificial anodes.

The distribution of higher-current anodes is complex and techniques are being evolved to provide mathematical modelling. This is not always applicable to present designs for these higher-current anodes such as 200 A conduit-mounted container anodes and 500–1000 A seabed-mounted remote anodes.

16.5.6 Cathodic protection monitoring

The cathodic protection design for a fixed offshore platform should always contain provision for monitoring.

These monitoring provisions should include fixed, permanent reference electrodes which are automatically scanned and may be addressed on demand. Such reference electrodes, which may be high-purity zinc, or zinc and silver/silver chloride dual electrodes with data transmitted by either cables or acoustic transmission through the water, provide the routine 'base line' performance monitoring information.

These electrodes should give routine structure/electrolyte potential data under all weather conditions. Surveys by divers and remote-controlled vehicles (RCV) may provide sound data for calm weather conditions. However, they will not necessarily provide suitable information relating the influence of wave height to the level of cathodic protection on offshore platforms. It is common for fixed systems to monitor not only potentials but also the current output of selected sacrificial anodes and/or all impressed current anodes. Theoretically, cathode current density can be monitored with permanent sensors but it is not known whether this method is being used in practice.

In addition to the routine monitoring procedures, it is advisable to survey the structure at suitable intervals in order to develop an overall picture of steel/sea potentials at points other than the fixed electrode locations. Such surveys are normally undertaken by RCVs on deep-water platforms and by divers on shallow platforms. It is, of course, essential that such surveys are carried out properly by well-trained, properly supervised crews. Routine fixed monitoring electrode scans with the data recorded and sent onshore for expert interpretation should probably occur every 24–36 hours. Potentials during selected heavy storms should be scanned on demand. If sufficient priority is given to the cathodic protection monitoring so that it is commissioned early in the life of a platform, intervals of some eight hours are possibly appropriate for recording data during the early months of operation. An initial survey by RCV or diver should be undertaken as soon after commissioning as practical. This should obtain representative steel/sea potentials for 100% of the structure.

Thereafter annual surveys may measure potentials on reduced percentages of the structure depending upon the results obtained initially. In practice, the Classification Society may dictate minimum levels of inspection but even for structures certified by the same Classification Society there are substantial variations in the detail and quality of cathodic protection monitoring data.

Any parts that are isolated from the main structure present problems in both monitoring and surveying. There have been reports of risers suffering from interaction current discharge with resultant corrosion. In the view of some authorities, electrical isolation from the structure should occur only when the riser and/or the pipeline require more negative potentials than the structure to overcome temperature or bacterial effects. Where risers are isolated they require special provision for monitoring.

Structure-mounted reference electrodes close to the minimum top and bottom of the riser may be used as permanent monitoring electrodes if cable connections for monitoring are connected to the riser on the seaward side of the isolating joint. A similar approach applies to electrically isolated caissons.

All other components of the total structure, including piles, pre-installed conductors etc, should be definitively bonded to the structure, preferably by welding, thereby ensuring adequate protection and effectiveness of monitoring.

16.6 CATHODIC PROTECTION OF HARBOURS, JETTIES AND RELATED FACILITIES

Much of the information given for the cathodic protection of fixed offshore structures is also relevant to the protection of harbours, jetties and related steel structures found in coastal and estuarine waters. These structures are generally based on tubular driven piles or sheet steel piles; sometimes a combination of both.

The tubular pile is widely used for jetty and causeway construction and is analogous to the steel tubular fixed offshore platform. Similar design requirements and types of anodes are applied to both forms of marine tubular construction. The sheet steel pile, often with interlinking piles as in the Larsen type, are widely used in wharf, harbour-wall and water-channel construction. Generally the inshore side of the sheet steel piling is filled with soil or rubble and often horizontal tie bars are used to retain the sheet steel, the tie bars themselves being anchored on the inshore side with further sheet steel piles (see Section 19.2 for details of protective systems for piles).

The environment, the scale and the specific structural aspects of these inshore constructions dictate the differences in cathodic protection requirements and design compared with the large fixed offshore structures.

The environment is likely to be far more variable than in the open sea. Estuarine sites in particular are liable to have considerable fluctuations in the electric resistivity of the water which must be accommodated in anode design and material selection. Increased water resistivity may not necessarily result in reduced cathodic protection current demand. High levels of oxygen, high flow rates and pollution can all result in estuarine current demands being higher than those presented in *Table 16.3*. The increased resistivity will, however, proportionally increase anode/electrolyte resistance and therefore reduce cathodic protection current output. Sacrificial anode materials are only effective in fairly low resistivities; for most practical applications aluminium anodes should be used in waters of up to 150 ohm-cm, zinc up to 500 ohm-cm and magnesium in waters and soils of up to 10 000 ohm-cm, although extruded or other high-surface-area/weight anode shapes can extend these limits. Some impressed current anodes, in particular lead, are unsuitable for applications in other than fairly pure sea water and even platinum-coated anodes show increases in consumption in diluted sea water.

One major advantage that the environment of harbours, jetties and related structures has, compared with fixed offshore platforms, is the relative ease of access. The costs of monitoring and repair of inshore cathodic protection systems is far lower than offshore equivalents. This is probably the most important single factor in favour of impressed current systems for use on inshore facilities. The impressed current system can be economically designed with the facility for routine maintenance. Such impressed current systems could have initial costs as low as one half the cost of a sacrificial system with the equivalent life and capacity, but would require substantially more maintenance with a continuing electrical power cost throughout the life of the structure.

Jetty and harbour facilities are often extensive but seldom of the same enormous scale as a deep-water fixed offshore platform. A significant proportion of tubular piled structures are coated with organic coatings, thus further reducing the required size of the cathodic protection system. However, the economics of this choice are even less beneficial with lower cost impressed current systems than in the comparison given in *Table 16.5* for fixed offshore structures. Whereas a large cathodic protection system for a fixed offshore platform may be of the order of 10 000 A, a large jetty or harbour system would often be one tenth of that. This reduction in scale may result in increased distribution difficulties with large numbers of tubular piles concentrated within a small volume of sea, introducing a somewhat similar problem to that found on offshore platform conductors. The proper sizing and distribution of anodes within groups of piles is essential.

The component parts of harbour, jetty and related facilities to which cathodic protection is applied may not be electrically continuous. Causeways and jetty approaches are often constructed from individual pile groups with steel or concrete deck structures that are not themselves either electrically continuous or assured (e.g. by welding) of continuity to the individual piles. Sheet steel piling, its associated tie bars and anchor piles, although in close mechanical contact, may not be sufficiently electrically continuous to ensure that all components are in the cathodic protection system with none isolated from the main structure. If isolated, they can act as collectors and dischargers of current, suffering 'interference' or 'interaction' effects from the cathodic protection system and enhance the corrosion rate at the point(s) of discharge.

In all cathodic protection schemes applied to these types of structure, it is essential to provide adequate low resistance and mechanically robust continuity bonds between all components of the construction. Normally such continuity is provided by welding directly between components, e.g. between adjacent sheet steel piles, to their tie bars and to their anchor piles or by welding continuity straps between components.

Practical inshore cathodic protection schemes may comprise flush sacrificial anodes welded to sheet steel piling. Smaller versions of the 'stand-off' anodes from offshore platforms welded to tubular piles with numerous variations (some of low reliability) may also be used to avoid the necessity for underwater welding of anodes to the piles after driving. Impressed current schemes generally utilise anodes rigidly mounted on the piles; flush-mounted with integral dielectric shields for both tubular and sheet steel piles. The more vulnerable cantilever anodes have been used successfully as have, in calm sheltered waters with suitable bottom conditions, simple impressed current anode arrays on the sea bed. Some systems have utilised anodes suspended from the jetty or similar structure, by cables or ropes; in areas of high marine growth and/or occasional heavy seas, these schemes have had poor reliability.

Some points relating to the design of cathodic protection systems for harbour and jetty facilities are worth noting. For sheet steel piling which has a sea water electrolyte on one face and a soil infill on the back face it is normally necessary to provide cathodic protection for both faces, not just the sea water exposed face. The infill will often be highly corrosive imported fill or alternatively may be quite innocuous coarse rubble, both of which may flood with sea water through leaks in the piled wall; the result will probably be a significant corrosion rate. To provide proper protection for the inshore side of such constructions may require additional onshore buried groundbeds, e.g. high silicon–iron anodes in a carbonaceous backfill. One approach for combined front and back face protection of sheet steel piling is to drive boreholes below the depth of the sheet steel piling, on the landward side of the piles, and construct semi-deep-well groundbeds positioned to distribute current to all faces of the piling.

Another frequent omission is the adequate coating of the splash zone, without which the corrosion rate at and above the air/water interface may proceed unhindered irrespective of how effective the cathodic protection is in preventing corrosion below the water. The monitoring of cathodic protection systems for harbours and jetties is generally limited to manual, steel/water or steel/mud potential measurement followed by adjustment of the cathodic protection output where appropriate. Some impressed current systems may be of the automatic potentiostatic type with a closed loop control circuit, sensing steel/electrolyte potential(s) from fixed reference electrode(s).

16.7 CATHODIC PROTECTION OF SUBMARINE PIPELINES AND OUTFALLS

Submarine and outfall pipelines are generally constructed from welded steel with a suitable protective coating (see Section 17.13). The coatings themselves do not usually provide full long-term protection for the steel. However, they reduce the current demand from the cathodic protection system, so allowing increased spacings between anode installations on the pipelines. Pipeline coatings may have areas of weakness, particularly at barge or field-formed joints. Furthermore, in service the pipelines may be subject to damage due to anchors, trawls, debris and surf. All these areas must be able to receive sufficient cathodic protection current from the installed system, even though the location of such potential defects will not be known at the time of the design. Many cathodic protection designs for submarine pipelines are based on an estimated single average coating defect percentage, often in the range 2.5–10%. Some designers are now estimating initial defect percentage, anticipated breakdown rate and the 'worst case' defect size at the maximum distance between anodes.

The greatest proportion of submarine pipelines are provided with sacrificial anodes, flush mounted on the pipe often in the form of bracelets and in direct electrical connection with the pipeline. These anodes are of zinc or aluminium, cast round steel cores. The cores are mechanical supports for the sacrificial anode material, particularly for the last years of the anode life, and they also provide suitable attachment points for mechanical and electrical connections.

Typical weights and spacings for such anodes are shown in *Table 16.10*. Impressed current cathodic protection, the predominant technique for the protection of lengthy buried pipelines, has been restricted in its application to submarine pipelines. This is due primarily to the requirements for electrical power supplies and suitable location for electrical equipment along the route of such pipelines. Therefore impressed current systems have normally been applied only to outfalls, to the shore ends of long submarine pipelines and, occasionally, to pipelines in the vicinity of fixed offshore structures.

Generally these land-based systems have comprised high-silicon–iron anodes in carbonaceous backfill forming buried groundbeds some 25–150 m remote from the pipeline near the shore with a transformer–

Table 16.10 Some characteristic data for cathodically protected pipelines in the North Sea

	Length (km)	Diameter	Average anode spacing (m)	Anode surface (m²)	Nominal area relationship anode:cathode	Anode weight (kg)	Comments (km)
Lehman–Indefatigable fields	64	0.813 m (32 in)	495	—	—	454	
Forties–Cruden Bay	224	0.813 m (32 in)	188	—	—	227	
Ekofisk–Teesside	349	0.864 m (43 in)	120	0.957	1:343	376	
Ekofisk–Emden	396.5	0.915 m (63 in)	134	1.417	1:272	454	
			134	0.609	1:632	454	ca 25
Piper–Flotta	ca 220	0.457 m (18 in)	283	—	—	109	
Frigg–St. Fergus	355	0.813 m (32 in)	168	1.932	1:222	610	
2 parallel			168	1.308	1:328	610	ca 33
			268	1.005	1:427	610	ca 3

rectifier as the source of d.c. power between the groundbed and the pipeline.

The design steps for submarine pipeline and outfall cathodic protection systems are similar to those for offshore platforms except that the distribution calculation is relatively simple and easy to define and almost all submarine pipelines are coated.

Table 16.11 and *Table 16.12* summarise the available information and recommendations regarding protective potentials and current densities for submarine pipelines and the effects of temperature on these parameters. The calculations for impressed current or sacrificial anode/ electrolyte resistance are undertaken precisely as for offshore platforms. For sacrificial anodes the size constraints of manufacture and the frequent requirement to have an anode thickness near that of the concrete weight coating will normally result in optimum anode spacings in the region 5–12 pipe joints (70–170 m) for pipelines in the 10-inch to 42-inch (254– 1067 mm) range using current design parameters. It is possible to model mathematically various sizes of coating defects between such anodes in order to assess the likelihood of large local defects being adequately protected.

For impressed current systems with semi-remote groundbeds it has long been the practice to estimate attenuation of protection along a pipeline using a standard formula[30].

Table 16.11 Submarine pipeline potential requirements

	Recommended pipeline steel/electrolyte potential (V wrt Ag/AgCl)				
	Aerobic		Anaerobic		
	+ve limit	−ve limit	+ve limit	−ve limit	Variation with temperature
DNV TNA 703 Rev 0	−0.80	−1.05	−0.90	−1.05	1 mV/°C more negative between 25 °C and 100 °C
NACE RP-06-75	−0.80 and another	—	Noted as abnormal require special action	—	Noted as abnormal require special action
NACE RP-01-69 (1976 Rev)	−0.80 and others				
Department of Energy UDC 620.193.27 OT-R-8292. Sept. 1982 'The effects of Heat Transfer on the External Corrosion of Submarine Pipelines and Risers'	−0.775 (ASTM sea water/potted sea water −0.800 sea water	—	−0.800	—	Not affected by temperature/heat transfer. Note: limited data
Norwegian Corrosion Control of Offshore Pipelines 11 NTNF Project 1830.5585. March 1982	−0.800	—	−0.800 to −0.950	—	Not affected by temperature/heat transfer. Note: limited data

It can be shown, for example, that a 16-inch pipeline with a wall thickness of 0.25 inches and a good-quality, well-inspected, coal-tar enamel coat and wrap system can be protected for in excess of 20 miles in each direction from one suitable impressed current system.

In using such a formula, it is assumed that the coating quality is uniform and it does not take into account the localised potential or current density that can be anticipated several miles from the groundbed if a section of coating is removed by mechanical damage. Some attempts are being made to estimate the effect of this by finite element analysis, both manually and by computer.

16.8 CATHODIC PROTECTION OF SHIPS

Many ships' hulls are cathodically protected. The method is also used for the internal surfaces of cargo/ballast tanks of crude oil tankers and bulk

Table 16.12 Submarine pipeline current density requirements

Origin	Recommended pipeline steel/electrolyte current density (mA/m^2)		
	Sea water	Mud	Variation with temperature
DNV TNA 703 Rev 0	Initial/mean/final 160/120/100	Initial/mean/final 50/40/30 pipeline 25/20/15 non-pipeline	Increase by $1\,mA/m^2/$ °C between 25 °C and 100 °C
NACE RP-06-75	86–130	11–22	—
NACE RP-01-69 (1976 Rev)	—	—	—
Department of Energy UDC 620.193.27 OT-R-8292. Sept. 1982 'The Effects of Heat Transfer on the External Corrosion of Submarine Pipelines and Risers'	Initial/equilibrium 320–4300/20–70 (polished, non-corroded specimens)	Initial/equilibrium 30–60/10–13 10 °C 70–240/10–60 65 °C	Suggests two fold increase in current density as safe design factor for steel at 70 °C in sea water, compared with 20 °C. Note: limited data
Norwegian Corrosion Control of Offshore Pipelines 11 NTNF Project 1833.5585. March 1982	Initial/equilibrium 3000/30–70	Equilibrium 20–45 at 5 °C 30–120 at 30 °C	1.5 to 3 fold increase in mud. No significant increase in sea water

carriers for corrosion control when sea water is carried in the tanks as ballast. All of the purpose-built drilling and multi-purpose support vessels involved in offshore oil and gas development are provided with cathodic protection for their external submerged surfaces and generally also for ballast tanks.

All ships' hulls are coated externally and this is reflected in the cathodic protection requirements (see Section 18.11 for coatings used on ships). Typical modern external hull cathodic protection designs will utilise average current densities of around $10\,mA/m^2$ implying a coating efficiency of some 90% in reducing bare steel current density demands from around $100\,mA/m^2$. For some vessels operating under particularly arduous conditions, such as ice breakers or oil platform supply vessels, average current densities of around $30\,mA/m^2$ are appropriate. The selected current density should reflect the intended service of the vessel and surface preparation, coating and inspection specifications for the external hull.

A ship's hull is a complex shape not amenable to simple surface area calculation. It is normal to use the formula

$$A = (1.8 \times LBP \times D) + (BC \times LBP \times B)$$

where A is the wetted surface area, LBP is the length between perpendiculars, D is the draft, BC is the block coefficient, B is the breadth.

Typical block coefficients for various classifications of vessel are

Naval vessels	0.55
Trawlers	0.55
Passenger vessels	0.60
Cargo vessels	0.75
Tankers	0.8 to 0.9

Using the above information, it is possible to calculate the average current demand for a particular vessel. This current may be provided with sacrificial anodes, generally of aluminium as owners appreciate the significant cost savings compared with zinc, or with impressed current. Magnesium anodes are sometimes used for vessels operating in fresh water but particular attention is required to avoid damage to the coating which may be affected by cathodic reactions adjacent to the anode.

The calculation of anode size, impressed current or sacrificial, is by the same procedure and formulae as for fixed offshore structures. One strange inconsistency that has continued in the shipping industry years after the deletion from other applications is the use of a notional 40% consumed anode for calculation of sacrificial anode current output. Anode current outputs are often calculated at 40% consumed and assumed to be this figure throughout life; this over simplification results in sacrificial anode designs for ships and drilling/support vessels that are incapable of providing the design current for the last half of the design life. Due to the requirements for dry docking ships, the sacrificial cathodic protection systems are often designed for quite short periods of two to four years, whereas the large semi-submersible vessels which are difficult, expensive or impossible to dry dock often have five- to ten-year systems.

Typical sacrificial anode external hull protection systems comprise relatively thin section, flush-mounted anodes in the range 5–100 kg in aluminium alloys, welded directly to the ship's hull. The distribution of anodes will reflect the higher current demand in the stern. This arises from the high water flow rates and agitation in this area. Furthermore, account must be taken of the galvanic effects of a copper-alloy propellor if present and fitted with a shaft/hull grounding system. Some 15% of anodes are located in the stern area. Anodes are installed parallel to lines of water flow over the hull to minimise hydrodynamic drag.

Impressed current systems for ships have had a mixed history of reliability but, in part due to the relative ease of access for repair compared with fixed offshore structures, the modern systems are generally considered adequate.

The impressed current systems are almost all of the automatic or potentiostatic type with a closed loop sensing/control circuit using permanently installed zinc or silver/silver chloride reference electrodes and electronic comparison/current output control to achieve protection within preset limits. This degree of control and the ability to obtain performance records of the steel/reference electrode measurements and transformer–rectifier output voltage and current are often taken to imply that a higher level of corrosion control is available from impressed current than with sacrificial anode systems. However, some such systems have the reference electrodes located too close to the impressed current anodes for the measured potentials to be representative. Although good experience has been obtained with many such systems, the use of 'stern only' impressed current systems with anodes and power sources all in the stern area, even on vessels up to 150 000 tonnes, may cause problems. Local coating damage to the hull in the forward part of the vessel may not be detected by the reference electrodes so allowing corrosion to proceed. For these 'stern only' systems, it is normal for the anodes to be long thin sections, bolted to doubler plates welded to the hull, with positive cables to the anode from the transformer–rectifiers passing through the hull via a coffer-dam arrangement. Typical individual anodes are rated at between 50 and 100 A and they are often installed in pairs.

For larger vessels, some proportion of the total cathodic protection system is provided forward, normally using recessed anodes of circular or elliptical shapes to reduce the risks of damage from debris and anchors. Some systems use such recessed anodes in both stern and forward locations.

Impressed current anodes on ships' hulls are generally provided with some form of dielectric shield or coating, applied to the hull for a distance of some metres around the anode, to improve current distribution and to limit the applied steel/sea potentials at the edge of the shield to a level that should not damage the normal hull coating. These shields may be fabricated glass-reinforced plastic assemblies glued or bolted to the hull or high-quality, often thick, two-pack coatings applied directly to the blast-cleaned steel of the hull.

The effects of a successful combined coating and cathodic protection system, impressed current or sacrificial, in reducing the increase of hull surface roughness between dry dockings produce significant fuel savings of some 5–10%[31]; these important cost savings can be expected to increase the percentage of vessels that are fitted with external cathodic protection. Internal cathodic protection of ships' cargo/ballast tanks has been particularly important with the growth in size and numbers of vessels carrying crude oil. These vessels generally travel between their point of cargo discharge to the oil-exporting countries in ballast with special ballast tanks or with some of the cargo tanks filled with sea water. During the 1960s, the massive growth in the numbers of this type of vessel was a stimulus to developments in the methods of corrosion control of these tanks. Classification Society regulations are now well developed and, if followed, allow 10–15% thickness reductions in the structural steels used for constructing such vessels.

Some of the relevant parts of the regulations can be summarised as follows:

(i) As a minimum, the top 1.5 m of any tank—including the deck head— shall be coated.
(ii) The remainder of the tank should be protected by coating all surfaces plus suitable cathodic protection, or by coating of selected structural members and more extensive cathodic protection.
(iii) Impressed current systems are not permitted in oil cargo tanks.
(iv) Magnesium anodes are not permitted in or adjacent to oil cargo tanks.
(v) Aluminium anodes are subject to certain positional restrictions to avoid sparking due to impact; in cargo tanks their potential energy is limited to 28 kg m and they should not be located under hatches or openings in the tanks. Zinc anodes are not similarly restricted, but it is normal to limit potential energy to 540 kg m.

A range of current densities are recommended for different cargo trades and different tanks:

Cargo/clean ballast tanks	86 mA/m^2
Lower wing tanks	86 mA/m^2
Ballast only and ballast/ white oil cargo tanks	108 mA/m^2
Fore and aft peak tanks	108 mA/m^2
Upper wing tanks	120 mA/m^2
Cargo/dirty ballast tanks	Dependent upon trade
Coated surfaces	5 mA/m^2

The normal cathodic protection systems for these tanks comprise combinations of zinc and aluminium anodes in the range 10–50 kg, complying with the above requirements. Systems are designed for a notional ballast factor and an intended period between anode replacement, variations in ballast factor and life of the anodes.

16.9 REFERENCES

(1) BRAND, J W L F, in L L Shreir (ed), *Corrosion*, Section 11.3, Butterworths, London (1976)
(2) FARRER, T W and WORMWELL, F, *Chemistry and Industry*, 108 (1952), 1444 (1954)

(3) BOOTH, G H and TILLER, A K, *Corrosion Science*, Vol 8 (1968), 583–600

(4) NTNF (*Norwegian Council for Scientific and Industrial Research*) *Project Report 1830.5585*, March (1982)

(5) ULANOVSKY, I B, *Zaschita Meallov*, **8**[2], 213–216 (1972)

(6) HARTT, W H and WOLFSON, *NACE Corrosion 80 Conference Paper 152* (1980), National Association of Corrosion Engineers, Houston, Texas, USA

(7) GRIGOREV, V P, *Zhur. Priklad. Khin* **34**, 186 (1961)

(8) DEPARTMENT OF ENERGY, *Project 340 Summary Report UDC 620.193.27 OT-R-8292*, September (1982)

(9) HARTT, W H, *NACE Paper 105* (1981), National Association of Corrosion Engineers, Houston, Texas, USA

(10) SCOTT, P M and SILVESTER, D R V, *Department of Energy, UK Offshore Steels Research Projects* (1977)

(11) WYATT, B S, BRITTAIN, and JONES, G R, *NACE 80 Conference Paper 148* (1980), National Association of Corrosion Engineers, Houston, Texas, USA

(12) WYATT, B S, private communication

(13) DEPARTMENT OF ENERGY, *UK Project Report 340*, September (1982)

(14) ARUP, H, *Eurotest Conference Safety of Offshore Structures* (1977)

(15) MIURA, C, *Corrosion Engineering*, **25**, 12, 733 (1976)

(16) DAVIS, J G, *OTC Paper 1461* (1971)

(17) LAQUE, F L, *2nd International Conference on Metallic Corrosion*, 789 (1963)

(18) WYATT, B S, *Anti-Corrosion*, April (1977) 5–10 and May (1977) 9–15

(19) SEAGER, R W, WARNE, M A and VENNETT, R M, *NACE Corrosion 82, Paper 162* (1982), National Association of Corrosion Engineers, Houston, Texas, USA

(20) WYATT, B S, *NACE Corrosion 79 Conference, Paper 253*, March (1979), National Association of Corrosion Engineers, Houston, Texas, USA

(21) BURGBACHER, J A, *Materials Protection*, **4**, 7, 26–29 (1968)

(22) WYATT, B S, *Proc. Cathodic Protection Theory and Practice, The Present Status*, Paper 8, Coventry (1982), Inst. Corr. Sci. Tech., London

(23) BROSS, H E and BURGBACHER, J A, *NACE 76 Conference, Paper 181* (1976), National Association of Corrosion Engineers, Houston, Texas, USA

(24) MOULTON, R J, *Cathodic Protection Theory and Practice, The Present Status*, Coventry (April 1982), Inst. Corr. Sci. Tech., London

(25) DWIGHT, H B, *Electrical Engineering* (1936) pp 1319–1328

(26) LLOYDS REGISTER OF SHIPPING, *Guidance Notes on Application of Cathodic Protection where Reduced Scantlings are to be Applied*

(27) WALDRON, L J and PETERSON, M H, *Naval Research Laboratory Report 4891*, February 26 (1957)

(28) McCOY, J E, *Trans. Inst. Marine Engineers*, **82,** No 6 (1970)

(29) SUNDE, E D, 'Earth Conduction Effects in Transmission Systems', Dover Publications, New York (1968)

(30) SCHWERDFEGER, W J, McDORMAN, O N, *T. Res. Nat. Bur. Stand.*, **47**, 104 (1951)

(31) JENSEN, F, *Proc. Cathodic Protection Theory and Practice, The Present Status*, Paper 7, Coventry (1982), Inst. Corr. Sci. Tech., London

17 Offshore structures for oil and gas production

Not all offshore structures are designed for oil and gas production. For example, an artificial island has been constructed 15 km offshore from Brazil to store and transfer to ships, sea salt produced by evaporating sea water[6]. In future, offshore structures might be expected to fulfil other similar rôles but most such structures are at present used for oil and gas extraction. As there are a number of problems specific to this particular usage, it is convenient to cover them in this chapter. Of course, the general problem of protecting structures from corrosion is the same no matter for which purpose they are being used.

There is no essential difference in the methods employed to protect offshore structures and other marine installations. The approach to the problems is, however, greatly influenced by the nature of the operation, in particular the limitations on space and the high value of the product: oil or gas. This leads to problems with maintenance and difficulties with replacement of defective parts. There are also problems specific to oil and gas production that might be experienced on shore but which are generally more serious in offshore situations.

The following aspects will be considered in relation to offshore structures:

(i) Economics
(ii) Environments and corrosion control
(iii) Maintenance
(iv) Design
(v) Fouling
(vi) Hot riser corrosion
(vii) Down hole corrosion
(viii) 'Sweet' and 'sour' corrosion
(ix) Sulphide stress cracking
(x) Corrosion fatigue

17.1 ECONOMICS

The costs involved in working on offshore structures are generally higher than those for similar work onshore. Often costs assume an importance that may over-ride virtually all other considerations. Any requirement that conflicts with the production of oil becomes a major problem. This is, of course, true for onshore situations but it is more acute when working some kilometres from the coast, particularly in locations where the weather is bad and there are difficulties not only in supplying materials but also in storing them. Accommodation is always likely to be in short supply and this further limits the numbers able to be working at any one time. Time itself is also important, especially in locations where the weather conditions suitable for maintenance painting are limited. This may sometimes result in requirements for painting at night which causes further problems.

The cost of maintenance is always higher in offshore situations for the reasons noted above and the actual additional costs of scaffolding etc, for access.

It follows that the planning and supervision of all operations is a matter of fundamental importance. To the points mentioned must be added the aggressive nature of the environment itself, which would make steel protection and corrosion control difficult anyway.

17.2 ENVIRONMENTS AND CORROSION CONTROL

The general environments have been discussed in Chapter 3. They are shown diagrammatically in *Figure 17.1* with an indication of the main problem areas and the methods of controlling corrosion. Clearly, the splash zone is the most aggressive and difficult to maintain because cathodic protection cannot be used and painting has to be carried out under poor conditions over comparatively limited periods.

There are many platforms in different regions of the world and the problems vary. The design of a platform must take into account many factors such as wave motion and wind speed and these may also influence the corrosion control methods chosen for a particular platform.

The choice of protective coatings for offshore platforms is based on cost, tests and—in particular—service experience. An important factor to be taken into account is coating maintenance, which has to be carried out at fairly frequent intervals.

Some coatings, such as two-pack epoxies, may require some abrasive treatment before overcoating, e.g. by light blast-cleaning. Other coatings, such as chlorinated rubber, can be overcoated without this

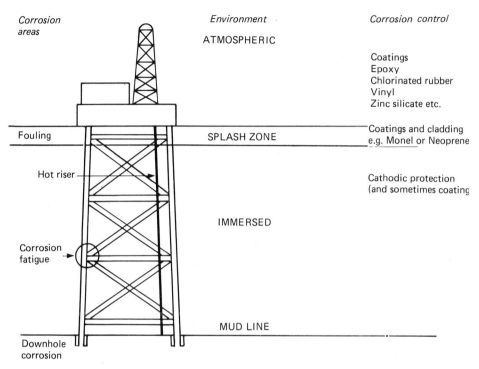

Figure 17.1 Offshore structure: areas of corrosion and types of control

additional preparation. In some parts of the world, urethane finishing coats are used to provide improved appearance and durability.

The long-term protection requirements for offshore platforms are not easy to achieve, but success is determined very largely by the degree of attention given to the guidelines in Chapters 9, 10 and 12.

The platform structure can, from the corrosion standpoint, be divided into three broad zones: submerged zone; splash zone; atmospheric zone. The essential factors that influence the choice of control methods in each zone are as follows.

17.2.1 Submerged zone

Cathodic protection (see Chapter 16) is the method invariably used in this zone. Generally, protective coatings are not used in conjunction with the cathodic protection but some organisations do, additionally, specify coatings for submerged areas. The decision is an economic one. Fewer anodes would be required for cathodic protection and they would require to be replaced less frequently during the design life of a platform and— where protective coatings are used—the anode life would be expected to be considerably longer. This must, of course, be balanced by the cost of protecting such a large area of steelwork.

The coating system would generally be of a two-pack coal-tar epoxy type.

17.2.2 Splash zone

This area, which is difficult to maintain, is normally protected by high-duty organic coatings but there are alternatives, particularly for risers; again the choice is determined by economic factors. The following coatings are used:

(a) Corrosion-resistant alloy sheathing, usually a high nickel alloy such as Monel 400 of reasonably thin gauge, about 1 mm thick, which is fitted round the tubular steelwork in the splash zone and welded into position to seal any gap between the sheathing and steelwork. Although used for risers, this method is not widely employed for North Sea platforms partly because of the possibility of tearing such thin material, particularly by impact, and because of the concern that if sea water enters the gap between the sheathing and steelwork, significant hidden corrosion could occur. Additionally, the cost of such sheathing is high.

(b) Thick rubber or neoprene coatings up to 13 mm in thickness.

(c) Silica-filled polymeric resin cladding and glass-flake filled polyesters.

(d) Increased thickness of steel in the splash zone may be used to allow for the loss arising from corrosion and wear. The steel is usually coated in the same way as the rest of the structure.

(e) Coatings are most commonly used to protect the splash zone. Many different types of coating systems are used. These are generally of the same type as those used for the atmospheric zone but are usually applied to a greater thickness. In NACE Standard RP-01-76[1], the view is expressed that organic coatings cannot be considered as permanent protection in this zone because of the impact damage that occurs.

17.2.3 Atmospheric zone

The approach to the protection of the structural parts exposed to the air is similar to that for other atmospheric marine conditions. To withstand the aggressive conditions encountered in this situation, chemical-resistant coatings are used. Basically, these are two-pack epoxides, chlorinated rubbers and vinyls. Typical systems are given below, although other systems not listed may also be used.

Paint systems for the atmospheric zone of offshore structures

The paint systems for offshore platforms are based on one-pack and two-pack chemical-resistant paints to provide a comparatively thick overall coating. The choice of system is based on the factors noted earlier, on test data and on service experience. Experience is gained continuously and this leads to variations on specifications for new platforms as compared with older ones. The specifications of coating systems for platforms is a specialised matter. In all cases, blast-cleaning to Swedish Standard Sa 2½ and Sa 3 with some primers would be considered the minimum requirement for new work and a blast primer of some type would probably be used during fabrication, although some specifiers prefer the primer to be applied directly to the blast-cleaned steel.

The following systems indicate the range of options.

	Thickness (μm)
Inorganic zinc silicate primer	75
Two-pack epoxy	125
Vinyl acrylic finishing coat	50
	250
Two coats of zinc phosphate pigmented chlorinated rubber primer	150
Chlorinated rubber high-build finish	150
	300
Zinc phosphate pigmented two-pack epoxy primer	75
Two coats: two-pack epoxy high build	225
	300
Vinyl system (3–4 coats)	200–250
Inorganic zinc silicate primer	75
Two coats two-pack epoxy	250
	325
Zinc silicate primer	75
Tie coat	25
Chlorinated rubber (pigmented with micaceous iron oxide (MIO))	100
Chlorinated rubber: high build	100
	300

17.3 MAINTENANCE

As already indicated, maintenance painting is particularly difficult under offshore conditions. The general guidelines in Chapter 14 are relevant but there are a number of specific points worth noting.

  (i) Sound specifications and a high level of quality control (inspection) are particularly important for maintenance painting.

 (ii) All equipment should be in proper order before work begins and suitable spare parts should be available to ensure continuation of working.

(iii) When blast-cleaning, it is necessary to ensure that abrasives do not damage equipment in the vicinity. Proper covers should be used to protect equipment and other steelwork. Where necessary, plugs and covers should also be used to prevent drainage areas becoming clogged.

(iv) It may be quicker and more economic to remove heavy scale and rust by pneumatic-driven chisels and other descaling tools before blast-cleaning. Care must be exercised to ensure that steelwork is not damaged by these methods.

 (v) It is often a requirement to measure wall thickness of pipes and critical structural elements. This is most usefully carried out immediately after blast-cleaning. The methods may be comparatively simple using pit-depth gauges. More complex methods may, however, be required to obtain accurate readings.

(vi) All abrasives should be removed by suitable methods such as vacuum cleaning once blasting has been completed.

(vii) Although there should be adequate access to all vulnerable parts, this is not always practicable, particularly as additions to equipment and structures are made in service. Where necessary, clamps and other supports should be removed to ensure proper cleaning and painting of steelwork.

(vii) Because of the difficulties involved with offshore maintenance, comparatively minor matters can have a considerable influence. For example, ease of mixing ratios for two-pack materials and clear marking of paint containers with all the relevant information; such marking, incidentally, should be weatherproofed. Anything that is likely to lead to smooth operations is advantageous in any situation, but particularly offshore.

17.3.1 Splash zone

The most difficult areas for coating maintenance are the lower structural members in the splash zone. The procedures will depend upon the methods used for coating and the amount of breakdown that has occurred. Where areas of coating have deteriorated to the stage where blast-cleaning is required, various procedures can be considered but all of them are less than ideal. Generally, the steel is blast-cleaned and painted during low tides, preferably during reasonably good weather conditions. It may, in some locations, be possible to clean off the old paintwork, rust, etc, and recoat with thick, quick-drying epoxy paints before the steelwork is covered by the rising tide. More often, it is necessary to carry out rough blasting one day and then to follow with a sweep blast and paint coats the next day.

In some situations, glass flake reinforced polyester coatings may be used and again rapid application techniques are required. It is possible to blast-clean and apply specially formulated paints under water. This technique may sometimes be worth consideration. Where Monel is used as a cladding, damaged areas should be repaired, and to prevent salt water seepage between the Monel sheeting and the structural steel, suitable sealing must be carried out. Damaged neoprene coatings are repaired by special techniques.

17.3.2 Atmospheric zone

The maintenance in this zone follows the general pattern of other marine installations. The maintenance systems will usually be similar to those chosen for the initial protection. Whatever the choice of systems for maintenance, they must be compatible with the original coatings applied to the structure. Epoxies—widely used to protect atmospherically exposed steelwork—generally require abrading or light blasting before the maintenance coats are applied. Some organisations, therefore, prefer to use a final coat of chlorinated rubber paint which does not require this treatment prior to repainting. Others use a final urethane coating over the epoxy, claiming that this both performs better and reduces maintenance painting.

Metal coatings, particularly hot-dip galvanising, can be used for coating a number of components and parts of the structure, e.g. handrails and gratings. It is advisable to paint these items before the zinc has been completely removed, leaving rusty steel that will require cleaning prior to painting.

17.3.3 Immersed zone

Generally, reliance is placed on cathodic protection in this area (see Chapter 16). Where protective coatings are applied as additional protection, they are not usually repainted during maintenance. If necessary, there are methods of cleaning and painting under water, but these are costly procedures.

17.4 SURVEYS FOR MAINTENANCE

Maintenance should not be planned on an ad hoc basis, particularly on offshore structures where there are limitations on time and access; it is advisable to plan ahead and this is generally done. There are a number of factors that will determine when recoating is to be carried out, e.g.

(i) Weather conditions.
(ii) Schedule of essential operations and whether maintenance coating will interfere with them.
(iii) Extent of corrosion in critical areas.
(iv) Additional costs likely to be incurred by delays.
(v) Appearance.

The structure should be periodically surveyed and a list of essential work reported. Where practicable, essential maintenance should be carried out without delay, even if other areas are given lower priority.

Quality control and inspection is particularly important and the guidelines in Chapter 12 should be followed.

17.5 DESIGN

Many of the comments in Chapter 7 are relevant to offshore structures but there are a few points worth emphasising.

The splash zone is the area where coating maintenance of steelwork is most difficult. It is, therefore, advantageous to minimise the surface area of steel in this zone. In particular, angles, struts, etc, which are difficult to paint should be avoided. Where this is not practicable, consideration should be given to enclosing them with steel plates. Additional steel in the form of thicker sections or weld wraps should—where appropriate—be considered.

Other general points include the following:

(i) Elimination or welding of crevices; skip welds should be avoided.
(ii) Where practicable, tubular members are preferable to other types of sections as they are easier to paint.
(iii) Boxing in of complex structural units is a method of reducing corrosion problems and easing maintenance painting.
(iv) Avoidance of bimetallic corrosion situations, particularly where aluminium alloys are used in conjunction with steel.

(v) Ensure that there is good drainage throughout the structure.
(vi) Although difficult in practice, it is beneficial to avoid clusters of equipment and pipes, which make maintenance, painting, repair and replacement difficult. It may be necessary to remove clamps in order to gain access to difficult areas. There are, therefore, advantages in designing them for easy removal.
(vii) Access to all parts is important. The provision of pad eyes assists with maintenance work.
(viii) The provision of corrosion allowances via wear plates may be useful but on the main structural elements caution should be exercised because of the additional loads that may arise from additional equipment or other requirements.

17.6 FOULING

Fouling is the term used to denote the attachment of plant and animal organisms to metallic surfaces, whether bare or coated.

As indicated in Chapter 18, fouling is a problem on ships because it affects their movement through water, so increasing the amount of fuel required to maintain velocity. This adds considerably to costs and specially formulated anti-fouling compositions are applied to ships' hulls. These control fouling for a year or so. The ship is then dry docked and, after suitable cleaning, further anti-fouling paints are applied.

This approach is less suitable for offshore structures because of the problems of cleaning and re-application of the anti-fouling compositions.

The effects of fouling on structures are different from those on ships because they are basically static.

The problems generally fall into the following categories:

(i) The weight of fouling can be considerable and may cause loads above the original design limits. This can lead to failures of bracing, etc, if the fouling is not removed.
(ii) They cause considerable problems with visual inspection and examination of paintwork and metal surfaces prior to maintenance. They have therefore to be removed before recoating.
(iii) Their effect on the breakdown of coatings and corrosion of alloys depends upon the type of coating and alloy and also upon the type of fouling.

17.6.1 Effects of fouling on corrosion

Pitting can be caused under barnacles, particularly with passive film-type alloys, e.g. stainless steels. In some cases the organisms themselves may produce corrosive substances, e.g. acids, if they are damaged. Even where they are inert from the corrosion standpoint, they can cause problems, acting as deposits or causing the development of concentration cells and providing stagnant situations where sulphate-reducing bacteria may flourish.

Fouling can also affect the flow round a structure. In condensers, heat exchangers, etc, fouling can cause blockages and erosion–corrosion effects leading to severe attack on alloys. Alloys vary in their resistance to fouling, depending on the velocity of the water and the nature of the fouling organisms. The resistance to pitting where barnacles and similar organisms become attached tends to follow the resistance pattern at crevices. Consequently, titanium, copper and cupro-nickels would be expected to be resistant and stainless steels and nickel the least resistant. In practice, however, other factors may operate to alter this general assessment of pitting resistance.

17.6.2 Removal of fouling

Anti-fouling paints are not generally used to control fouling on offshore platforms. The methods commonly employed for structures are hand or mechanically driven tools similar to those used for rust removal, e.g. brushes, needle guns and rotating discs and various forms of water jetting.

Chemical methods—sometimes in conjunction with plastic pigs—are used to clean pipes and tubes, for example in heat exchangers (see Section 8.54). This method is, however, difficult to apply to structures and is not generally used. Where chemical methods are used on pipes, considerable control must be exercised to ensure that the solutions, often oxychlorides, are not present in concentrations sufficient to attack the tube materials.

17.7 HOT-RISER CORROSION

All steel immersed in sea water is susceptible to corrosion even where it is coated. Coatings may have defects, become damaged or—over a period of time—deteriorate. This always causes problems but a particularly difficult situation can arise on riser pipes. If these perforate through corrosion or fracture then considerable loss of oil can occur. Furthermore, there are safety hazards in such a situation. Therefore corrosion of riser pipes is a particularly important matter which has to be prevented.

Failures have occurred in various parts of the world and a considerable amount of work has been carried out in this particular field of corrosion. The British Department of Energy has issued a report on hot-riser corrosion and this covers matters in some detail.

As this particular type of corrosion is somewhat specialised, albeit very important to those having to deal with it, only a summary of the problems will be considered here. Those directly involved are advised to maintain contact with the work in progress, which will provide the most useful current information and data.

A number of factors have to be taken into account when considering the corrosion of hot risers:

(i) The temperature of the oil or gas being carried in risers is high (between about 50–100 °C).
(ii) The oil is transported through a series of pipes of varying diameters and great internal pressures can build up.
(iii) Externally the pipe is subjected to the corrosion profile shown in *Figure 3.1* (p 39), so at the splash zone the conditions are particularly aggressive; immersed parts are often covered with concrete.
(iv) There will be a difference between the internal temperature of the pipe carrying hot oil and the external temperature in contact with sea water. The temperature of the sea water will vary depending upon factors such as location but in the North Sea can be below zero centigrade. Consequently there is a steep temperature gradient leading to heat transfer problems with increased corrosion.

In view of the aggressive nature of the environment, sound methods of protection must be adopted. Organic coatings such as epoxies have not always performed as well as might have been anticipated, in particular blistering has occurred. This may have been caused by lack of adhesion arising from the expansion of the pipe under internal pressure or from stray currents or lack of cathodic protection arising from insulation of the pipes from the main steelwork or the design of the system. Clearly the environment is very severe and it is becoming common practice at the splash zone to either sheath the riser pipes with Monel or to use highly-resistant neoprene coatings which can be applied at works.

17.8 DOWNHOLE CORROSION

Downhole corrosion cannot strictly be considered purely in the context of marine corrosion because it occurs as a result of the methods used for oil production, whether or not it is offshore. It is not therefore confined to offshore operations but, because of the saline nature of the water, it may lead to a more corrosive situation than would be expected on land. In most production operations, water from the formation layer under the crude oil eventually mixes with the oil. In marine situations the salinity of the water varies but as it does not usually contain oxygen it is not necessarily highly corrosive. However, other gases, particularly carbon dioxide (CO_2) and hydrogen sulphide (H_2S), may be present, and these can lead to serious corrosion.

Inhibitors are widely used as a control method for wet sour gas gathering systems. Duncan has provided information on gas wells in Bahrein[7] and Gatlin has published the results of a programme of evaluation of inhibitors[8].

17.8.1 'Sweet' corrosion

This term is often used to denote corrosion arising from the presence of water containing dissolved carbon dioxide. The tubing of gas condensate wells when affected by CO_2 usually corrodes by deep pitting that may lead to perforation of the wall over a comparatively short period. The pitting arises from the attack by carbonic acid at areas of droplet condensation on the tubing wall. This pitting may be quite localised with areas where there is no condensate being virtually unattacked.

Salt water and organic acids tend to increase the attack on steel as does oxygen, if it is introduced in some way. (A useful reference source published by NACE is 'CO_2 corrosion in oil and gas production', compiled by NACE Task Group T-1-3.)

17.8.2 'Sour' corrosion

This term is commonly used to denote corrosion resulting from the presence of hydrogen sulphide (H_2S). H_2S is very soluble in water and forms weak acids which can lead to pitting of steel and other alloys, partly because of the acidity formed and partly because a sulphide film is formed from sulphide ions (HS^- and S^{2-}) which is cathodic to most alloys. H_2S can also arise from the presence of sulphate-reducing bacteria in the presence of sulphates.

Hydrogen sulphide may cause problems other than pitting, and a number of different terms have been coined to describe some of these forms of attack. This can be confusing, but the two main types of corrosion are (i) hydrogen induced cracking (HIC), (ii) sulphide stress corrosion (SCC). The term 'stepwise cracking' is based on the terminology used by NACE Task Group T-IF-20 in their work on 'Stepwise Cracking of Pipeline Steels' and refers to cracks arising from initial hydrogen blistering.

17.9 HYDROGEN-INDUCED CRACKING AND BLISTERING

In sour gas service, the cathodic reaction is not the one commonly experienced when steel corrodes in aerated solutions, i.e.

$$O_2 + 2H_2O + 4e \rightarrow 4OH^-$$

Instead of the production of hydroxyl ions, hydrogen is produced at the cathode. This can be denoted by the following equation:

$$H_2S + Fe + H_2O \rightarrow FeS + H_2$$

This does not explain all the steps in the process, in particular the formation of adsorbed nascent hydrogen at the steel surface. This nascent hydrogen may combine to form gaseous molecular hydrogen (H_2), but in the presence of H_2S the atomic hydrogen can diffuse into the steel instead of combining in this way. The hydrogen atoms may, however, combine inside the steel at various discontinuities (traps) formed by, for example, inclusions such as sulphides. The hydrogen gas having formed internally cannot escape so may form internal blisters. These are formed under pressure and may cause problems with the steel. Whether or not this reaction occurs will depend upon factors such as the moisture content of the gas, the partial pressure of the H_2S and the temperature. Sour gas situations are concerned with certain levels of H_2S where these reactions are likely to occur. Hydrogen blistering which indicates the high pressure exerted internally by the gas does not usually lead to brittle fracture but can cause rupture of the steel.

Where a number of blisters are formed in the steel and are close together, these may result in cracking in steps, i.e. 'stepwise cracking' (SWC). This is shown diagrammatically in *Figure 17.2*. This form of attack does not depend upon the presence of applied or internal stresses. The severity of hydrogen cracking will be determined by factors such as:

  (i) Concentration of hydrogen sulphide.
 (ii) Shape of internal defects where hydrogen forms, e.g. a long narrow crack is likely to produce stress concentrations at the end.
(iii) Amount of hydrogen diffusing into the steel.
(iv) The environment.
 (v) The amount of inclusions and laminations in the steel, i.e. whether it is 'clean'.

A considerable number of laboratory investigations into this phenomenon have been carried out in attempts to determine the relative importance of the variables concerned. At present the results of such work have not provided sufficient useful data to establish clear guidelines for its control. There are, however, a number of useful points that have arisen from the investigations and a detailed review has been published by Biefer[2].

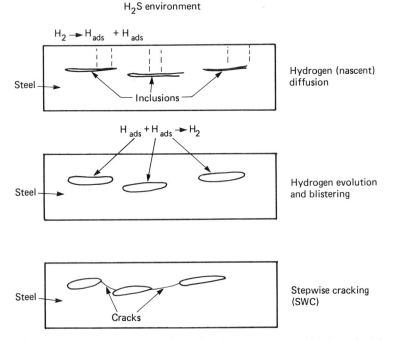

Figure 17.2 Stepwise cracking (SWC). H_{ads} represents a hydrogen atom adsorbed onto the steel surface

17.9.1 Test methods

Two test methods are commonly used: (i) the 'BP Test' and (ii) NACE (National Association of Corrosion Engineers) Method TM01-77; both are immersion tests. In the 'BP Test' specimens 20 mm × 100 mm are cut from the circumference of the pipe at different points. They are immersed in a de-aerated synthetic sea water solution saturated with H_2S. At the end of the test period, the specimens are removed, polished and microscopically examined.

The NACE method uses a more aggressive acidified chloride solution. Other test methods are also employed; sometimes the actual pipe is tested. The work of NACE Task Group T-IF-20, which is considering test methods, was reviewed in a presentation by Bruno and Hill, at the NACE Annual Conference in 1980 (since published as NACE Standard TM-02-84. Test Method 'Evaluation of pipeline steels for resistance to stepwise cracking').

17.9.2 Summary of work

Many of the investigations have produced conflicting results, but the following general points can be made:

(i) The effects are generally worse with sea water than other forms of water.
(ii) The combination of CO_2 and H_2S is particularly severe.
(iii) The type of steel has an influence on the probability and nature of attack. This results from the effects of different processes on the segregation and form of inclusions, e.g. controlled rolling produces elongated sulphide inclusions.
(iv) The additions of rare earth metals and about 0.01% Ti are considered to improve the shape of the inclusions, i.e. to produce a more rounded form.
(v) Copper and other additions, e.g. nickel and chromium, may—under certain circumstances—reduce the absorption of hydrogen into the steel.
(vi) Low pH tends to increase susceptibility.
(vii) Temperature influences susceptibility but does not vary much in the range 15–35 °C.
(viii) Increased concentration of H_2S leads to increased susceptibility to attack. There is evidence of a threshold value below which attack does not occur.

17.9.3 Control of hydrogen-induced cracking

The control methods will be based on the background knowledge concerning the causes of HIC. These can be summarised as follows:

(i) Control of the environmental situation that causes the formation and absorption of hydrogen into the steel. There are practical difficulties in achieving this but methods such as coatings, inhibitors, raising the pH, removing moisture or hydrogen sulphide would effect such control.
(ii) Use of steels less susceptible to this form of attack. This again leads to practical difficulties. Furthermore, much of the data indicating the cause of susceptibility is based on laboratory tests and such results may not be generally applicable. Changes to reduce or alter the shape of inclusions, alloying steel with, say, copper, and general improvement in steel quality all appear to be the possible approaches in this area.

17.10 SULPHIDE STRESS CRACKING (SSC)

The type of attack considered above can reasonably be considered as hydrogen blistering. Although cracks may appear where there are stress concentrations at the ends of sharp inclusions, basically stress is not a requirement for this type of attack. There are, however, two forms of attack that can occur where steel is stressed. A range of terminology has been used to describe these forms of corrosion, e.g. 'hydrogen sulphide cracking', 'sulphide stress corrosion cracking', 'sulphide cracking' and 'sulphide stress cracking', the term used here. Irrespective of the terminology, the phenomenon is essentially hydrogen cracking stimulated by sulphide, which acts as a 'promoter' for hydrogen absorption. The two types of attack differ in the form of cracking that occurs.

In the first type the application of stress may lead to cracks appearing in the steel parallel to the line of blisters that can occur in the absence of stress. This is stress cracking in the sense that failures which would probably occur anyway become more acute in the presence of stress.

The other type, though, is more akin to stress corrosion cracking or hydrogen embrittlement (see Section 4.6) and cracks are at right angles to the applied stress.

The factors that lead to hydrogen absorption, e.g. the presence of hydrogen sulphide and traces of moisture, are a necessary requirement for this type of cracking. However, only steels of high strength under tensile stress, applied or residual, are affected. The strength levels at which cracking can occur are also related to the microstructure so may depend on the steel composition and production route.

There appears to be a progression from hydrogen-induced cracking to sulphide stress cracking. For example, Ohki *et al*[2] have shown with stressed specimens that by changing the shape of inclusions stepwise cracking was eliminated and resistance to failure was improved but sulphide stress cracking occurred. However, it is by no means certain that measures to improve stepwise cracking (SWC) will necessarily improve resistance to SSC. Furthermore, the level of stress may influence the form of failure, i.e. hydrogen-induced cracking or sulphide stress cracking. It is probably sufficient to accept that in the presence of hydrogen sulphide and moisture many steels will be affected by blistering and under stress may crack.

The situation at welds is likely to be a critical area for this form of attack and the advantages of methods such as stress relief would be expected to influence the nature and severity of any attack.

17.11 CORROSION FATIGUE

This is covered in Section 4.7. By the nature of offshore structures situated in positions where they have to withstand cyclic stressing from the action of winds and waves, corrosion fatigue is a potential problem, particularly at weld areas. A good deal of investigational work is being carried out and information is being collected continually. It is, however, a complex form of corrosion and only limited design data is available. Some of this is conflicting; for example, the influence of cathodic protection. Furthermore, corrosion fatigue data tends to have an inherent scatter.

As noted in Section 4.7, the general ways of dealing with the problems can be summarised as follows:

(i) Reduction of stress concentrations wherever practicable.
(ii) Materials selection to avoid potential problems.
(iii) High quality control of all stages, especially of fabrication and welding.
(iv) Use of safety factors in design.
(v) Monitoring of performance.

Comments on corrosion fatigue of offshore structures are given in References 10, 11 and 12 in Chapter 4.

17.12 SUBMARINE PIPELINES

The term 'submarine pipeline' is used to indicate a pipeline below the sea, generally used to carry oil and gas from offshore production but also employed for other purposes, e.g. effluent and sewage disposal. They may vary in length from 1000 m or less to many kilometres. Throughout the world there are several hundreds of kilometres of such pipelines and the protection of the exterior of the pipes against attack by sea water has, over the last few years, been a matter of considerable interest and development. The techniques for producing and laying the pipes will not be considered here in any detail. The methods have, however, an influence on the form of coatings used, so it is worth noting the most common ones. These are:

(i) Pulling method, where the pipeline is constructed in lengths and pulled into position on the sea bed by winches fixed on anchored ships at sea. This method is generally used only for comparatively short pipelines, and requires a large construction site near the sea.
(ii) Lay barge method, in which the pipe is assembled and joined on a special barge then lowered to the sea bed from one end of the craft. Pipework in suitable lengths is supplied to the barge.
(iii) Float and sink method, in which the pipeline is constructed on the sea shore or river bank and floated, using buoyancy tanks, then towed to position, where it is lowered into place from pontoons. The

pipe sections are joined on a special barge close to the pontoons. This method is not generally used for long pipelines laid under the sea, mainly because of the problem of handling large floating lengths of pipeline under the conditions prevailing some distance from the coast.

(iv) Reel barge, a method in which the pipeline is reeled in long lengths from special craft has been developed and is now widely used. Pipelines may rest on the sea bed or be buried in trenches formed by dredging, ploughing or jetting. Although this is not a matter for consideration here, the nature of the sea bed clearly has some relevance to the abrasion-resistance requirements of the protective coatings used on the pipeline.

The majority of pipelines are of steel, although for some purposes plastics materials and non-ferrous alloys have been used.

17.13 EXTERNAL COATINGS FOR SUBMARINE PIPELINES

A wide range of coatings is available for the external protection of pipelines immersed in sea water. A number of factors have to be taken into account when choosing a coating but one, specially applicable to submarine pipelines, is the use of concrete weight coatings to ensure negative buoyancy. The adhesion of concrete to organic coatings is not always easy to achieve by simple methods. Generally, submarine pipelines are protected by both coatings and cathodic protection. Consequently, depending on the type of coating chosen and the method of fixing the anodes to the pipeline, it may be necessary to remove small areas of coating to enable the welding of steel plates for fixing the anodes. This is followed by local repairs to the coating. Coatings will also have to be applied to field welded areas of pipes that have been coated in works.

(i) *Factors for coating selection*
The following factors have to be considered in selecting a suitable protective coating for the external surfaces of submarine pipelines:

Resistance to sea water
Ease of application
Impact resistance
Long-term stability
Adhesion to steel pipe
Resistance to bacteria and marine organisms

Abrasion resistance
Electrical resistance
Flexibility
Resistance to cathodic disbondment
Adhesion of concrete to coating
Temperature resistance
Effects of stress
Cost

(ii) *Coatings*

The coatings most widely used for submarine pipelines are as follows:

Bitumen 'coat and wrap'. At one time this was extensively used but new coating systems, particularly thin film types developed in recent years, have tended to replace it for many situations.

The steel surface is cleaned, preferably by blast-cleaning, although sometimes manual methods such as wirebrushing have been used. For cold applications, the pipe is primed, usually with a quick-drying primer based on chlorinated rubber or a similar polymer. This is usually followed by two coats of bitumen or coal tar with suitable fillers and reinforced with a suitable material such as glassfibre. Finally the coating is wrapped, e.g. with felt. For hot application one method described is as follows[3]: the steel is blast-cleaned and preheated to 90°C. While the pipe is still hot a coal-tar-type primer is applied followed by a flood coating with enamel. Simultaneously an inner wrap of glassfibre is pulled into the enamel. Finally a coal-tar-impregnated thermoglass wrapping is applied.

Fusion-bonded epoxy coatings. A comparatively recent development, fusion-bonded epoxy coatings have been used for a number of pipelines but, because the coating is usually only about 0.3 mm thick, their effectiveness depends very much upon the quality of the coating procedures. Although these may vary with different coating applications, the following indicate the nature of the procedures.

(a) Pre-heat pipe.
(b) Blast-clean to Swedish Standard Sa$2\frac{1}{2}$ or Sa3.
(c) Heat pipe.
(d) Application of epoxy powder by electrostatic methods as the pipe rotates and moves through coating chambers.
(e) Quenched in water.
(f) Final inspection.

Thorough inspection of the coating procedures at all stages is carried out to check the presence of small pores or holes ('holidays') by suitable methods, usually by spark testing. Small defects are repaired, often by using a melt stick of epoxy at local points. Where lengths of coated pipe are welded together, these areas are usually treated by blast-cleaning followed by induction heating and application of epoxy powder to produce a high-quality coating. Special equipment has been designed for cleaning and coating welds. These fit over the pipe and the coating operation takes 3–5 minutes at each weld, depending on the size of pipe. Special equipment is also available for coating the internal welds of pipes, where the fusion banded coating is also applied internally.

Polyethylene coatings. Polyethylene materials can be applied in a number of ways—by tape, from fluidised beds or by extrusion methods. There is a range of polyethylene materials, but they are not all suitable for pipeline coatings. Low-density polyethylene used for fluidised bed coatings has a high melt flow index and this leads to stress-cracking difficulties. This material is suitable for smooth decorative finishes but where low-melt flow indexes can be used this reduces the stress-corrosion problems. Nevertheless, coatings applied by fluidised-bed processes followed by sintering have not proved completely satisfactory for underwater coatings. Polyethylene tapes have been widely used for underground pipelines but for submarine pipelines the tendency is to use high-density polyethylene coatings applied by extrusion methods, e.g. the flat die process. In this method, the polyethylene is extruded on to the pipe as a wrapping sheet as part of the following general process to provide a coating 3–4 mm in thickness:

    Blast-cleaning
    Pre-heating
    Application of adhesive
    'Spiral' application of polyethylene coating
    Cooling
    Cutting of coating at ends

One problem with this method, compared with fusion bonding of epoxy is that the welds cannot be protected by this process in the field. Other methods such as cold wrapping tapes or heat shrink wraps have to be used. The latter method requires care and some expertise to ensure satisfactory protection. Both methods may provide a level of protection somewhat lower than the main coating. As environmental stress cracking resistance (ESCR) is important with polyethylene materials, suitable checks must be made, e.g. tests to ASTM D1698-70.

General. Other materials are available for coating the exterior surfaces of submarine pipelines, e.g. Neoprene, Hypalon, nylon and solventless urethanes, but the three noted above are commonly used at present (1983).

The technology of coating pipelines will no doubt be developed with the application of new techniques, and possibly new coating materials will be used. Cathodic protection is generally applied to ensure adequate corrosion protection for long periods. Furthermore, regular checks on corrosion are carried out by suitable methods. The internal coatings for the pipes will not be considered here because these will be determined by the material being transported in the pipes and this will not be affected by whether the pipe is buried in soil, immersed in water or exposed above ground. Submarine pipelines are widely used in the oil industry and for many purposes no internal coating is required. With modern methods of reel-laying of pipelines, problems may arise with the application and repair of coatings at welds and due attention should be paid to this aspect.

(iii) *Concrete weight coating*

The use of concrete coating to achieve negative buoyancy can lead to problems with bonding between the concrete and the protective coating. A number of approaches have been considered, including mechanical bonding with spirals on the protective coating and special polymer cements. The concrete coating provides additional protection provided there is no slippage. Where concrete coatings are used, the field joint should be made up to the concrete level with suitable mastic materials.

17.14 QUALITY CONTROL OF SUBMARINE PIPELINE COATINGS

Quality control of all coating applications is important (see Chapter 12) but it is particularly so for submarine pipeline coatings. Some specific areas that require particular attention with fusion-bonded coatings are noted below.

(i) The surface quality of pipes is important, particularly for the thinner fusion-bonded coatings. All defects should be carefully ground before blast-cleaning.

(ii) Thorough cleaning of the steel surface is essential. Preferably pipes should not be allowed to rust in stock to a situation where they become pitted, as this makes cleaning more difficult.

(iii) Careful checks should be made on the quality of the powders used.

(iv) Temperature control of a high order is necessary during the coating process, as is careful control of the quench cure.

(v) High standards of 'holiday' detection over the whole surface is an essential requirement for these coatings.

Other tests are also carried out to check the following requirements:

(i) Cathodic disbondment as a check that the cathodic protection process does not result in loss of bond between the coating and the steel. Although test methods are available, e.g. ASTMG8 Method, some authorities have developed their own, e.g. British Gas, and, for a particular pipeline, the methods specified by the client have to be adopted.
(ii) Stress-corrosion cracking.
(iii) Repairs of the coating and protection at joints.
(iv) Resistance to flow.
(v) Resistance to impact.

These and other tests are discussed in papers by Hankins[4] and Bayliss[5].

17.14.1 Stress corrosion

Stress-corrosion cracking is recognised as a cause of failure in buried cathodically protected on-shore pipelines. It is generally accepted that the cracking is caused by the alkalinity at the pipe wall arising from the cathodic protection process and polarisation into a critical potential range where stress-corrosion cracking is likely to occur.

Although no such failures have been reported with submarine pipelines, recently reported investigations[9] indicate that the possibilities of such failures cannot be discounted. Because attack on-shore occurs at defects in coatings, e.g. disbondment, the higher standards of surface preparation used for submarine pipelines would be expected to reduce the likelihood of attack. Furthermore, calcareous deposits formed in sea water would be expected to protect steel from this form of attack.

17.14.2 Summary

The protection of submarine pipelines for long-term durability provides a most challenging task. Although monitoring of the pipelines is carried out, maintenance work is obviously difficult and any serious corrosion of the pipe would be a matter of concern. High-quality coatings with cathodic protection should provide the required durability, but there is only a limited indication of the long-term performance of many of the coatings at present being used or considered for use. Clearly a high level of quality control is essential and undoubtedly further developments in this field are to be anticipated.

17.15 REFERENCES

(1) BIEFER, G J, *Materials Performance*, **21**, 6 (1982) p 19

(2) OHKI, T, *et al, ASTM*, STP610 (1976)

(3) KIERNAN, E G, *First International Conference on the Internal and External Protection of Pipelines*, BHRA (1975)

(4) HANKINS, E E, *Third International Conference on the Internal and External Protection of Pipelines*, BHRA (1979)

(5) BAYLISS, D A, *Corrosion 82*, Paper 121 (unpublished paper), National Association of Corrosion Engineers, Houston, Texas, USA

(6) DUTRA, A C, *Materials Performance*, **16**, 6 (1977)

(7) DUNCAN, R N, *Materials Performance*, **19**, 7 (1980) p 45

(8) GATLIN, L W, *Materials Performance*, **17**, 5 (1978) p 9

(9) MOLLAN, R *et al*, *Materials Performance*, **21**, 8 (1982) p 50

18 Protective coating of ships

The general principles for the protection of steelwork outlined in Chapters 9–12 are relevant to steel ships. Surface preparation of the steelwork and coating application require the same standards and degree of control for ships as for other structures exposed to marine conditions. Ships are, however, different from static structures in that they can be dry-docked for maintenance painting. On the other hand, their speed and fuel costs are markedly affected by any marine fouling that collects on the hull.

There are a number of fairly clearly defined areas of steelwork on ships as indicated in *Figure 18.1*, and the requirements for each differ in a number of ways. These areas are broadly as follows:

   (i) Underwater plating— ships' bottoms
  (ii) Boot-top plating
 (iii) Topside and superstructure
 (iv) Cargo holds
  (v) Cargo and ballast tanks
 (vi) Internal accommodation
(vii) Machinery, etc

These will be considered in more detail later but a number of general points are relevant to all the steelwork of a ship, in particular surface preparation before painting.

The surface preparation of steelwork has been considered in detail in Chapter 9, and for sound performance from coatings, a high quality of cleanliness of steel surface is essential. For the initial painting of ship plating this should not be a problem. A common procedure is to blast clean the ship plate in automatic plants before fabrication and to coat it with a thin 'holding' or prefabrication primer, which is quick drying and capable of being welded through during fabrication. It is essential to apply the prefabrication primer before any rusting of the steel occurs. This will depend upon the conditions of temperature and relative humidity in the shop, but a maximum period of four hours after blast-cleaning is often specified.

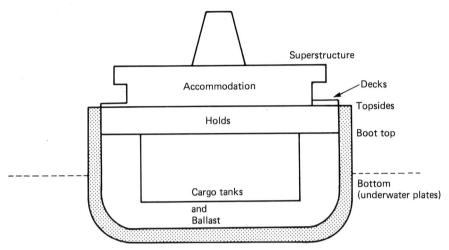

Figure 18.1 Areas of protection on ships

It is advantageous to use new steel from the mill with as little rust as possible. Although the blast-cleaning will remove all scale and surface rust, if the steel has pitted to any extent during weathering after rolling, some salts may remain on the steelwork leading to problems as discussed in Chapter 9. The automatic plant uses centrifugal shot-blasting equipment with a number of impellers. The abrasive fed to each impeller can be controlled individually to take account of the dimensions of the steel plates or sections. Plates are fed through horizontally on a roller conveyor with both sides being cleaned in one pass with the abrasive being recovered after screening. The plates then pass through an automatic spraying unit for the application of the prefabrication primer. They are then removed to a storage area. Some plants have a preheater unit to warm the plates before they enter the blast-cleaning plant. It is claimed that this assists adhesion and is beneficial in cold weather because it ensures that the plates are dry when cleaned and primed.

The standard of blast cleaning for most ship plate will be Swedish Standard Sa $2\frac{1}{2}$ or the equivalent if another standard is specified, e.g. Second Quality Finish (BS 4232). For some coatings, e.g. zinc silicate or sprayed aluminium metal, a higher standard may be specified. The surface profile will depend upon a number of factors but will generally be a maximum of 100 μm. These standards can be achieved on new plate in automatic plant by proper control of the blasting conditions and abrasive.

During fabrication the primer will be damaged, either deliberately, e.g. through welding, or accidentally. It may be necessary to touch up certain areas to protect the steelwork before the final coating is applied.

Touching up should not, however, be carried out indiscriminately. Advice from the paint suppliers should be sought; for some coatings it may be necessary to blast-clean rusted areas before recoating them.

Before the full coating system is applied, the steelwork should be thoroughly cleaned to remove grease and dirt. This is an operation that does not always receive sufficient attention. During fabrication, grease, dust and general dirt inevitably reach the surfaces to be painted and they must be properly removed if the full potential of the coating system is to be achieved. All grease should be removed by solvent cleaning and dust removed completely, not transferred from one part of the surface to another. Washing down should be carried out with clean fresh water and if a zinc primer has been used any corrosion products should be thoroughly brushed off. Effective washing is particularly important on ships, which are constructed near the coast where sea salt is likely to be deposited on the steel.

18.1 PROTECTIVE COATINGS FOR NEW CONSTRUCTION

The coatings for the different areas shown in *Figure 18.1* will be considered individually.

18.1.1 Ships' bottoms—underwater plating

Paint systems for underwater plating of ships have to resist conditions of high conductivity sea water and, as most ships' hulls are cathodically protected, they must be non-saponifiable. Consequently, the conventional paints used for many structures are not suitable. Typical systems for ships' bottoms are based on the following binders:

Coal-tar epoxy
Chlorinated rubber
Vinyl-tar
Bitumen or pitch

The coating system is usually applied to provide a high dry film thickness and the bitumen and tar paints are generally pigmented with aluminium flake. Modern chemically curing coatings, such as epoxies, can be applied in thick single coats, although it is generally advisable to apply at least two coats to overcome any defects in a single coat, e.g. pinholes or 'holidays'.

Some authorities consider that zinc-pigmented primers should not be used for underwater service whereas others consider that they are suitable. Advice should be sought from paint suppliers where such primers are included in an underwater protective system.

Apart from the anti-corrosion compositions, anti-fouling coatings also have to be applied to ships' bottoms.

18.1.2 Anti-fouling coatings

Although fouling by the attachment and growth of plants and other organisms occurs on many marine structures, it is of particular concern on ships because it can impair their efficiency. This arises because of the friction between the hull with the attached fouling and the water. This either reduces the speed of the ship or involves the use of more fuel to maintain the required speed, although in some cases this is not possible if fouling is too severe. Furthermore, the requirement to remove fouling limits the time a ship can usefully be operating at sea.

Fouling takes place mainly when ships are in port and the incidence and characteristics of the fouling depends on the particular port, the time of the year, the temperature of the water and the time spent in the port. A report on fouling distribution in different parts of the world has been prepared by the British Ship Research Association [1]. The general way in which fouling occurs on painted steel immersed in sea water has been explained in various publications [2]. The surface first becomes covered with slime, weeds then grow from the spores entrapped in the slime and finally barnacles, tube worms and other animals become attached.

Most anti-fouling coatings incorporate biocides, such as copper and its compounds, to control fouling by poisoning the growths, but other biocides such as lead, mercury and arsenic are still used. The copper compounds used are expensive and other substances such as organo-tin and organo-lead compounds have been investigated and, in some cases, e.g. tributyl tin oxide, have been reported to have been successful on a limited scale. Anti-fouling paints work essentially by maintaining in the water in close proximity to the painted surface a concentration of poison that will kill all fouling organisms. Paints must, therefore, be formulated to contain sufficient of the toxin to allow a reasonable life for the anti-fouling paint and to allow it to leach out at an appropriate rate. There are two main types of paint used:

(i) The 'continuous contact' type in which the paint matrix remains unattacked and the toxin particles are leached out at the required rate.
(ii) The 'soluble matrix' or 'self-polishing' type in which layers of paint containing the toxin are removed to provide the required amount of poison. This method has an advantage because by the use of suitable

coloured layers it is possible to provide an indication of when the anti-fouling paint requires replacing.

Tests on an oil exuding anti-fouling coating have been reported [3]. These incorporate an exuding fluid in a silicone rubber matrix and have the advantage of being non-toxic. These have been developed primarily for offshore structures but at present are not as resistant to mechanical damage as conventional anti-fouling coatings and are less easy to apply.

Anti-fouling paints containing copper or mercury should never be applied directly to steel because of the possibility of galvanic attack. The performance of anti-fouling compositions can be affected by the type of anti-corrosive or tie coat over which they are applied. It is therefore advisable to consult the paint manufacturer before choosing an anti-fouling composition and preferable to purchase all the paints, including the anti-fouling, from one manufacturer. It is usual practice when replacing anti-corrosion composition, to apply a barrier coat over the old anti-corrosion composition, e.g. an aluminium-bitumen paint.

The choice of anti-fouling will be determined by the operating conditions of the ship, the time between dockings and cost. The anti-fouling coats have a limited life because the toxins are released and used until a stage is reached where they are no longer effective in preventing the attachment of marine organisms. The life of the coating will depend upon the type, thickness and operating conditions and will vary from one to three years. Typical systems for ships' bottoms are shown below. All require an additional anti-fouling coat. Cathodic protection is generally used for this area of the ship. This is considered in Chapter 16.

Protective system	Number of coats	Total thickness (μm)
Aluminium pigmented vinyl tar	3	225
Aluminium pigmented bituminous	2	200
Coal-tar epoxy	2	250
Chlorinated rubber	3	225

The following areas require special treatments as outlined in the 'Recommended Practice for the Protection and Painting of Ships' (British Ship Research Association) [2]:

(i) Keel plates should be fully painted with an additional thickness of 50–100 μm over the specified dry film thickness for the bottom plating. This should be applied before or shortly after the keel is laid.

(ii) Special provision should be made to ensure adequate protection of areas that will become inaccessible for painting on the stocks. The complete coating system, possibly including the anti-fouling composition, should be applied to such areas before the fabricated parts are placed on the stocks or before the launching cradle is fitted.

(iii) High-duty coatings are desirable on stabiliser fins, shaft brackets, rudders and stern posts and wherever the water speeds are sufficient to cause damage to the coatings. Thick epoxide coatings (500 μm) and solvent-free epoxides have been used for this purpose. Such coatings may, with advantage, be reinforced with glassfibre on the leading edge.

(iv) Where there is a likelihood of considerable erosion or cavitation, tough coatings such as neoprene may be considered, although it may be difficult to obtain satisfactory adhesion.

(v) Non-ferrous fittings such as sea water inlet and discharge pipes may be in contact with underwater plating and the bimetallic effects can result in corrosion of the steel. This effect will generally be suppressed where cathodic protection is used. Otherwise, local cathodic protection with sacrificial anodes should be considered.

18.1.3 Boot topping

This is the area equivalent to the splash zone of marine structures and cathodic protection is not effective for such conditions. The boot topping is potentially the area liable to maximum corrosion, so good protection is essential. The decorative aspect may be important, so coloured finishes may be required. The choice depends on the type of ship and the service conditions under which it will operate. In some cases the same system is used for both the underwater plates and the boot topping. Typical systems include chlorinated rubber and two-pack epoxy, both applied to a total film thickness of 230–250 μm.

18.1.4 Topsides and superstructures

These areas are typical of structural steel exposed to atmospheric marine conditions. Colour and gloss are often important, particularly for passenger ships, but there is generally a wide choice of paints. These include systems based on the following binders:

Alkyd gloss
Vinyl
Chlorinated rubber
Two-pack epoxy

The systems are generally of three or four coats including suitable primers and a total dry film thickness of 175–200 μm.

Light alloys are also used in the atmospheric zone of ships. Aluminium should be anodised but, where paint coatings are required, suitable degreasing and the application of a wash primer is necessary to ensure good paint adhesion.

18.1.5 Internal accommodation

Conventional alkyd systems or two-pack epoxy or polyurethane systems are used for these areas, which can be considered as being similar to the internal steelwork of buildings exposed to marine conditions.

18.1.6 Cargo and ballast tanks

Severe corrosion may be experienced in cargo and ballast tanks of ships, particularly oil tankers. This results from the corrosive nature of sea water and other cargoes that may be carried. This is further aggravated by the washing of the tanks with sea water. Refined oil products provide little protection, although crude oil leaves a waxy type film on the surfaces of the tanks, which provides a measure of protection. This protection is not, however, continuous, so any salt water ballast carried later may result in local attack and pitting on unprotected areas. Some crude oils contain sulphur compounds, which may react with moisture to form sulphuric acid with serious corrosive effects on tanks which contain them.

A number of methods other than coatings are used to control corrosion in tanks. These include:

(i) Cathodic protection.
(ii) Injection of inert gases to remove oxygen.
(iii) Dehumidification to remove moisture in empty tanks or above oil cargoes.
(iv) Use of inhibitors.
(v) Float coats, e.g. an oil film on the surface of the water.
(vi) Wax and grease coatings.

The selection of the coating system will depend on the cargoes to be carried and advice will obviously be sought from coating manufacturers before a choice is made.

For ballast tanks that are permanently in salt water ballast, cathodic protection alone is suitable. Generally, however, where a coating either alone or in conjunction with cathodic protection is used, a coal-tar epoxy paint with a dry film thickness of about 250 μm can be used. For tanks that are out of ballast for reasonably long periods, a combination of cathodic protection with a coating system such as coal-tar epoxy is likely to provide the best results. Such coatings are necessary because the cathodic protection will operate only in the parts of the tank containing sea water.

Where tanks are used for cargo/ballast, then the coatings selection must be suitable for the liquid cargoes to be carried. Coal-tar epoxies are suitable for some cargoes, but for others a range of coating materials is available as follows:

System	Number of coats	Total thickness (μm)
Zinc silicate	2	120
Coal-tar epoxy	2	250
Solvent-free coal-tar epoxy	1 or 2	300
Polyurethane	2	250
Two-pack epoxy	2	250
Solvent-free epoxy	1 or 2	250

18.1.7 Fresh water tanks

For non-potable water, protection can be achieved with tar or bituminous paints applied to a thickness of 150–200 μm. For drinking water, however, the coating must be non-tainting and paints based on natural or artificial bitumen are available, e.g. to British Standard 3416 Type II. Newly painted tanks should be rinsed out with fresh water before filling with fresh drinking water. Other coating systems for fresh water tanks include epoxy-polyamide paint and solvent-free epoxies.

High standards of surface preparation and coating applications are required for tanks and full quality control procedures should be adopted, e.g. testing for pores or 'holidays'. Suitable precautions such as adequate ventilation to remove solvents and, where appropriate, humidity control, are advisable when painting the interiors of tanks.

18.1.8 Deck plating

Deck plating is difficult to protect because of the combined effects of corrosion, abrasion and wear. Furthermore, for safety reasons, the coating must have non-slip qualities.

The non-slip requirements are usually obtained by using a specially formulated paint containing grit for the weather coat or by adding special grit to the finishing coat. Sprayed metal coatings have also been used for weather decks. Composite coatings with zinc applied to the steel followed by aluminium and a number of coatings of epoxide paint have been used on naval vessels.

All the following paint coatings have been used for weatherdecks: alkyds, bitumen, vinyls, chlorinated rubber, two-pack epoxies and zinc silicates. Zinc silicate coatings of 125 μm thickness are reported to have performed well in service. They are particularly useful as abrasion-resistant coatings where cargoes are not of an acid or alkaline nature. For acidic or alkaline conditions, epoxy systems are to be preferred. Bituminous compositions are not satisfactory for vessels exposed to high temperatures, e.g. in the tropics.

Where appropriate, different systems can be used on different parts of the deck, depending upon the actual service requirements.

18.1.9 Deck machinery, pipes, etc

It is difficult to generalise on deck machinery, hatch covers and pipes. Often machinery is supplied already coated or with a protective priming coat. In the latter case suitable systems such as those used for the other areas exposed to the atmosphere can be used. Generally, careful attention must be paid to the machinery because of its shape and the various components and fittings that may be bolted to it.

The design is important and often suitable covers will reduce corrosion at inaccessible areas.

Where appropriate, pipes should be hot-dip galvanised before painting. In some situations, suitable wrapping tapes can be used to protect pipes.

18.2 MAINTENANCE OF SHIPS' COATING SYSTEMS

Maintenance of ships can be considered in relation to (i) the underwater steelwork and (ii) steelwork not permanently immersed in sea water.

Although methods have been developed for cleaning and painting under water, ships' bottoms are generally recoated in dry dock. The above-water steelwork can be repainted without dry-docking.

18.2.1 Underwater steel

The preparation for maintenance painting in dry dock is essentially different from that for new work. Generally, maintenance consists of applying new paint over old apart from damaged areas or where premature failure of the coating system has occurred. The sides of the ship are usually scrubbed down with scrapers and brooms. High-pressure water may also be used to clean the steelwork. The aim is to remove the fouling and loose anti-fouling paint from the under-water parts of the ship. Any defective paintwork, e.g. where it has blistered, is treated with scrapers, wire brushes, etc and, where appropriate, blast-cleaning is carried out, either with dry abrasives or by water-blasting. All salt deposits and salts in the pits of any rusted steel should be removed and it may be necessary to control the size of abrasive to ensure thorough cleaning of the pits. Conventional anti-fouling coatings are usually sealed before recoating with a sealer chosen to take account of both the anti-fouling coating and the type of anti-corrosive composition. When coatings that have broken down are removed before repainting, care must be exercised in choosing the maintenance coating because at the edge of the area that has been cleaned there may be a number of coats present from previous maintenance paintings and the effects of applying certain types of coating must be checked. For example, a coal-tar epoxy may have been coated with a vinyl anti-fouling coating and, at the edge of the repaired area, the anti-fouling could be sandwiched between two coats of coal-tar epoxide. This could well lead to an adhesion failure during service. To avoid this possibility, a vinyl tar might be preferred to the coal-tar epoxy. Again, changes in the type of anti-fouling during maintenance should not be made without seeking specialist advice.

18.2.2 Boot topping

The environmental conditions at the boot topping are generally more severe than those on the area above, i.e. the topsides. Often, the two areas are painted the same colour but during maintenance it is advisable to examine the coatings carefully to ensure that the boot topping is adequately protected. If appropriate, the protective coatings for the boot topping should be extended to the topsides. The reverse, i.e. extending the protective systems appropriate to the topside to the lower areas, may cause problems. The boot topping coatings should be suitable for cathodic protection as they will be immersed in sea water for periods during service. Furthermore, anti-fouling may be carried into this area and a suitable coating must be selected to withstand intermittent immersion and atmospheric exposure.

18.2.3 Superstructures

Superstructures are often maintained by the ships' crews and conventional coatings are often used for these areas. If, for some reason, the maintenance is not being carried out adequately, or where conventional paints do not provide sufficient colour-retention and gloss, other coating systems, e.g. polyurethanes, may be considered, but advice should be sought before overcoating conventional paints with two-pack materials.

18.2.4 Tanks

Maintenance of tanks is usually carried out when the ship is dry docked. The tanks should be carefully examined and thoroughly cleaned before any recoating is carried out. Where appropriate, blast-cleaning should be used to clean parts, or even the whole, of the tank, before recoating.

18.2.5 Summary

The protection of ships by coatings follows the same principles as those outlined in Chapters 9–14 for marine structures. There are some basic differences and the main points to be considered are:

(i) Anti-fouling requirements for underwater areas and selection of coatings that are compatible with anti-corrosive coatings. This may be of particular importance when considering maintenance.
(ii) The selection of systems for underwater plate that are compatible with cathodic protection, i.e. non-saponifiable.
(iii) High-quality surface preparation and application techniques combined with sound selection procedures for ballast, water and cargo tanks. Sound ventilation procedures during coating operations.
(iv) Careful attention to design, particularly where steel is used in conjunction with light alloys or timber, e.g. decks.
(v) A good deal of the maintenance painting of ships is carried out during service, but the underwater plating and tanks are usually dealt with at dry docking. Often, in the limited time available, it is not possible to carry this out under ideal conditions and to the standard required. It is, therefore, advisable to examine underwater plating carefully so that the areas requiring immediate attention are properly dealt with.

The following chapters should be considered in relation to this particular chapter:

Chapters 7, 9, 10, 11, 16.

18.3 REFERENCES

(1) 'Fouling Distribution in the Major Oceans', BSRA Tech. Mem., No 397 (1971)
(2) 'Recommended Practice for the Protection and Painting of Ships', British Ship Research Association
(3) WILSON, R W, *Proc. UK National Corrosion Conference*, p vii (1982), Inst. Corr. Sci. Tech., London

19 Steel piling

There are two broad types of steel piling: driven piles used for foundations or supporting structures and sheet piles used for retaining purposes.

An investigation into the underground corrosion of steel piles driven into undisturbed soils has led to the conclusion that it is negligible, irrespective of the soil type and characteristics[1]. Pitting may occur in the water table zone but this is not generally regarded as affecting the structural integrity of the piling, although this may not be so in some marine sediments[2]. Investigations of piles that have been in service for 25 years show that corrosion varies from nil to 30 μm/y, with a mean of about 10 μm/y.

Generally, therefore, corrosion of buried piles is not excessive in ordinary soils; attention should, however, be given to situations where the water may contain excessive amounts of chlorides.

Piles also have to withstand the same conditions as other steelwork in marine environments: immersion, spray and atmospheric conditions. The corrosion rates follow the same pattern as for other steelwork but, for economic reasons, the approach to protection is different.

19.1 CORROSION RATES

The actual corrosion rates of steel piles depend upon the specific location, but there are broadly four zones: mud, fully immersed in sea water, intertidal zone and spray-air.

(i) Beach groynes

Piles used to stabilise beaches on the coast are often called 'groynes' and, depending on the nature of the beach, they may suffer considerable attack just above the beach line. This arises from abrasion by sand or pebbles under the influence of movements of the tide. Measurements carried out on piles at Felixstowe on the East Coast of England after 32

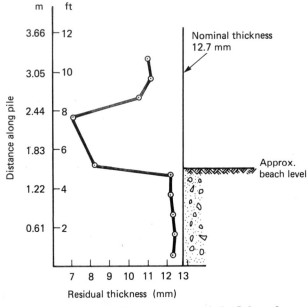

Figure 19.1 Residual web thicknesses on an extracted pile (Reference 3)

years' service demonstrate the effect of abrasion as shown in *Figure 19.1*. Although the corrosion below beach level was low, just above it the loss of steel by corrosion and abrasion was much higher. On other piles at Felixstowe even greater losses of steel were recorded just above the beach line, allowing only six or seven years' life for a pile of 12.7 mm thickness[3]. In the absence of abrasion from beach shingle, lives of over 20 years have been obtained. Piles manufactured from steels of higher abrasion resistance have been developed for such situations, but limited information on service performance is available.

(ii) *Immersed zone*
Corrosion of bare steel in the immersed zone tends to be high initially but over a period of time it decreases, partly because of the attachments of various organisms and marine growths. Tests have shown that corrosion tends to be linear with time with a rate of about 80 µm/y for sheet steel piling immersed on one side[1].

(iii) *Intertidal zone*
This zone between the mean high- and mean low-water levels tends to accumulate dense barnacle growths, which provide appreciable protection to bare steel by limiting access of oxygen and possibly by providing

some protection from wave action. The rusts formed in this area have been shown to contain up to 15% by weight of lime. This arises from barnacle secretions, and assists in reducing corrosion to a value similar to that in the immersion zone, i.e. about 80 μm/y[1]. Just below the mean low-water level the corrosion rate tends to be somewhat higher—about 100 μm/y, possibly as a result of bacterial activity.

(iv) *Splash and atmospheric zone*

These areas do not become covered with marine fouling and so do not receive the same protection as other areas of piling. Furthermore, the splash zone is continually wetted and dried, so the corrosion rates tend to be highest in this area—from 100–250 μm/y. The corrosion in the atmospheric zone depends upon the height above the splash zone and is within the range 50–100 μm/y.

19.2 PROTECTIVE SYSTEMS

Although in principle they can be coated in the same ways as other marine structures, there are a number of considerations specific to piles. There is no real requirement for protective coatings in the parts of the pile that are driven below ground level, except where they are exposed to high-chloride-containing material, particularly under conditions of virtual waterlogging. Again, where there is evidence of bacterial activity, some form of corrosion control may be necessary, possibly cathodic protection. Piling is also susceptible to mechanical damage, particularly where it is used on beaches or in harbour installations.

The decision whether to use protective coatings and, if so, the type and thickness, will be determined by economic factors. Generally, the choice is the same as for other marine structures, with the same problems of maintenance in all but the atmospheric zone.

A considerable number of tests have been carried out on coatings for steel pilings but it is difficult to draw conclusions, because, in many cases, the performance of coatings in one series of tests differs from that in another. This may arise from factors other than the coatings, e.g. application and surface preparation.

The results from three series of tests are of particular interest:

(i) The Sea Action Committee of the Institution of Civil Engineers (United Kingdom) began a series of tests in 1962 to assess nine different types of paint applied to blast-cleaned steel piles in Portsmouth, England, and in Belfast, Northern Ireland[4]. The results were somewhat disappointing. The only coatings that provided sound protection at both the intertidal and spray zones after four years at Portsmouth were a polyurethane pitch (350–480 μm) and a Neoprene coating. A solventless epoxy (500–660 μm) was in good condition but showing signs of embrittlement. The other coatings, which included the coatings listed below, were not satisfactory and suffered from defects such as lack of adhesion, intercoat adhesion problems, blistering and penetration by barnacles.

Two-pack coal-tar epoxy	(500–750 μm)
Zinc silicate-chlorinated rubber	(500 μm)
Hot tar enamels	(3000 μm)
Coal-tar pitches	(500 μm)
Vinyl copolymer	(75 μm)

The results at Belfast, covering fewer coatings, produced similar results.

(ii) An extensive series of tests designed to continue for 15 years has been carried out at Dam Neck, Virginia, USA, and a report on the first eight years has been published[5]. The tests were carried out by coating 8 in × 8 in (0.20 m × 0.20 m) mild steel H piles. Additionally, some 8 in (0.20 m) diameter pile pipes were also included. Some low-alloy (Mariner) steels were also tested.

Twenty-three protective coating systems were tested, with two of them cathodically protected. The coatings covered both polyamide and amine cured coal-tar epoxies, aluminium pigmented amine cured coal-tar epoxies, zinc silicates, flake glass filled polyester and zinc and aluminium metal coatings. All the coatings were applied to blast-cleaned steel. The piles were placed in position with 5.7 m of pile sunk into the sea bed. On some piles, the bottom 3.6 m of the piles were left uncoated. A group of piles was removed after six years and a thorough examination was carried out.

The following broad conclusions can be drawn:

(a) Coal-tar epoxies performed well when applied over a zinc silicate primer but were much less effective when applied directly to the steel, mainly because of their poor performance in the atmospheric zone.

(b) Aluminium pigmented coal-tar epoxy was superior to non-pigmented coal-tar epoxies.

(c) Coal-tar epoxies applied over an organic zinc-rich primer were much less effective than those applied over the zinc-silicate primer.

(d) Vinly-sealed sprayed aluminium metal coatings performed well.

(e) Hot-dip galvanised coatings were reasonably satisfactory.

The good results obtained with sprayed aluminium coatings are confirmed by the American Welding Society tests [7] (see Chapter 11).

A review of the corrosion and protection of steel pipes has been carried out by L L Watkins [6]. He concludes that damage from marine organisms may render some coal-tar and asphalt coatings unsuitable for use in sea water. The modern coal-tar epoxies, however, have proved to be durable. Other points made include the following:

(i) Polyamide-cured epoxies appeared to be less effective than those that were amine-cured.
(ii) Coal-tar epoxies appeared to provide good service over a seven-year period.
(iii) Neoprene coatings have given excellent service.
(iv) Chlorinated-rubber paints have not provided sound protection.

19.2.1 Summary

Because of variations in performance of the coatings in different test programmes, it is difficult to draw firm conclusions regarding the protection of steel piles. Costs are important and coatings such as neoprene are probably not likely to be widely used. It has been reported that isocyanate-cured pitch epoxies are considered by one piling manufacturer to represent the best coating system for piles. However, for many parts of the pile, that corrosion rate is such that protective coatings may not be essential. The following overall possibilities have been proposed [3]:

(i) Underground: no protection required.
(ii) Sea water immersion: protective coatings become damaged and, as maintenance is a problem, bare steel is recommended. Cathodic protection may be considered.
(iii) Intertidal zone: protective coatings such as coal-tar epoxies may be considered or, alternatively, an addition of a corrosion allowance to the pile.
(iv) Splash zone: the use of concrete jackets can be considered as extensions from the coping, where the tidal range is negligible. A minimum cover of 100 mm is required.

For high tidal ranges, the use of an isocyanate-cured pitch epoxy system (400 µm) applied to steel blast-cleaned to 1 m below mean high-water level is considered to be suitable.

19.3 REFERENCES

(1) MORLEY, J, *BSC Report* T/CS/1114/1/78/C
(2) BJERRUM, L, *Geotechnique*, **7**, 73 (1957)
(3) MORLEY, J, *BSC Report* T/CS/906/6/78/C
(4) Institution of Civil Engineers, Sea Action Committee, Technical Reports on 'Steel Box Piles at Portsmouth'
(5) NBS Monograph 158, 'Corrosion and Protection of Steel Piles in a Natural Sea Water Environment', US Department of Commerce, National Bureau of Standards
(6) WATKINS, L L, *Tech. Mem. No. 27*, May 1969, US Army Corps of Engineers, Washington, DC
(7) 'Corrosion Tests of Flame-Sprayed Coated Steel, 18-Year Report', Amer. Welding Soc., Miami, Florida, USA

20 Reinforced concrete, wood and plastics

20.1 REINFORCED CONCRETE

The two most commonly used structural materials for marine environments are reinforced concrete (RIC) and carbon steel. Concrete may deteriorate in a number of ways, e.g. by alkali aggregate reactions[1]. These forms of deterioration are not related to corrosion and will not be discussed here. However, a serious form of concrete breakdown arises directly from the corrosion of steel reinforcements and this will be considered below.

Concrete is strong in compression but comparatively weak in tension. In tension, it is no stronger than the bond between the cement and the aggregate, so steel is used to provide the tensile strength required. The steel is generally used in the form of bars or mesh so that the tensile stresses are transferred to it and the material is called *reinforced concrete*. Steel can also be used in another way—it can be pre-tensioned so that the surrounding concrete when loaded does not have to withstand any tensile stresses. This is termed *prestressed concrete*.

Concrete produces an alkaline environment for steel which passivates it, so preventing corrosion. Provided this passivity is maintained, no problems should arise with steel reinforcements. The alkalinity can, however, be affected if gases, such as carbon dioxide or sulphur dioxide, permeate the concrete and react with alkali hydroxides forming other compounds such as carbonates and sulphates, with a reduction in pH. If this carbonated layer extends to the steel reinforcement then the environment surrounding the steel is no longer sufficiently alkaline to provide the conditions for passivation. A degree of carbonation generally occurs in concrete but this is usually a surface effect and does not lead to corrosion of the steel reinforcements. Clearly, the depth of cover of the concrete must be greater than any potential depth of carbonation. As various codes of practice recommend, a cover of about 25 mm compared with a normal carbonated layer of 6–7 mm, then there should be no problems.

However, the quality of concrete has an influence on the permeability of carbon dioxide and moisture and also on the likelihood of deep cracks occurring. Cracks in concrete do not necessarily lead to problems of steel corrosion but they may do so. This will depend upon the size and type of crack and whether it becomes blocked with products arising from reaction within the concrete. In immersed sea water tests, cracks below 0.5 mm in width did not lead to corrosion [2]. However, it is possible in air for carbonation to proceed from the crack, so causing loss of alkalinity at the steel reinforcement. If the crack itself develops sufficiently then corrosive species such as oxygen, moisture and atmospheric contaminants may react with the steel, causing corrosion. The process discussed above can lead to corrosion of steel reinforcement in industrial atmospheres, although this is not generally a serious problem except where the concrete cover is insufficient. However, in marine situations, the presence of chlorides aggravates the problems.

20.1.1 Effects of chlorides

Chlorides can locally destroy very protective passive films formed on alloys such as austenitic stainless steels and this causes pitting. Equally, they can cause breakdown of passivity of reinforcement in concrete even though the general environment is alkaline. There is some reaction between chlorides and the concrete itself which can convert the chlorides to non-reactive compounds but the presence of carbonation may lead to release of active chlorides. Furthermore, the concentration of chlorides will be important in determining how quickly the steel corrodes. The quality and depth of cover of the concrete are important factors in determining the likelihood of attack by chlorides, but in many situations problems have arisen.

Once corrosion has occurred, the rust that forms will have a greater volume than the steel from which it is formed and this will exert sufficient pressure on the concrete to cause cracking or spalling. Chlorides can penetrate even dense concrete and depassivate steel under appropriate conditions. Clearly, reinforced concrete made with constituents containing chlorides may be susceptible to this form of attack. The use of sea water or chemical agents that contain chlorides may well lead to rapid depassivation of the reinforcements. It seems probable that the concentration of chloride necessary to cause problems with concrete that has cured is less than for chlorides used in the concrete mixing process itself. Although the concentrations of chloride required for depassivation of the steel will depend upon a number of factors—not least the quality of the concrete—it appears from investigational work that 0.5–1.0% chloride in the concrete is sufficient to cause the conditions necessary for corrosion.

Although not directly related to marine environments, the use of salts containing high proportions of chlorides for de-icing operations on concrete bridge decks has provided useful information on the levels of chloride likely to cause problems and in one large-scale investigation it was found that 0.4% by weight of chloride in concrete caused depassivation of the steel reinforcement [3]. The time for the permeation of the chloride through the concrete has been studied under immersed conditions and this has shown that chloride can penetrate the usual concrete cover employed in marine situations (50–60 mm) in about two years, depending on the conditions of test.

In other investigations a chloride concentration of about 1.18 kg/m³ in concrete was enough to promote corrosion [4].

20.1.2 Corrosion of reinforcements

The breakdown of passivity will lead to localised corrosion and where this is sufficient to cause the formation of a large volume of corrosion products some form of concrete failure will occur. The exact nature will vary. Sometimes cracks appear which penetrate to the steel. If these do not eventually lead to spalling it must be assumed that corrosion has been stifled in some way, but generally spalling of some sort is likely to occur.

It would be expected that local breakdown of passivity would lead to a galvanic couple being formed between the passivated and corroding steel because of the potential difference arising from the two situations. This has been confirmed by investigational work; consequently, once the steel is depassivated locally, provided that the concrete contains sufficient moisture to produce an electrolyte of adequate conductivity and there is sufficient permeability of oxygen, corrosion is likely to be accelerated by the cathodic action of the large area of passivated steel.

Where the anode/cathode ratio is very small, considerable local corrosion may occur. It has been shown that with a $1:10^5$ anode:cathode ratio the corrosion of bare steel connected to passivated steel in concrete can reach 1.75 mm/y under sea water conditions, compared with the usual rate of about .0125 mm/y [5].

20.1.3 Avoidance of corrosion

To avoid or decrease the rate of corrosion of steel reinforcements in concrete exposed in aggressive marine situations, there are a number of straightforward actions that can be taken. These are indicated below, but it should be emphasised that the composition and structure of concrete as a material is important and this has not been considered in this chapter. There are codes of practice covering the requirements for concrete and these are updated as additional information and experience are obtained.

The most aggressive conditions are in the splash zone area for offshore structures, but any structure or building on the coast must be regarded as a potential problem area unless care is exercised in the design. Factors to be considered include:

   (i) High-quality concrete that will provide a high level of protection to the steel with low permeability. Clearly, the incorporation of chlorides into the concrete mix may lead to problems.

 (ii) Adequate concrete cover over the steel reinforcement. This should take into account the aggressiveness of the conditions and would not necessarily be the same for all situations.

(iii) Careful attention to the positioning of reinforcements before casting the concrete to ensure that a proper thickness of cover is obtained.

(iv) Sound quality control procedures.

 (v) Adequate sealing of any joints that could lead to the collection of moisture and chlorides on the concrete surface.

A number of investigations into the corrosion of steel reinforcements in concrete bridge decks has led to methods of treating the concrete to either reduce permeability or to improve the corrosion resistance of the reinforcement bars. All the methods would not necessarily be appropriate to marine structures. However, as a matter of interest, some of these will be considered.

Kilareski in a survey of different methods of treating concrete for bridge decks [6] includes methods such as cathodic protection but also some rather less conventional approaches of an experimental nature.

   (i) The introduction of wax beads into the concrete during mixing. After the concrete has cured, heat is applied to it. This results in the melting of the wax which seals the voids to reduce the permeability. It is clearly an expensive method and no information on its practical efficacy is given.

 (ii) Experiments have been carried out with methods of electrochemical removal of chlorides from concrete. An ion exchange resin was used in an electrolyte on a concrete deck. Current was applied to a platinized titanium anode and the top reinforcement mat. This caused the chloride ions to migrate to the surface where they were collected by the ion exchange regin. According to the report this method resulted in the complete removal of the chloride ions. It seems doubtful whether the economics of the process would, at present, be acceptable for marine structures.

(iii) A method similar to the wax method has also been used with monomers which are forced into the concrete after it has been heated to a temperature of over 350 °C for a substantial period of time. The moisture is expelled and the liquid monomer is polymerised to a solid form to seal the concrete. Again, this is an experimental and expensive method.

A number of methods are, however, used in practice, including the following:

(i) The use of stainless steel for reinforcing bars. This has been employed on a limited scale for marine situations. Recently, stainless steel or nickel cladding of reinforcing bars has been used—again to a limited extent. This method may, however, prove to be economic for some particularly difficult situations. As stainless steel and other resistant coats will be cathodic to the steel reinforcements, a sufficient thickness of cladding must be applied to ensure that any pits that form on the cladding do not extend to the carbon steel. This could lead to a dangerous situation where a large cathode/anode ratio would lead to pitting of the steel.

(ii) Metallic coatings, generally galvanised zinc, have been widely used for steel reinforcements. Different views on the effectiveness of such methods have been expressed. Reports from the various bodies concerned with marketing zinc have shown the advantages[7] but other reports have cast some doubts on this form of protection. Arup[8] has expressed the view, based on Scandinavian experience, that zinc is not necessarily a satisfactory method for protecting steel in concrete, particularly under immersed conditions or in the splash zone. The different views expressed may arise from the nature and aggressivity of the environment and the particular type of structure being considered.

(iii) A method that is being employed in America for bridge decks is the protection of reinforcements with a powder epoxy coating. The reports on performance in practice indicate that this is an effective method. Clearly, the quality of the coating and avoidance of damage are important factors in achieving success.

(iv) Additional protection of the concrete exterior by the use of organic coatings such as epoxies or chlorinated rubber is also employed to reduce the ingress of corrosion species. Only non-saponifiable binders can be used because of the alkalinity of concrete.

(v) Cathodic protection can be applied to reinforcements in some situations, although in practice this may be difficult to achieve.

20.1.4 Detecting and monitoring corrosion in reinforced concrete

The most common method of monitoring the performance of reinforced concrete is by visual inspection of the exterior of the structure for rust stain, cracks etc. This is not completely satisfactory because when there are visible signs, some attack on the reinforcements has already occurred and it provides no indication of the rate of attack. A number of methods of detecting and monitoring corrosion of reinforcements have been developed. They are basically the same as those discussed in Chapter 21. As, however, there are a number of minor variations in the use of these methods for concrete, they will be briefly reviewed here.

(i) Detection of corrosion using potential measurements is based on the change in potential that occurs when steel reinforcement changes from the passive to the active state. The reference electrode used is normally the copper/saturated copper sulphate electrode (CSE). The potential (CSE) for passive steel is about 0.20 V but 0.35 V for corroding steel[9] which provides an indication of whether or not corrosion is occurring. However, potential readings between 0.20 and 0.35 V do not provide a clear indication of the condition at the steel surface. The results are usually plotted as potential contours which indicate the corroding areas.

   Potential measurements do not provide a measure of the rate or the amount of corrosion, nor can the method be used for concrete immersed in sea water.

   Arup has expressed the view that the $Cu/CuSO_4$ half cell should not be used for concrete. In testing old concrete structures he found that the half cell gave erratic results and considered the standard calomel electrode to be preferable[9a].

(ii) Other electrochemical techniques are being used either in laboratory investigation or to a limited extent in the field:

   (a) Detection using polarisation methods. Briefly, this consists of applying a small fixed anodic current to the reinforcement and observing the change of potential with time. The level of potential attained will indicate whether there are passive or active conditions at the steel surface.

   (b) Use of the electrical resistance probe[10] to determine the loss of steel in the reinforcement. This method does provide the opportunity of calculating the corrosion rate by making a series of measurements over a given time period.

   (c) Use of the polarisation resistance probe[11]. This method requires a suitably conducting electrolyte which may not be present in all concretes.

   These are monitoring rather than detection methods and are based on embedding the probes in the concrete. Generally, the applications of such methods should be considered as part of the design of the structure.

(iii) Special tubes may be embedded in the concrete. These can then be used to monitor loss of steel by ultrasonic probes moved along the tube.

(iv) A new type of probe has been developed[12] and has been described[13]. The probes are embedded in the concrete in areas considered to be critical for the particular structure. Each probe consists of one or more separate electrodes of the same steel as used in the reinforcing bars. The electrodes are enclosed in a stainless steel casing with a reference electrode. The probes are placed in the concrete so that they are located at different levels below the concrete surface. It is claimed that this method is suitable for large offshore concrete platforms and can provide information on the corrosiveness of the concrete, on potentials, resistivities and corrosion rates of the steel.

Other methods are also used to examine concrete; these include simple measurements of the thickness of the concrete cover using standard meters and the removal of concrete cores for laboratory examination and analysis. The core technique is particularly useful for determining the degree of carbonation and chloride penetration.

20.1.5 Remedial measures

On marine structures, remedial measures should be undertaken as soon as corrosion is detected. The general approach is to take suitable action to remove, so far as practicable, the source of the problem and to repair the concrete. Where corrosion has occurred, any chloride should be removed and the rust cleaned from the steel—preferably by blast-cleaning. Where corrosion has been severe, it will be necessary to assess the structural integrity and, where appropriate, to take suitable action.

Various methods of repair with different materials can be employed. These include the spraying of concrete on the affected areas or the application of epoxy mortars. The disadvantage of many mortars of this type is that they do not provide suitably alkaline conditions at the steel surface. There are, however, materials available for application to the steel that are claimed to have alkaline properties. Such materials should be applied directly to the cleaned steel reinforcements.

20.2 CORROSION BY WOODS

When wood and metals are in contact, particularly in marine environments, degradation of one or both of the materials may occur. Timber hulls are still used for some ships and boats and a good deal of decking is made from wood. The different types of wood vary in their corrosive effects on metals.

20.2.1 Corrosion of metals by wood

Wood can absorb moisture and in marine situations this may contain a substantial quantity of chloride. Where damp wood is in contact with a metal, particularly steel, corrosion may well occur. Apart from salt-containing moisture, woods contain various corrosive substances such as acetic acid. Because of their volatility, these may cause problems in wooden cases used for packaging. Generally, damp wood will be more corrosive than when dry, but certain types such as oak can produce acetic acid vapours even when dry and these are a potential corrosion hazard.

Salt—up to 4% by weight—can be introduced into wood by some seasoning processes and can lead to the corrosion of screws and other components in contact with the wood.

Other treatments for flame-retardant purposes and to prevent attack by various insects and fungi may also result in the addition of corrosive compounds to wood. In suitable circumstances, such treatments can accelerate the corrosion of fasteners in contact with them. The main problem with metals in contact with wood arises in situations where the area of contact is rather greater than with screws. Damp wood can act as an electrolytic path, especially if it contains chlorides. This can lead to a deposit-type of attack on steel. It causes concentration cells to be set up and may even result in bimetallic corrosion between different metals in contact and embedded in the wood.

20.2.2 Avoidance of corrosion by wood

The choice of components of corrosion-resistant materials rather than of steel or galvanised steel is one method of avoiding corrosion. Generally, however, the methods are based on insulation of the metal from the wood by means of coatings of suitable film thickness. Ordinary paints are not usually adequate for this purpose and chemical-resistant types such as bitumen or epoxies are used. The paints should be applied to the surface of both the metal and the wood.

20.2.3 Deterioration of wood in contact with metals

The nature of corrosion, with alkali produced at the cathodic areas and iron salts at the anodic areas, leads to the deterioration of wood. Similar problems of alkali attack on wood can occur where timber is in contact with cathodically protected steel.

A useful guide to the topic has been produced by the Department of Industry (UK)[14].

20.3 CORROSION BY PLASTICS

Plastics coatings have been considered in Chapter 10, but plastics materials are also used as alternatives to metals. Although these serve an important role in some areas of marine engineering, they will not be discussed in this book, which is concerned with the corrosion of metals. Plastics tend to be looked upon as inert with almost indefinite resistance to deterioration, but this is not completely correct. Plastics do deteriorate and, under some conditions, may cause metals to corrode either by direct contact or as a result of vapours produced in an enclosed space. These effects are not common in everyday usage but many plastics materials are liable to produce volatile compounds of a corrosive nature at higher temperatures and some even at ambient temperatures. A useful review on the corrosive effects of various plastics materials has been published[15].

20.4 REFERENCES

(1) *BRE Digest* 258:1982
(2) VENNESLAND, O and GJORV, O E, *Materials Protection*, **20**, No 8, 48 (1981)
(3) VON DAVEER, J R, *ACI Journal*, December (1966) p 697
(4) CLEAR, K C and HAY, R E, FHWA, RD-73-32 and 33, April (1973)
(5) VRABLE, J B, *Materials Performance*, **21**, No 3, 51 (1982)
(6) KILARESKI, W P, *Materials Performance*, **19**, No 3, 48 (1980)
(7) *Galvanised Reinforcements for Concrete*, US Zinc Institute (1970)
(8) ARUP, H, *Materials Performance*, **10**, No 4, 41 (1979)
(9) AMSI/ASTM C876-77
(9a) ARUP, H, *Materials Performance*, **20**, No 6, 26 (1981)
(10) KELDSEN, J T, NACE Conference 1978 Paper 123, National Association of Corrosion Engineers, Houston, Texas, USA
(11) ANDRADE, C and GONZALES, J A, *Inst. Ed. Torraja de la Construction y del Cemento*, Report No 33, Madrid (1978)
(12) Norwegian Patent No 139504
(13) GJORV, O E and VENNESLAND, O, *Materials Performance*, **21**, No 1, 33 (1982)
(14) The corrosion of metals by wood, No 2 of *Guides to Practice in Corrosion Control*, UK Department of Industry (1981)
(15) DONOVAN, P D, in L L Shreir (ed), *Corrosion*, Vol 2, Chap 19.8, Butterworths, London (1976)

Other references

'The corrosion of steel and its monitoring in concrete', No 7 of *Guides to Practice in Corrosion Control*, UK Department of Industry (1981)

BROWNE, R D, 'The performance of concrete structures in the marine environment', *Proc. Conf. Inst. Mn. Eng.*, London (1973)

Solving Rebar Corrosion Problems in Concrete (1963) National Association of Corrosion Engineers, Houston, Texas, USA

EVERETT, L H and TREADAWAY, K W J, 'Deterioration due to corrosion in reinforced concrete', *Building Research Establishment Information Paper 1*, p 12/80

British Standard CP110, *The Structural Use of Concrete*, Part 1

21 Testing and monitoring

Corrosion testing and corrosion monitoring are—with service experience and research investigations—the ways in which data and information are obtained concerning corrosion processes and methods of controlling these processes. Service experience is particularly important and while a good deal of information is made available in published papers and in seminars and conferences, even more could, with advantage, be published. The report of actual performance in service of protective coatings and alloys is of considerable value. The reporting of failures, in particular where the cause has been established, is clearly of the greatest value in assisting in the avoidance of similar failures. In this chapter, consideration will be given to corrosion testing and monitoring. There is not always a clear demarcation line between the two, nor is there a clear definition of testing and the same equipment or apparatus may be used for testing as is used for research. In the context of this book, testing is considered as the determination of data rather than the study in detail of mechanisms of corrosion and protection.

Corrosion monitoring may be considered as a particular form of testing. The definition in the UK Department of Industry's Committee on Corrosion booklet on monitoring[1] seems, however, to be reasonable——'The systematic measurement of the corrosion or degradation of an item of equipment with the aim of assisting the understanding of the corrosion process and/or obtaining information for use in controlling corrosion or its consequences'. The term 'equipment' presumably is taken to include structures and the aim of assisting in the understanding of the process should, perhaps, refer specifically to the problem in hand, so far as this chapter is concerned.

Corrosion monitoring originates from plant inspection techniques and from techniques developed for corrosion testing. Although it often employs reasonably complex apparatus and equipment, this is not necessarily the case. The simplest of all corrosion tests, i.e. placing alloy coupons in the atmosphere to determine their corrosion loss over a given period of time, is a perfectly valid method of monitoring the performance of a structure, provided that it gives real data relevant to the structure itself. This is not always a simple matter to achieve and may require a specialist knowledge of the corrosion processes concerned. This is a fundamental point to be appreciated with testing and monitoring. The value of the particular method of testing or monitoring is not determined by the complexity of the equipment used but by the validity of the method in producing the required data.

21.1 CORROSION TESTING

In the limited space available here it is only possible to provide a summary of testing with some pointers to the approach to be adopted. Books have been written on this subject[1a),(2),(3),(4),(16)] and there is a multitude of published papers on the topic. References to some of the most relevant are given at the end of the chapter. It can be said with some truth that it is better to do any corrosion testing rather than none at all provided the aim of the test is clear and the conclusions drawn are valid. Too many tests are carried out without a clear aim and the conclusions drawn are often related to conditions far removed from those of the test. Furthermore, the statistical basis of many tests is very doubtful so the results may not be representative of those that would be obtained if a larger sample had been used. This is clearly a problem because, as the number of specimens involved in a test rises, the time and cost increase considerably.

Nevertheless, there is no point in carrying out tests if the results are going to be irrelevant or misleading. If the costs for carrying out a series of tests properly are considered to be too high and there is no way of providing genuinely useful data in some other less costly way, then it may be better to abandon the idea of testing. Having said this, it must be added that in the hands of specialists all tests will provide some useful information if they are correctly interpreted. This, though, is a different matter from carrying out a programme of routine tests under conditions completely different from those that will be experienced in practice or where the statistical significance of the results is so low that they may be quite random and of little value.

A particular problem arises with tests on coatings because the performance of a coating system is determined by many factors other than the system under test. Misleading results can be obtained if tests are carried out on substrates markedly different from those that will arise in practical situations. This may not matter if the limits of the test are fully appreciated, but often there is a tendency to ignore these limitations. An obvious example is the salt-spray test, of which there are a number of variations. It is a comparatively simple test, so, particularly for information relevant to marine conditions, it is often specified. Because it uses sodium chloride it is deceptively easy to imagine that it will reproduce the conditions of marine environments. This is not necessarily so, and many workers will have experience of tests carried out simultaneously in a salt-spray test and at a marine exposure site, where the results within the same groups of test specimens are quite different. This does not mean that salt-spray tests are of no value; rather that they must be used to obtain information and data relevant to the test conditions.

21.2 TYPES OF TEST

The different types of corrosion tests can be grouped as follows:

Laboratory tests
Field tests
Service tests

21.3 LABORATORY TESTS

These may be standardised test methods such as salt spray, or special tests developed for specific purposes. Many of the tests are called 'accelerated' because they cause corrosion or breakdown of coatings more rapidly than occurs in service. Generally, though, the acceleration factor cannot be expressed numerically (e.g. × 10) and usually varies with the particular alloy or coating being tested.

Some laboratory tests are 'simulated' in the sense that an effort is made, so far as is practicable, to simulate the actual service conditions. Generally, such tests are carried out by research organisations and complex apparatus may be required; often they are more concerned with research than testing.

Another group of tests carried out in the laboratory on protective coatings should be mentioned. Although they are not corrosion tests, the results obtained are related to the performance of the coatings. These include physical tests for resistance to abrasion, scratch and impact, and tests to determine application qualities such as drying. These are all useful for quality control purposes but are not always easily interpreted in relation to corrosion protection requirements. For example, there often appears to be little evidence to support the relationship between, say, flaking of a paint coating and adhesion test data. The problems with these physical tests arises in part from the changes in properties that may occur over a period of time.

Many different tests are carried out under laboratory conditions, often with standardised procedures and apparatus. Some of the more commonly specified tests will be briefly considered later but, first, attention will be given to a number of comparatively simple tests that require little except the standard apparatus found in most laboratories.

21.3.1 Immersion

Simple immersion tests are rarely carried out to obtain straightforward data for marine situations. They are widely carried out to determine the effects of chemicals and acids on alloys and sometimes on coatings. They may, also, be used to determine galvanic effects on certain assemblies in sea water or other solutions. Although the tests are simple and therefore comparatively easy to carry out, their limitations must be appreciated; in particular the effects of corrosion products which remain in an enclosed and limited solution. Additionally, the effects of temperature and aeration must be considered carefully when carrying out such tests. Temperature control can be achieved reasonably easily at lower temperatures by the use of a water bath. However, the possibility of salts concentrating in the solution because of evaporation must be taken into account and it may be necessary to fit a reflux condenser. Although air or oxygen can be bubbled through the solution, it is not always appreciated that a considerable amount may be required to ensure that realistic corrosion rates are obtained. Air or oxygen should be passed into the solution by means of a sintered glass disc and the bubbles should not directly impinge on to the test piece. ASTM G31 provides recommendations for immersion testing[5].

Partially immersed or water-line tests can be carried out by placing a specimen in a fixed vertical position in a suitable solution. The solution should be maintained at a constant level by the addition of distilled water.

Other tests such as alternating immersion are easily carried out but must be carefully controlled if valid comparisons are to be made. This often calls for apparatus that is rather more complex than the simple beaker test.

21.3.2 Condensation-type humidity and spray tests

These condensation-type tests require sufficient humidity to ensure that either water vapour which has condensed on the specimen or sprayed salt droplets, remain in contact with the alloy surface or protective coating for a reasonable length of time. Such tests are useful when standardised to provide comparative testing conditions. Many humidity tests are arranged to prevent condensation of moisture while allowing droplets of test solution to remain moist.

The limitations of salt-spray tests have been mentioned earlier, but there are many such standard tests, e.g. BS 1391:1952 'Performance tests for protective schemes used in the protection of light-gauge steel and wrought iron against corrosion', ASTM B 117:1973 'Test for salt spray (fog) testing', DIN 50907-1952 'Resistance to marine climate and sea water'.

Variations of the straightforward salt-spray tests have been developed, in particular the CASS (copper accelerate acetic acid salt spray) Test: ASTM B 368:1968. This has been used widely for testing nickel-chromium coatings on steel. Various concentrations of NaCl solution have also been specified, although 3% NaCl has been widely used.

21.3.3 Electrochemical tests

Electrochemical tests are widely used for research and monitoring but less generally for laboratory testing, although some forms of potential measurement are used for field tests. Their main use is for galvanic or bimetallic tests where potentials of alloys are measured relative to a standard electrode to produce results similar to the 'Galvanic Series in Sea Water' and for specialised tests for crevices, pitting, etc, as discussed later. Electrochemical tests appear to be an attractive basis for determining data in a way much more rapid than with conventional tests and they have been used to some extent for tests on coatings and for alloys.

Instruments such as potentiostats and zero-resistance ammeters are widely available. Although they are not used for routine testing, there are situations where such techniques may, in the future, be used more widely to obtain fairly routine data.

21.3.4 Special tests

A wide range of special tests has been developed to obtain specific information on stress-corrosion cracking, intergranular corrosion, cavitation, etc. Details of these tests are to be found in published standards and books on corrosion testing. They are generally carried out by laboratories specialising in these particular forms of corrosion and in some cases the cost of the apparatus itself would preclude its purchase for occasional tests. Some of the specialised tests are, however, reasonably straightforward and require little or no specialised apparatus but often the interpretation of the results is of greater significance than the results themselves.

An indication of the general nature of some of these tests is given below.

(a) *Velocity tests*

The importance of velocity in influencing the corrosion performance of alloys has been discussed in Chapter 4. Tests have accordingly been devised to provide data on the effects of velocity.

Simple rotor tests where specimens are rotated in synthetic or natural sea water have been used fairly widely. However, they are difficult to control satisfactorily because there is a tendency for the solution to follow the specimens through the water so that the relative velocity is reduced. This can be overcome to some extent with baffles but, irrespective of whether the specimens are moved through the solution or the specimen remains stationary and the solution is made to pass over them, it is difficult to determine the correct velocity or to reproduce practical conditions. Nevertheless, apparatus based on rotors is useful for sorting tests before carrying out more expensive field or service trials.

Rotors using discs as the specimens have also been used, particularly for higher velocity tests to determine the 'critical velocities' for different alloys in sea water, i.e. the velocity below or above which protective films break down and corrosion occurs. The 'critical velocities' may, however, be specific to the particular test method.

More complex apparatus has been devised to obtain data on the effects of velocity. These include special rigs containing pipes and sometimes other equipment such as condensers[3]. A method developed by the British Non-Ferrous Metals Research Association is widely used to assess the effects of impingement attack on alloys employed for heat exchanger tubes[6]. Other special tests for erosion and cavitation effects have been developed, but generally they are used in specialist laboratories and details can be obtained from the references at the end of this section.

General advice is given in NACE publication TM-02-70 'Methods of Conducting Controlled Velocity Laboratory Corrosion Tests'.

(b) *Crevice corrosion tests*

Various electrochemical methods have been used to determine the critical pitting potential of alloys. However, simple tests are often used with crevices formed from the materials under investigation. Some indication of resistance to crevice attack can be gained by placing sand on the surface of the alloy and testing it under suitable conditions. Methods using plastics washers bolted to the material under investigation are used to provide the small gap necessary for crevice attack and other simple tests can be used (*Figure 21.1*).

(c) *Stress-corrosion cracking*

Many test methods have been developed to assess the susceptibility of

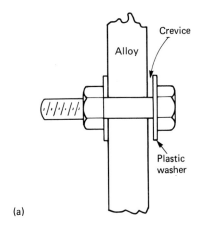

(a)

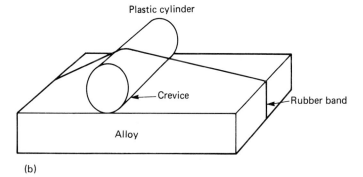

(b)

Figure 21.1 Simple crevice corrosion tests (a) Plastic washer with gap (b) Plastic cylinder

alloys to this form of attack. The stressing of test specimens is usually achieved by one of the following methods: (i) constant deflection, or (ii) constant load (*Figure 21.2*).

Of these, the constant-deflection method is usually the simplest to carry out, particularly where U-bend specimens are used. Other methods of constant-deflection testing have made use of two- or four-point bending techniques and all have been widely employed. Constant-load tests allow for accurate determination of stress but rather large and heavy pieces of testing equipment are required.

The specimens suitably stressed are then exposed to the type of environment under consideration, e.g. sodium-chloride solutions for marine situations. For stainless steels, 42% boiling $MgCl_2$ is frequently used to assess susceptibility of an alloy to stress-corrosion cracking. This solution is, of course, far more aggressive than normal marine environments and it does not follow that failures in such a test will be repeated in

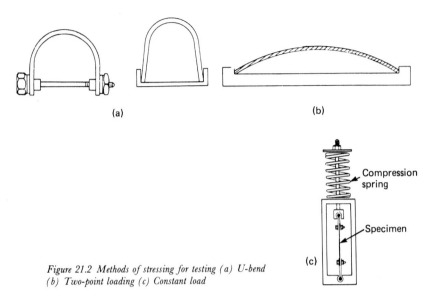

(a)   (b)

Figure 21.2 Methods of stressing for testing (a) U-bend
(b) Two-point loading (c) Constant load

service: 3.5% NaCl is frequently used for aluminium alloys. As with other tests, variations in the test solutions can influence results, e.g. aeration and pH. There are similarities between stress-corrosion cracking and cracking arising from hydrogen embrittlement in high-strength steels. Hydrogen can be introduced into the steel either during manufacture or treatment, e.g. by pickling or electroplating, or by corrosion processes during service. The test methods reflect these two ways of introducing hydrogen and they may be carried out in either the presence or absence of a corrosive environment, depending upon the circumstances. For example, the susceptibility of high-strength steels to hydrogen cracking after Cd plating is evaluated by subjecting plated specimens to a sustained tensile stress, the specimens being exposed to a laboratory atmosphere. In other tests, the specimens are immersed in dilute acid or neutral solutions containing a hydrogen promoter such as As and subjected to the tensile stress. Methods somewhat similar to those used for stress-corrosion testing are generally employed. Although there are standard laboratory tests for stress corrosion, they must be carried out with care, and specialist advice should be sought before relating results to practical situations.

(d) *Corrosion fatigue*

Testing for corrosion fatigue is usually carried out on the same types of apparatus as used for ordinary fatigue testing. A conventional rotating cantilever beam apparatus such as the Wöhler is used with the corrodent being brought into contact with the specimen by means of a wick or a small cell. Other test methods have been used, e.g. push-pull tests, and some variations in results have been obtained with the different methods. It is difficult to reproduce service conditions in any laboratory corrosion-fatigue test and such tests require specialist knowledge, particularly for the interpretation of the results.

The requirements for corrosion-fatigue testing have increased with the developments of offshore technology, with special reference to welds. Special apparatus has been designed to study the performance of welds in situations such as the North Sea, where wave motion is an important factor.

The results from tests are presented in the form of S–N curves and compared with curves obtained in the absence of a corrodent, such as sea water. This provides an indication of the effect of the corrosive environment on fatigue performance.

Test methods for corrosion fatigue generally fall into the field of investigation rather than testing.

(e) *Intergranular tests*
Stainless steels are susceptible to intergranular attack resulting from the precipitation of chromium carbides when the alloy is heated within a specific range of temperatures. Many different tests have been developed to check the susceptibility of such alloys, although most such tests are far more aggressive than the conditions to which the alloys will usually be exposed. The best-known such test is the Huey test using nitric acid as the corrodent, and named after the author of the paper in which it was first described. Another test, also named after the worker who first described it, is the Strauss test which employs saturated $CuSO_4$ solution containing H_2SO_4 and Cu turnings.

These tests are useful in evaluating the properties of unstabilised alloys containing more than about 0.03% C and to check whether welding operations are likely to make the alloys susceptible to this form of attack. The tests are carried out for periods of up to 5 days and the effects are measured in various ways including loss of weight, appearance and metallographic examination. ASTM A 262:1970 'Recommended Practice for Detecting Susceptibility to Intergranular Attack in Stainless Steel' provides details of the procedures generally considered to be acceptable for this type of test.

(f) *Galvanic tests*
Useful tables of potentials can be prepared, such as the Galvanic Series in Sea Water. These provide a broad indication of the outcome of

connecting different alloys together in the presence of sea water. Such series do not, however, take into account factors other than potential differences so various tests have been devised to provide practical data on the effects of anode–cathode area ratios and polarisation effects. Such tests can be carried out without the requirements for complex apparatus, and the following methods are commonly used for atmospheric testing:

(i) A bolted galvanic couple of the type shown in *Figure 21.3*. This is based on the recommendations of Subcommittee VIII of ASTM Committee B-3[7].

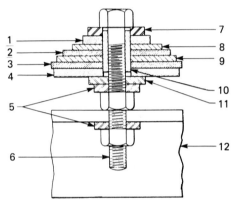

Figure 21.3 Galvanic tests—bolted assembly (1) Bakelite washer 19.0 × 3.2 mm (2) Metal B disc 30 × 1.6 mm (3) Metal B disc 36.6 × 1.6 mm (4) Bakelite washer 35.5 × 3.2 mm (5) Stainless steel lock washer (6) Stainless steel bolt 4.8 × 38.1 mm (7) Stainless steel washer 15.9 mm o.d. (8) Metal A disc 25.4 × 1.6 mm (9) Metal A disc 35.5 × 1.6 mm (10) 11.1 mm Bakelite bushing 5.2 mm i.d. × 7.9 mm o.d. (11) Stainless steel washer 15.9 mm o.d. (12) Galvanised angle support

(ii) A bolt and wire test couple as shown in *Figure 21.4*. This was originally developed by the Bell Telephone Laboratories but has been widely used by other testing organisations. The more noble metal is used for the bolt and wire of the less noble metal is wound round it.

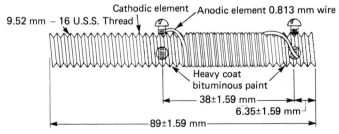

Figure 21.4 Galvanic tests—bolt and wire

(iii) To determine the effects of fasteners, simple specimens with the appropriate nuts, bolts and washers are fixed to the metal in question through a hole using an ordinary bolted connection.

(iv) Two plates of the alloys under investigation can be bolted together. This is a simple method of test but it is qualitative in the sense that the actual areas in contact cannot be determined.

Under immersed conditions, specimens of the two alloys can be connected in a cell and by measuring the current an indication of the corrosion rate will be obtained. If there are doubts concerning the relative potentials of the two metals, these can be measured with a suitable reference electrode. The current measurements should be made with a zero resistance ammeter or other suitable method.

21.3.5 General comments on laboratory tests

The more specialised tests such as those for stress-corrosion cracking or intergranular corrosion are generally carried out by specialist organisations. They are widely used in developing alloys and for obtaining broad data. This is useful as an indication of the likelihood of attack on specific alloys but it has to be used with caution as design data.

The more straightforward types of test such as salt spray are widely used and may form parts of the requirements for some coating standards. Their limits should, however, be appreciated. The variation in results that can be obtained by different laboratories on the same series of coatings is well illustrated in the reports of the Methods of Testing Sub-Committee of the former British Iron and Steel Research Association[8].

A symposium on accelerated tests for marine coatings organised by the Western Region Conference of NACE in 1966 and published as part of the NACE Coatings and Linings Handbook illustrated some of the approaches made to the laboratory testing of coatings. The topics covered include the following:

(i) Accelerated tests for coatings exposed to cold and high humidities.
(ii) Accelerated tests for zinc-rich topside coatings.
(iii) Use of an environmental test room in coating evaluation.
(iv) Equipment and methods for evaluating erosion and cavitation resistance of coatings.
(v) A laboratory method for evaluating anti-fouling paints.

This illustrates the wide-ranging attempts to devise methods for the rapid evaluation of coating systems. Although, as already indicated, in the hands of specialists the results are useful, they must be critically assessed before applying them to practical situations.

21.4 FIELD TESTS

These are tests carried out in the atmosphere or under immersed conditions, usually at test sites specially prepared for this purpose. Although such tests relate more closely to service conditions, they do not necessarily replicate them. It is quite common, for example, for large organisations to have test sites representing different types of environment, e.g. marine, industrial and rural atmospheres and rafts moored in bays for various immersion tests. A single site cannot, however, provide results that accurately relate to a wide range of other conditions. Nevertheless such tests provide more realistic data than do laboratory tests for many situations.

Field tests have been widely used to provide data on corrosion rates of most alloys in a variety of environments. They are also commonly used to obtain information on the performance of protective coatings.

In some ways, such tests are deceptively easy to carry out compared with some of the complex laboratory tests. Tests must, however, be carried out in accordance with certain requirements if they are to provide useful data and some of these requirements will be considered below in relation to marine testing, which covers atmospheric and immersion conditions.

21.4.1 Atmospheric tests

The general procedures for atmospheric tests are reasonably well standardised. Specimens of suitable size are exposed on racks or stands at suitable angles to the horizontal and facing the direction required. In the USA, tests on bare alloys tend to be exposed at 30° to the horizontal, whereas coated specimens are generally exposed at 45°. The specimens are usually placed facing the main direction of the sun, although at coastal sites they are commonly exposed facing the sea. Small specimens are often used, 13.4 cm × 10.1 cm (6 in × 4 in) being a popular size; they are generally mounted on a rack and separated with shaped porcelain insulators. There have been variations in the methods of carrying out the tests. The British Iron and Steel Research Association which carried out large programmes of exposure tests in the United Kingdom on steel specimens, bare and coated, preferred to use larger specimens, about 26.8 cm × 23.5 cm (12 in × 10 in), and these were generally exposed on steel stands by means of insulated nuts and bolts.

(a) *Alloys*
The exposure angle and direction affect the corrosion rate of alloys in

various ways, although for many purposes it is not important, because the information obtained is not vital from the design viewpoint. There are, however, exceptions, e.g. specimens that may be exposed to obtain data on metallic coatings. Generally, the corrosion data obtained is an average for the two sides of the specimen, but sometimes one side of the specimen is coated so that it does not corrode and a corrosion rate on one side only can be obtained.

The usual way of measuring corrosion rates is by taking the difference in the initial weight of the specimen before exposure and the weight obtained after removal of the corrosion products at the end of the exposure period (see 'Removal of Corrosion Products', p 391). For many corrosion-resistant alloys, descriptive data, e.g. amount and distribution of pitting and staining, may be of more value than weight-loss measurements.

The criterion used in the extensive test programme carried out by ASTM Committee A-5 on long-term tests of bare and zinc-coated sheets was visible perforation of the sheet. This may be reasonable to determine the performance of sheets but generally such criteria of failure provides limited data and is affected by random occurrences. Sometimes tests are conducted for one year only and this provides useful comparative data, particularly if such tests are carried out at a range of sites. Of more value are longer-term tests where the data can be used to construct a corrosion–time curve, such as that shown in *Figure 21.7*. A number of specimens are initially exposed and groups are removed at regular intervals to plot the corrosion rate after a particular period of exposure.

The exact nature of the data obtained from atmospheric exposure tests will be determined by the type of test. Corrosion-fatigue and stress-corrosion tests can be carried out at test sites, as can other types of test such as intergranular attack. In fact, most of the apparatus and equipment discussed under laboratory tests can be modified for field tests in the atmosphere but, with the exception of a few specialist research organisations that conduct tests on different forms of corrosion under atmospheric conditions, most field tests are carried out to determine straightforward corrosion data, e.g. rate of metal loss or pitting, and to determine the performance of coatings.

Although atmospheric marine tests can be carried out at any suitable site near the sea, precautions have to be taken to ensure that the specimens remain secure during the tests and generally that the site is monitored in some way so that the site conditions can be related to the corrosion data obtained. Routine measurements—much as temperature, rainfall, relative humidity and hours of sunshine—are frequently taken. Additional measurements of sulphur dioxide and of salt collected on surfaces are also taken where appropriate.

Automatic equipment, which can be left to record much of this information, has reduced the staff requirements for test sites and, with suitable computer programs, the data can be processed comparatively easily. In marine testing, the measurement of the salt (NaCl) content of the atmosphere is generally important to determine the nature of the test site. Although there are various ways of measuring it, none is completely satisfactory. Chlorides collected in deposit gauges from rainfall provide a measure, but variations of the method devised by Ambler and Bain are still used [9]. In essence, the test consists of trapping salt particles on a piece of muslin wrapped round a test tube with an extension into a reservoir of water in which the salt accumulates. The quantity of salt collected on the muslin and in the water over a period of time provides a measure of the amount of salt in the air. This test method does not necessarily provide an accurate measurement of all the salt in the air, but is useful for comparing salt concentrations at different parts of a site or even at different sites.

(b) *Protective coatings*

The atmospheric testing of protective coatings in coastal and marine areas is not basically different from tests carried out at inland sites but it is more difficult to define one marine site relative to another. The nature of the prevailing winds, the degree of shelter and the amount of sunshine— all influence the rate of breakdown of organic coatings and test results may vary markedly between sites.

Large-scale atmospheric tests on protective coatings have been carried out at marine sites throughout the world. A typical example is that of the American Welding Society series of tests on flame-sprayed metal-coated steel [10], the results of which are discussed in Chapter 11.

Tests were carried out at four atmospheric marine sites, as well as others at atmospheric industrial sites. Additionally, sea-immersion tests were also carried out.

These tests illustrate the scale of organisation required for tests of this magnitude. In all, some 1600 specimens were exposed at the four marine sites in this co-operative research programme, which has continued for over 20 years.

The conditions in marine locations vary in different parts of the world and these should be taken into account when planning test programmes. For example, the Arabian Gulf has become an important area of industrial activity in the last decade. The conditions on the coastal areas are clearly different from those in Europe. Periods of very high humidity may be experienced, particularly during the late summer and early autumn. The dew point is often exceeded and condensation occurs on all surfaces when their temperature approaches that of the ambient conditions. This differs from European conditions where, in coastal areas, rain usually falls before the air at ground level becomes saturated with moisture to an extent where the dew point is exceeded[11]. Such conditions can clearly influence test results.

The testing of organic coatings at marine sites is carried out as a matter of routine by many coatings' manufacturers and suppliers. Tests are also carried out by independent groups of users, but there can be problems in this area of investigation.

Generally, tests are required on new types of paint formulation which are often developed to provide increased protection. They are also developed for other purposes, e.g. improved application or drying, but from the standpoint of atmospheric corrosion testing, the aim is usually to determine the improved protective performance of such coatings. To assess such coatings realistically would require some years of test and this is generally not practicable. Attempts to reduce this period by, for example, testing only one or two coats of a system may provide misleading data. Methods such as using standardised scratches to investigate spread of rust from damaged areas may, however, be advantageous. Probably the main problems with field tests of protective coatings arise from the difficulties of providing test specimens that realistically represent the conditions that coating systems will have to withstand in practice. It is almost inevitable that specimens prepared under laboratory conditions will perform better than those in which the coating is applied under actual service conditions.

21.4.2 Immersion tests

Immersion tests for alloys and for protective coatings are carried out from some fixed situation such as a raft moored in the sea or from the underside of a pier projecting into the sea. Sometimes test jetties are used at special sites. Tests under immersed conditions are subject to variations in environmental factors but provide a reasonable method of assessing the corrosion of alloys or the performance of protective coatings. There are a number of points worth considering in relation to such tests:

(i) Specimens should be suspended at the same depth for direct comparison.

(ii) To allow for ease of examination, they should be attached to the test rig or raft so that they can be removed without the requirement to unbolt them from the rig.

(iii) The specimens should be arranged so that none is sheltered by others. Failure to do this may lead to increased attack on the 'front row' of specimens. Preferably they should be placed parallel to the flow of water.

21.4.3 General comments on field tests

The conduct of field tests is comparatively straightforward. This may lead to the carrying out of tests with insufficient regard to the requirements for valid results. Some of the problems that arise, particularly on tests of coating systems, are noted below and are worth reviewing before undertaking such work.

(i) No single site will provide results appropriate to every situation. The results obtained at one marine site cannot necessarily be applied to others, although 'orders of merit' are likely to be similar at different locations.

(ii) The order of merit, i.e. the relative performance of coating systems, is likely to be more reliable than the actual performance of individual specimens.

(iii) The surface preparation prior to coating has an important influence on the results. Generally it should be similar to that to be used in practice and, for example, cold reduced steel specimens should not be used for tests on coating systems to be used on structural steel. Even where blast-cleaned steel panels are used, it should be appreciated that it is easier to clean small panels to a high level, so there is always a strong probability that coatings will last longer on such panels. Often test panels are obtained directly from the steel mill. Such specimens may be of a much higher standard, e.g. no pitting or rusting, than steel used in practice.

(iv) The thickness of the protective system also has an important influence on performance. Some authorities prefer to compare systems at the same dry film thickness. This can be achieved by preparing coatings with a suitable doctor blade, which provides a coating of known thickness. This may, however, be unrealistic because of the method of application, e.g. brushing or spraying is likely to produce a coating that is different from one produced by application with a blade. Furthermore, some coatings are high build and tests should be carried out on coatings with a natural film thickness. It should, however, be borne in mind that variations in thickness will influence the test results, so the dry film thickness of all coatings on the test specimens should be recorded and, where appropriate, compared when assessing the results.

(v) In practice, breakdown on coated structures often tends to originate at welds, on edges and at specific features such as water traps. So, although flat specimens are the most suitable for direct comparison between different coating systems, the results are generally rather better than will occur in practice.

Channels are preferred by some authorities and special fabricated

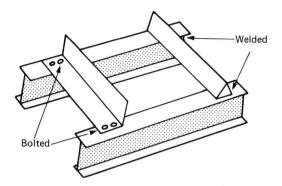

Figure 21.5 Fabricated test specimen

specimens have also been used[12] (*Figure 21.5*) to provide more realistic test specimens. Such attempts have not always been completely successful for various reasons, e.g. the difficulty of applying the coating evenly on such specimens. It is, however, worth considering slightly more complex specimens; at the least a scratch through the coating to the steel surface will provide information on undercutting at damaged areas (*Figure 21.6*).

(vi) The angle of exposure is important and may have a significant influence on the corrosion of metals and the performance of coatings. This is not particularly important for comparisons between coatings but the results obtained in tests carried out under different conditions cannot be directly compared. For tests on alloys, the angle and orientation of exposure may have an important bearing on the data obtained. In tests on steel specimens the difference in corrosion between those exposed facing north was 30% greater than

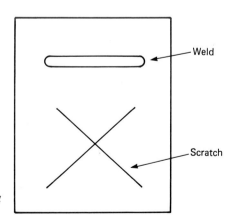

Figure 21.6 Specimen with weld and coating damaged by scratching

those facing south [13]. Although vertical exposure is in many ways advantageous, it is difficult to ensure that specimens are fixed in a vertical position and slight deviations may influence the results.

21.5 SERVICE TRIALS

Service trials are a special form of field test, carried out on actual plant or structures. The advantage of such tests compared with field tests is that the environment is the one where the coatings will actually be used and it is possible to take into account problem areas, such as welds, in the tests. Basically, the structure is divided into test areas of approximately equal size and of similar orientation. Where tests are being carried out for future maintenance repainting, as they often are, the general state of the surface to be repainted should be similar for the different test areas.

Trials such as this are more closely related to service conditions than are field tests, where the specimens are often prepared and coated in the laboratory. On the other hand, control of cleaning, paint application and coating thickness is more difficult. Often, painters—realising that they are painting test areas—apply the paint with more care than they would under the normal conditions of coating a structure. All dry paint film thicknesses should be measured carefully. In a series of trials on a gasholder, there were wide variations in the coating thickness applied to different areas (see *Table 21.1*). Clearly the performance of the different paints are related to their thickness and it is necessary where more than one painter is being used to ensure that he does not apply only one of the systems to be tested.

Other points to note are:

(i) Service trials should be clearly marked to avoid the loss of all information by the overpainting of test areas during treatment of other parts of the structure.

(ii) Care should be taken in selecting test areas to ensure that variations in the environment, e.g. prevailing winds, are taken into account.

Table 21.1 Variations in paint thickness in service trial

| Painter | *Average paint film thickness* (μm) | | | |
	Paint 1	*Paint 2*	*Paint 3*	*Paint 4*
A	43	47	59	39
B	31	37	39	32
C	35	35	45	23
D	45	45	44	—

This is particularly important where they may be moving salt spray on to the surface.

(iii) Areas should be selected so that the systems under test are divided equally between the different orientations.

(iv) The test area should be photographed or mapped prior to repainting trials to indicate whether premature breakdown has arisen from previous rusting.

Ideally, service trials should not be necessary if the performance of coating systems is regularly logged. The only exception would be where new maintenance methods and coatings were being tested.

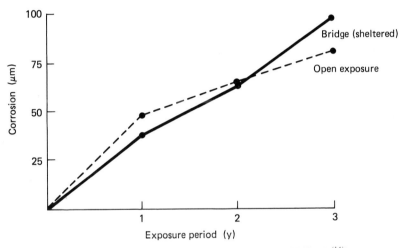

Figure 21.7 Corrosion monitoring of steel in a marine environment (after McKenzie [14] *)*

21.6 REMOVAL OF CORROSION PRODUCTS

Corrosion data on alloys exposed to marine atmospheres or immersed in the sea is often obtained from weight loss measurements. Specimens are cleaned and weighed before the test and then, after an appropriate period, are removed, cleaned to remove corrosion products, and reweighed. The difference in weight can then be related to the area and the density of the alloy, to provide a calculation of the corrosion rate in suitable units, e.g. µm/y.

Clearly this method will only give an accurate assessment of the corrosion rate provided all the corrosion products are removed. Further, in removing them, it is necessary to avoid removal of uncorroded metal from the specimen itself, as this will lead to errors in the calculation of metal loss. In practice, even with the use of suitable corrosion inhibitors, it may be difficult to achieve the removal of corrosion products without some effect on the alloy itself. It is, therefore, advisable before removing corrosion products to weigh clean unexposed specimens and then to treat them with the solutions to be used for the test specimens. After reweighing, any weight loss should be taken as a correction factor. If, however, this factor is too great, a new solution should be prepared and the procedure repeated. The correction factor should be calculated as weight loss/time and the correction used for the appropriate time taken to remove all the corrosion products.

ASTMGI:1972(31) 'Recommended Practice for Preparing Cleaning and Evaluating Corrosion Test Specimens' provides a guide to the methods used for the removal of corrosion products from metal specimens. Some of the more commonly used treatments are as follows:

(i) *Steel*
Immerse in Clark's Solution, which is made up from hydrochloric acid using Sb and Sn as inhibitors:

HCl (SG 1.19), 1 litre
Sb_2O_3 (antimony oxide), 20 g
$SnCl_2$ (stannous chloride), 50 g

The solution should be vigorously stirred before use and agitated during use. The time for removal of rust will depend upon the amount and type present. All loose rust should be scraped off first.

Other methods such as treatment in molten sodium hydride are more accurate but special equipment is needed, and safety precautions of a high standard are required.

(ii) *Copper and nickel alloys*
Dip for one to three minutes in 1:1 hydrochloric acid or 1:10 sulphuric acid at room temperature. Scrub lightly with a bristle brush under running water using a fine scouring powder if necessary.

(iii) *Aluminium alloys*
Make up an aqueous solution containing 2% by weight of chromic acid and 5% by volume of orthophosphoric acid. Maintain at 80 °C. Dip for five to ten minutes in the solution whilst agitating; then rinse in water to remove acid, brush lightly with a soft bristle brush to remove loose products and rinse again. If a film remains, immerse for one minute in concentrated nitric acid and repeat the procedure.

21.7 STANDARDS FOR TESTING

There are many national and international standards and these are listed in the appropriate handbooks and year books issued by the national standards bodies. Other standards are issued by organisations such as the National Association of Corrosion Engineers (NACE) and the American Society for Testing Materials (ASTM). These are not national standards but often have an equivalent authority and are often eventually accepted as such.

A selected number of standards is provided below, but this is by no means exhaustive.

British Standards

BS 1391:1952	Performance tests for protective schemes used in the protection of light-gauge steel and wrought iron against corrosion
BS 1747	'Methods for measurement of air pollution'. Issued in five parts, covering determination of sulphur dioxide, etc
BS 3745:1970	'The calculation of results of accelerated corrosion tests on metallic coatings'
BS 3900:1971	'Methods of test for paints'. In a number of parts, issued separately, covering physical tests, thickness, measurement, etc
BS 4351:1971	'Methods of detection of copper corrosion from petroleum products by the copper strip tarnish test' (identical with ASTM D 130)

ASTM Standards

A 262	† Detecting susceptibility to intergranular attack in stainless steels
A 279	Total immersion corrosion test of stainless steels
A 309	* Weight and composition of coating on long terne sheets by the triple spot test
A 393	† Conducting acidified copper sulphate test for intergranular attack in austenitic stainless steel
B 117	* Salt spray (fog) testing (method is intended for ferrous and non-ferrous metals with or without inorganic or organic coatings)
B 154	Standard method for mercurous nitrate test for copper and copper alloys (stress-corrosion cracking)

B 287	Method for acetic acid-salt-spray testing (this standard is intended for the same purpose as B 117 with differences)
B 368	Standard method of test for copper accelerated acetic acid-salt-spray (fog) testing
B 380	Corrosion testing of decorative chromium plating by the Corrodkote procedure
D 130	Copper corrosion from petroleum products by the copper strip test
D 610	Evaluating degree of rusting on painted steel surfaces
D 659	Evaluating degree of resistance to chalking of exterior paints
D 660	Evaluating the degree of resistance to checking of exterior paints
D 661	Evaluating degree of resistance to cracking of exterior paints
D 662	Evaluating degree of resistance to erosion of exterior paints
D 714	Evaluating degree of blistering of paints
D 772	Evaluating degree of resistance to flaking (scaling) of exterior paints
D 1014	Conducting exterior exposure tests of paint on steel
D 1654	Evaluation of painted or coated specimens subjected to corrosive enrivonments
F 64	* Corrosive and adhesive effect of gasket materials on metal surfaces
G 1	† Preparing, cleaning and evaluating corrosion test specimens
G 4	† Conducting plant corrosion tests
G 5	† Standard reference method for making potentiostatic and potentiodynamic anodic polarisation measurements
G 7	† Atmospheric environmental exposure testing of non-metallic materials
G 28	* Detecting susceptibility to intergranular attack in wrought nickel-rich chromium-bearing alloys
G 30	† Making and using U-bend stress-corrosion test specimens
G 31	† Laboratory immersion corrosion testing of metals
G 32	Vibratory cavitation erosion test
G 33	† Recording data from atmospheric corrosion test of metallic-coated steel specimens

G 34	Exfoliation corrosion susceptibility in 7 XXX series copper-containing aluminium alloys
G 36	† Performing stress-corrosion cracking tests in a boiling magnesium chloride solution
G 37	† Mattson's solution of pH 7.2 to evaluate the stress-corrosion cracking susceptibility of Cu-Zn alloys

NACE Standards

TM-01-69	Laboratory corrosion testing of metals for the process industries
TM-02-70	Method of conducting controlled velocity laboratory corrosion tests
TM-01-74	Laboratory methods for the evaluation of protective coatings used as lining materials in immersion services
TM-03-74	Laboratory screening tests to determine the ability of scale inhibitors to prevent the precipitation of calcium sulfate and calcium carbonate from solution

* Indicates a test method.
† Indicates the standard is a recommended practice.

21.8 MONITORING

Many of the techniques and instruments used for testing are also used for corrosion monitoring and there is not always a clear distinction between the two. Monitoring techniques may be used to investigate processes or to obtain direct information and data on corrosion that is occurring in a plant or on a structure. Until about 20 years ago, most monitoring was really a form of service test in which specimens or coupons were placed in suitable parts of the plant or structure and one or more periodically removed to provide information on corrosion over a known period. These methods are still used but have some disadvantages (and also advantages) over more refined forms of monitoring using instruments and probes.

Monitoring has developed because, by its nature, corrosion and control methods are not completely predictable. It is therefore advantageous to be able to maintain a check on the corrosion process and, in particular, the corrosion damage that is occurring. In this way a warning of problems is obtained and necessary action can be taken in advance of any really serious—or even catastrophic—situation arising.

Corrosion monitoring in the sense in which the term is generally used developed mainly to assist in the assessment or prediction of the corrosion behaviour of plant and equipment between shutdowns. Monitoring of the plant in use allowed improved diagnosis of any problems that might occur leading to improved plant operation. It also provided data for future design work on other plant or equipment. It should be noted, however, that monitoring has also been applied to situations other than plant for many years. Protective systems for ships, structures and vehicles have all been monitored over periods of time and, although this has often been described as testing, it really is monitoring where it is carried out as a continuous process over the life of an actual construction.

Monitoring is, of course, particularly advantageous in 'hidden situations, e.g. inside plant or pipes, where it is difficult to readily check corrosion. The developments in monitoring have arisen because of the improvements in equipment, instruments and techniques which have provided a much sounder basis for the understanding of corrosion processes and the ability to make 'instantaneous measurements' of corrosion.

In marine environments, there are a number of situations where monitoring would prove beneficial, for example, in pipes, cathodic protection and condenser systems. There are also many areas where the problems, whilst not specifically related to marine situations, nevertheless are accentuated by either the use of sea water for cooling or the construction of structures and plant on salt-containing soils and sand.

21.8.1 Monitoring techniques

Many techniques are now available for monitoring and they can be divided into four broad groups.

(i) Use of coupons of specific material.
(ii) Electrical techniques within the plant.
(iii) Analytical measurements.
(iv) Direct measurements on the plant itself.

Although modern instruments and apparatus have been designed so that they require little expertise to operate, it must be pointed out that the readings obtained will require expert interpretation by somebody who understands the operation of the plant or the problem areas of structures. Furthermore, expertise is required in choosing the most appropriate techniques. As with testing, the aim of the work must be clear and the limitations of any technique must be taken into account. A simple example to demonstrate the point concerns pipework, where the main problems may arise at joints and bends whereas other parts may not be seriously affected by corrosion; it follows that monitoring programmes must take this into account.

(i) *Use of coupons*

This method has already been considered under the section on 'Testing' and many of the comments are relevant to monitoring. It is a comparatively cheap and simple method to use but generally requires a considerable time to obtain data. It provides only average information on corrosion rates but does show any localised effects provided the coupons have been properly sited. Careful removal of corrosion products is necessary, and if this is not carried out correctly false ihformation will be obtained. Because it is basically a simple technique, it may be used by personnel with little or no appreciation of corrosion or of the varying corrosion influences on different parts of a structure or plant. In such situations, the information may be correct for the particular situation but misleading so far as the actual data required is concerned. An example of the way coupon testing is carried out—in this particular case to determine the influence of atmospheric and climatic factors on the corrosion of COR-TEN 'B' steel, a low-alloy weathering steel, is given in a paper by McKenzie[14]. Part of the monitoring was carried out at a marine site at Eastney on a narrow promontory in Langstone harbour and on the underside of a bridge at Shoreham-by-Sea in Sussex. The exercise was concerned with the determination of the difference in performance of the steel under freely exposed conditions and sheltered conditions, e.g. under a bridge. The test continued for five years; in *Figure 21.7*, the results over a three-year period are shown and this demonstrates some points of interest.

After one year's exposure the sheltered specimens were corroding at a lower rate than those openly exposed. After one year, however, the slopes of the corrosion-time curves for the two types of exposure changed, indicating that at some period they would probably cross, as in fact they did after about two years' exposure, when the sheltered specimens began to exhibit a higher corrosion rate than those on open exposure.

This illustrates one of the problems with coupon exposure, i.e. the time to obtain results. Furthermore, until the three-year exposure period, there could be no certainty that the relative rates of corrosion would continue in the same way. Nevertheless, this method of monitoring is useful for many situations. Apart from the corrosion rate data, there is also an opportunity to examine the type of corrosion and the composition of corrosion products when specimens are removed.

(ii) *Electrical tests*

The most commonly used tests in this category are polarisation resistance and electrical resistance. They are well-tried techniques and are not difficult to carry out, although they should not be used without some understanding of their limitations.

(a) *Polarisation resistance.* If a small voltage (about 20 mV) is applied to a corroding metal a current is produced. Special probes of the metal under consideration are made to carry out the technique, which can be used only in liquids with certain conductive requirements. The theory of polarisation-resistance measurements has been well described for readers who wish to study the matter further[15]. The method gives a plot of potential versus current for any linear polarisation and, by the use of suitable formulae, the corrosion current can be calculated.

Stern and Geary[15] showed that for a corroding metal there is a linear relationship between potential and current density, provided the potential is not displaced significantly (about 10 mV) from the corrosion potential E_{corr}. This can be expressed by

$$\frac{1}{R_p} = \left(\frac{\Delta i}{\Delta E}\right)_{E_{corr}} = 2.3\left(\frac{b_a + |b_c|}{b_a|b_c|}\right) i_{corr}$$

where R_p is the polarisation resistance, b_a and b_c are the anodic and cathodic Tafel slopes and i_{corr} is the corrosion current density since for a given system b_a and b_c are constant, then

$$\frac{1}{R_p} = \text{const } i_{corr}$$

A variety of commercially produced probes is available including automatic control systems suitable for use in multi-stage desalination plants as well as many other industrial applications.

(b) *Electrical resistance.* This method of monitoring is usually carried out with a probe, which consists of wire or strip of the metal being monitored. As the metal corrodes and the cross-section decreases, there is a change in resistance, which is measured by a suitable bridge circuit. This allows continuous monitoring of the corrosion of the alloy. Temperature compensation for resistance changes is incorporated ihto the probe by reference electrodes protected from the environment in which the monitoring is being carried out. The data obtained is an average over the time of monitoring. This method has been widely used in oil-refineries and multi-stage flash distillation plants. Often both electrical resistance and polarisation resistance techniques are used in plant to provide both instantaneous and overall data on corrosion.

(c) *Potential monitoring.* Other methods are not so widely used as the two described above but may have some advantages in certain situations. These include the use of potential measurements to indicate changes in corrosion, particularly breakdown of passivity, pitting, etc. This method is used to monitor the corrosion of reinforcing bars in concrete. Generally, a copper/saturated copper sulphate reference electrode is used. At potentials of 0.20 V (CSE) the steel is passive, whereas at 0.35 V (CSE) it is active. This is particularly important for marine environments, where chlorides may cause active corrosion of steel reinforcements.

(iii) *Analytical*
Automatic instrumentation is available to measure a number of variables which can influence corrosion and can be used as part of the overall monitoring process. These variables include pH, oxygen concentration, temperature and conductivity. Monitoring can also be achieved by determining the concentration of metal ion of metal salts produced by the corroding metal.

(iv) *Direct measurements*
Probably the most widely used method in this category is the Sentinel or 'tell-tale' hole. A small hole is drilled into the vessel or pipe wall to a depth equal to the design pressure thickness. The depth of wall not drilled is virtually the corrosion allowance and when this has corroded away a small leak develops, indicating that action is required. Although a simple and widely used test, it has the distinct disadvantage that process liquors, which may be dangerous, leak from the plant, albeit in small quantities. Other direct measurements include non-destructive techniques such as ultrasonic probes to measure the thickness of the alloy remaining in a piece of equipment or plant. Provided the original thickness is known or a series of tests is carried out over a period of time, an indication of loss of metal by corrosion (or possibly wear) can be established. There are a number of problems with this method, not least the coupling of the probe to the metal being examined. With some probes, it is necessary to remove the rust from the outside of the steel before carrying out measurements. There are, however, probes available that can measure through rusts of limited thickness so they do not have to be removed; probes for use at temperatures up to 500 °C are available. Ultrasonics involves the transmission of very high frequency sound waves through a metal and the time to traverse the thickness and return is measured. The probes are suitably calibrated for steel but, except where special probes are used, deposit or coatings on the steel may give false readings.

Eddy-current measurements, radiography and infra-red spectra are all methods that can be used for monitoring purposes. A simple caliper is also a useful tool for many purposes.

21.8.2 Selection of techniques

The selection of suitable techniques will be based on a number of factors, e.g. costs and actual requirements. Corrosion monitoring is a useful method of checking the course of corrosion, but it is not a replacement for inspection and common-sense. Problems can arise with instruments and, even where they are operating correctly, the readings and data will be those at the places where the monitoring is being carried out. Often, simple methods will be cheaper and as satisfactory as more complex methods.

Some of the factors to be considered before choosing a method include the following:

(i) Is the monitoring requirement for a new process or for one that is well established? If the latter, then a good deal of data will probably be available and the choice of method will be wide because it can usually be rapidly established whether the monitoring system is working correctly. If the former, then specialist advice may be necessary, particularly to interpret the results obtained.

(ii) What data is required? This is a fundamental point to be considered. If basically the requirement is to determine whether a structural member is safe, then monitoring is not required, only the measurement of the remaining thickness of the steel. If at this stage, however, there is some concern at the probable life of the structure, suitable monitoring—probably using ultrasonics—is required. It will not be practicable to measure every section, so the critical ones should be chosen. On a pipe, the same type of data on thickness may be required and similar methods could be used. It may, however, be that the real requirement is to determine the broad corrosion rate to decide whether inhibitors should be used, in which case a different technique might be chosen.

(iii) How quickly is the data required? If the data is required over a reasonably long term then coupon exposure can be considered, but otherwise a probe method will have to be chosen.

(iv) Is the data required of a general nature or specific to one part of a plant? Many of the probe techniques cannot take into account specific plant design details because the methods—particularly using the electrical resistance probe—provide data on the effect of the environment on the wire or strip of material in the probe.

This gives useful information on the corrosiveness of the environment to the selected alloy in a general way, but may not provide the specific data required.

(v) Is suitable technical expertise available? Although most of the monitoring methods are straightforward, it is necessary to have staff with a general understanding of the operation of the instruments and some ability to interpret the results. If such staff is not available it is advisable to use less sophisticated methods.

(vi) Is cost important? Cost is obviously relative to the value and importance of the plant or construction to be monitored, but the cost of equipment varies. For example, the cost of instruments for potential monitoring is probably about a quarter of the cost of other electrical methods. Furthermore, apart from instrumentation, manpower is required. The total costs are not negligible and the question to be asked is 'Do we require monitoring for this plant or structure or are visual inspections likely to be quite satisfactory?'

A useful book on corrosion monitoring is 'The Handbook of Industrial Corrosion Monitoring' published by HMSO (London).

21.9 REFERENCES

(1) 'Monitoring', No 6 of *Controlling Corrosion*, UK Department of Industry's Committee on Corrosion (1977)
(1a) CHAMPION, F A, *Corrosion Testing Procedures*, Chapman & Hall, London (1964)
(2) AILOR, W H (ed), *Handbook on Corrosion Testing and Evaluation*, John Wiley, New York (1971)
(3) LaQUE, F L, *Marine Corrosion*, Chap 3, John Wiley, New York (1975)
(4) SHREIR, L L (ed), *Corrosion*, 2nd ed, Vol 2, Chap 20, Butterworths, London (1976)
(5) *Recommended Practice for Laboratory Immersion Corrosion Testing of Metals*, ASTM G 31 (1972)
(6) CAMPBELL, H S, MP577, BNFMRA (1973)
(7) GORMAN, L J, *Proc. Amer. Soc. Test. Mat.*, 48:167 (1948)
(8) 'First Report of the Methods of Testing (Corrosion) Sub-Committee', *J. Iron Steel Inst.*, **158**, 463, April (1948)
(9) AMBLER, H R and BAIN, A A, *J. Appl. Chem.*, September (1955)
(10) 'Corrosion Tests of Flame-Sprayed Coated Steel, 19-year Report', AWS C2.14-74, Amer. Welding Soc., Miami, Florida, USA
(11) TEMPERLEY, T G, *Corrosion Science*, **5**, 581–589 (1965)
(12) 'Sixth Report of The Corrosion Committee, Special Report No 66', Iron Steel Inst., London (1959)
(13) LaQUE, F L, *Materials Performance*, **21**, No 4, 13 (1982)
(14) McKENZIE, M, 'Corrosion Testing and Monitoring for Highway Bridge Steelwork', *Proc. Conference on Corrosion Testing and Monitoring* (1977), Inst. Corr. Sci. Tech., London
(15) STERN, M and GEARY, A C, *J. Electrochem. Soc.*, **104**, 56 (1957)
(16) CARTER, V E, *Corrosion Testing in Metal Finishing*, Butterworths, London (1982)

Appendix

METHODS OF EXPRESSING CONCENTRATIONS OF SOLUTIONS

The concentration of a solid solute (e.g. NaCl) in a liquid solvent (e.g. water) is usually expressed in units of mass/unit volume and the accepted SI units are kg dm^{-3}. The litre, 10^3 cm^3, is an acceptable unit of volume although not SI. In chemistry the *molar mass* rather than the mass is used, and is defined as the amount of the substance that contains as many elementary units as there are in 12×10^{-3} kg of C-12 (the carbon isotope with mass number 12), i.e. 6.022 169 $\times 10^{23}$ specified elementary units, and this is a constant known as *Avogadro's number*, which is the number of elementary units in the molar mass, or the *mole*. Thus 1 mol (abbreviation for mole) of atoms, molecules, electrons, etc, will contain 6.022 169 $\times 10^{23}$ elementary units, and the unit of concentration is the mol dm^{-3}, which corresponds numerically with the unit used formerly, i.e. the molar concentration M or mol l^{-1}. It should be noted also that the molar mass (units, kg) replaces the atomic and molecular weights (units, g) used formerly.

Examples:
  1 mol Na atoms = 22.991 $\times 10^{-3}$ kg; 1 mol of Cl$_2$ gas = 70.91 $\times 10^{-3}$ kg (the molar mass of Cl atoms is 35.457 $\times 10^{-3}$ kg)
  1 mol Cl$^-$ ions = 35.457 $\times 10^{-3}$ kg (since the mass of the electron is 9.1055 $\times 10^{-31}$ kg, the mass of 1 mol is negligible)
  1 mol of NaCl crystals = (22.991 + 35.457) $\times 10^{-3}$ kg $\sim$ 58.45 $\times 10^{-3}$ kg (molar mass of Na atoms is 22.991 $\times 10^{-3}$ kg)

LAW OF MASS ACTION

The general equation for an equilibrium may be written in the form

$$aA + bB + \ldots \rightleftharpoons cC + dD + \ldots \tag{A1.1}$$

where A and B are the reactants, C and D the products, a, b, c and d are the number of each of the species involved in the equilibrium, and the reverse arrows $\rightleftharpoons$ indicate that the system is at equilibrium.

The probability of A and B colliding and combining to form C and D will be proportional to the product of their concentrations raised to the power of a and b, respectively, i.e. if there are two molecules of A and one molecule of B the probability of collision will be $c_A^2 \times c_B$. Thus the velocity of the forward reaction will be

$$v_1 = k_1 c_A^a \times c_B^b \qquad (A1.2)$$

where k_1 is the specific rate constant for the forward reaction. This will be very rapid initially, but will decrease with time as A and B react to form C and D. Conversely, the rate of the reverse reaction will be very slow initially owing to the very low concentrations of C and D, but will increase with time, and may be expressed as

$$v_{-1} = k_{-1} c_C^c \times c_D^d \qquad (A1.3)$$

REFERENCE ELECTRODES

There are numerous practical reference electrodes that are used for determining the potential of a metal/solution interface, and the choice of the one that is most appropriate is governed by a number of factors such as the nature of the solution in which the metal is immersed, reproducibility over prolonged periods of time, mechanical stability, etc. In order to minimise the liquid-junction potential at the interface between the two solutions, the reference electrode is selected so that the solution is similar or identical to the one in which the metal to be studied is immersed. Examples of widely used reference electrodes are given below.

Electrode and electrode equilibrium	Potential at 25 °C vs SHE(V)	Applications
Calomel $(Hg/Hg_2Cl_2, Cl^-)$ $Hg_2Cl_2 + 2e^- \leftrightarrows 2Hg + 2Cl^-$	$E_H = 0.2677 - 0.059 \log a_{Cl^-}$ 0.1 mol dm^{-3} KCl, $E_H = 0.334$ 1.0 mol dm^{-3} KCl, $E_H = 0.280$ Satd. KCl, $E_H = 0.241$	Laboratory studies of metals immersed in chloride solutions etc
Mercury/mercurous sulphate $(Hg/HgSO_4, SO_4^{2-})$ $HgSO_4 + 2e^- \leftrightarrows Hg + SO_4^{2-}$	$E_H = 0.6151 - 0.0295 \log a_{SO_4^{2-}}$	Laboratory studies of metals immersed in sulphate solutions
Copper/copper sulphate $(Cu/CuSO_4, Cu^{2+})$ $Cu^{2+} + 2e^- \rightleftharpoons Cu$	$E_H = 0.340 + 0.0295 \log a_{Cu^{2+}}$ For a saturated CuSO$_4$ solution, $E_H \simeq 0.30$	Monitoring potentials of cathodically protected underground steel structure
Silver/silver chloride $(Ag/AgCl, Cl^-)$ $AgCl + e^- \rightleftharpoons Ag + Cl^-$	$E = 0.2224 - 0.0591 \log a_{Cl^-}$ 0.1 mol dm^{-3} KCl, $E_H = 0.2881$ 1.0 mol dm^{-3} KCl, $E_H = 0.2224$ Sea water, $E_H \simeq 0.25$	Laboratory studies of metals immersed in chloride solutions, particularly at elevated temperature. Monitoring potentials of cathodically protected marine structures
Metallic zinc The potential of Zn in chloride solutions in a mixed or corrosion potential, which approximates to the reversible potential of Zn^{2+} (aq)/Zn, i.e. -0.76 V	In sea water, $E_H \simeq -0.80$	A robust electrode consisting of a rod of pure $Zn \simeq 25$ mm $\times$ 12 mm diam., which is used for monitoring potentials of cathodically protected steel structure immersed in sea water

Index